Simultaneous Inference in Regression

MONOGRAPHS ON STATISTICS AND APPLIED PROBABILITY

General Editors

F. Bunea, V. Isham, N. Keiding, T. Louis, R. L. Smith, and H. Tong

1 Stochastic Population Models in Ecology and Epidemiology *M.S. Barlett* (1960)
2 Queues *D.R. Cox and W.L. Smith* (1961)
3 Monte Carlo Methods *J.M. Hammersley and D.C. Handscomb* (1964)
4 The Statistical Analysis of Series of Events *D.R. Cox and P.A.W. Lewis* (1966)
5 Population Genetics *W.J. Ewens* (1969)
6 Probability, Statistics and Time *M.S. Barlett* (1975)
7 Statistical Inference *S.D. Silvey* (1975)
8 The Analysis of Contingency Tables *B.S. Everitt* (1977)
9 Multivariate Analysis in Behavioural Research *A.E. Maxwell* (1977)
10 Stochastic Abundance Models *S. Engen* (1978)
11 Some Basic Theory for Statistical Inference *E.J.G. Pitman* (1979)
12 Point Processes *D.R. Cox and V. Isham* (1980)
13 Identification of Outliers *D.M. Hawkins* (1980)
14 Optimal Design *S.D. Silvey* (1980)
15 Finite Mixture Distributions *B.S. Everitt and D.J. Hand* (1981)
16 Classification *A.D. Gordon* (1981)
17 Distribution-Free Statistical Methods, 2nd edition *J.S. Maritz* (1995)
18 Residuals and Influence in Regression *R.D. Cook and S. Weisberg* (1982)
19 Applications of Queueing Theory, 2nd edition *G.F. Newell* (1982)
20 Risk Theory, 3rd edition *R.E. Beard, T. Pentikäinen and E. Pesonen* (1984)
21 Analysis of Survival Data *D.R. Cox and D. Oakes* (1984)
22 An Introduction to Latent Variable Models *B.S. Everitt* (1984)
23 Bandit Problems *D.A. Berry and B. Fristedt* (1985)
24 Stochastic Modelling and Control *M.H.A. Davis and R. Vinter* (1985)
25 The Statistical Analysis of Composition Data *J. Aitchison* (1986)
26 Density Estimation for Statistics and Data Analysis *B.W. Silverman* (1986)
27 Regression Analysis with Applications *G.B. Wetherill* (1986)
28 Sequential Methods in Statistics, 3rd edition
G.B. Wetherill and K.D. Glazebrook (1986)
29 Tensor Methods in Statistics *P. McCullagh* (1987)
30 Transformation and Weighting in Regression
R.J. Carroll and D. Ruppert (1988)
31 Asymptotic Techniques for Use in Statistics
O.E. Bandorff-Nielsen and D.R. Cox (1989)
32 Analysis of Binary Data, 2nd edition *D.R. Cox and E.J. Snell* (1989)
33 Analysis of Infectious Disease Data *N.G. Becker* (1989)
34 Design and Analysis of Cross-Over Trials *B. Jones and M.G. Kenward* (1989)
35 Empirical Bayes Methods, 2nd edition *J.S. Maritz and T. Lwin* (1989)
36 Symmetric Multivariate and Related Distributions
K.T. Fang, S. Kotz and K.W. Ng (1990)
37 Generalized Linear Models, 2nd edition *P. McCullagh and J.A. Nelder* (1989)
38 Cyclic and Computer Generated Designs, 2nd edition
J.A. John and E.R. Williams (1995)
39 Analog Estimation Methods in Econometrics *C.F. Manski* (1988)
40 Subset Selection in Regression *A.J. Miller* (1990)
41 Analysis of Repeated Measures *M.J. Crowder and D.J. Hand* (1990)
42 Statistical Reasoning with Imprecise Probabilities *P. Walley* (1991)
43 Generalized Additive Models *T.J. Hastie and R.J. Tibshirani* (1990)
44 Inspection Errors for Attributes in Quality Control
N.L. Johnson, S. Kotz and X. Wu (1991)

45 The Analysis of Contingency Tables, 2nd edition *B.S. Everitt* (1992)
46 The Analysis of Quantal Response Data *B.J.T. Morgan* (1992)
47 Longitudinal Data with Serial Correlation—A State-Space Approach *R.H. Jones* (1993)
48 Differential Geometry and Statistics *M.K. Murray and J.W. Rice* (1993)
49 Markov Models and Optimization *M.H.A. Davis* (1993)
50 Networks and Chaos—Statistical and Probabilistic Aspects *O.E. Barndorff-Nielsen, J.L. Jensen and W.S. Kendall* (1993)
51 Number-Theoretic Methods in Statistics *K.-T. Fang and Y. Wang* (1994)
52 Inference and Asymptotics *O.E. Barndorff-Nielsen and D.R. Cox* (1994)
53 Practical Risk Theory for Actuaries *C.D. Daykin, T. Pentikäinen and M. Pesonen* (1994)
54 Biplots *J.C. Gower and D.J. Hand* (1996)
55 Predictive Inference—An Introduction *S. Geisser* (1993)
56 Model-Free Curve Estimation *M.E. Tarter and M.D. Lock* (1993)
57 An Introduction to the Bootstrap *B. Efron and R.J. Tibshirani* (1993)
58 Nonparametric Regression and Generalized Linear Models *P.J. Green and B.W. Silverman* (1994)
59 Multidimensional Scaling *T.F. Cox and M.A.A. Cox* (1994)
60 Kernel Smoothing *M.P. Wand and M.C. Jones* (1995)
61 Statistics for Long Memory Processes *J. Beran* (1995)
62 Nonlinear Models for Repeated Measurement Data *M. Davidian and D.M. Giltinan* (1995)
63 Measurement Error in Nonlinear Models *R.J. Carroll, D. Rupert and L.A. Stefanski* (1995)
64 Analyzing and Modeling Rank Data *J.J. Marden* (1995)
65 Time Series Models—In Econometrics, Finance and Other Fields *D.R. Cox, D.V. Hinkley and O.E. Barndorff-Nielsen* (1996)
66 Local Polynomial Modeling and its Applications *J. Fan and I. Gijbels* (1996)
67 Multivariate Dependencies—Models, Analysis and Interpretation *D.R. Cox and N. Wermuth* (1996)
68 Statistical Inference—Based on the Likelihood *A. Azzalini* (1996)
69 Bayes and Empirical Bayes Methods for Data Analysis *B.P. Carlin and T.A Louis* (1996)
70 Hidden Markov and Other Models for Discrete-Valued Time Series *I.L. MacDonald and W. Zucchini* (1997)
71 Statistical Evidence—A Likelihood Paradigm *R. Royall* (1997)
72 Analysis of Incomplete Multivariate Data *J.L. Schafer* (1997)
73 Multivariate Models and Dependence Concepts *H. Joe* (1997)
74 Theory of Sample Surveys *M.E. Thompson* (1997)
75 Retrial Queues *G. Falin and J.G.C. Templeton* (1997)
76 Theory of Dispersion Models *B. Jørgensen* (1997)
77 Mixed Poisson Processes *J. Grandell* (1997)
78 Variance Components Estimation—Mixed Models, Methodologies and Applications *P.S.R.S. Rao* (1997)
79 Bayesian Methods for Finite Population Sampling *G. Meeden and M. Ghosh* (1997)
80 Stochastic Geometry—Likelihood and computation *O.E. Barndorff-Nielsen, W.S. Kendall and M.N.M. van Lieshout* (1998)
81 Computer-Assisted Analysis of Mixtures and Applications—Meta-analysis, Disease Mapping and Others *D. Böhning* (1999)
82 Classification, 2nd edition *A.D. Gordon* (1999)
83 Semimartingales and their Statistical Inference *B.L.S. Prakasa Rao* (1999)
84 Statistical Aspects of BSE and vCJD—Models for Epidemics *C.A. Donnelly and N.M. Ferguson* (1999)
85 Set-Indexed Martingales *G. Ivanoff and E. Merzbach* (2000)

86 The Theory of the Design of Experiments *D.R. Cox and N. Reid* (2000)
87 Complex Stochastic Systems
O.E. Barndorff-Nielsen, D.R. Cox and C. Klüppelberg (2001)
88 Multidimensional Scaling, 2nd edition *T.F. Cox and M.A.A. Cox* (2001)
89 Algebraic Statistics—Computational Commutative Algebra in Statistics
G. Pistone, E. Riccomagno and H.P. Wynn (2001)
90 Analysis of Time Series Structure—SSA and Related Techniques
N. Golyandina, V. Nekrutkin and A.A. Zhigljavsky (2001)
91 Subjective Probability Models for Lifetimes
Fabio Spizzichino (2001)
92 Empirical Likelihood *Art B. Owen* (2001)
93 Statistics in the 21st Century
Adrian E. Raftery, Martin A. Tanner, and Martin T. Wells (2001)
94 Accelerated Life Models: Modeling and Statistical Analysis
Vilijandas Bagdonavicius and Mikhail Nikulin (2001)
95 Subset Selection in Regression, Second Edition *Alan Miller* (2002)
96 Topics in Modelling of Clustered Data
Marc Aerts, Helena Geys, Geert Molenberghs, and Louise M. Ryan (2002)
97 Components of Variance *D.R. Cox and P.J. Solomon* (2002)
98 Design and Analysis of Cross-Over Trials, 2nd Edition
Byron Jones and Michael G. Kenward (2003)
99 Extreme Values in Finance, Telecommunications, and the Environment
Bärbel Finkenstädt and Holger Rootzén (2003)
100 Statistical Inference and Simulation for Spatial Point Processes
Jesper Møller and Rasmus Plenge Waagepetersen (2004)
101 Hierarchical Modeling and Analysis for Spatial Data
Sudipto Banerjee, Bradley P. Carlin, and Alan E. Gelfand (2004)
102 Diagnostic Checks in Time Series *Wai Keung Li* (2004)
103 Stereology for Statisticians *Adrian Baddeley and Eva B. Vedel Jensen* (2004)
104 Gaussian Markov Random Fields: Theory and Applications
Håvard Rue and Leonhard Held (2005)
105 Measurement Error in Nonlinear Models: A Modern Perspective, Second Edition
Raymond J. Carroll, David Ruppert, Leonard A. Stefanski,
and Ciprian M. Crainiceanu (2006)
106 Generalized Linear Models with Random Effects: Unified Analysis via H-likelihood
Youngjo Lee, John A. Nelder, and Yudi Pawitan (2006)
107 Statistical Methods for Spatio-Temporal Systems
Bärbel Finkenstädt, Leonhard Held, and Valerie Isham (2007)
108 Nonlinear Time Series: Semiparametric and Nonparametric Methods
Jiti Gao (2007)
109 Missing Data in Longitudinal Studies: Strategies for Bayesian Modeling and Sensitivity Analysis
Michael J. Daniels and Joseph W. Hogan (2008)
110 Hidden Markov Models for Time Series: An Introduction Using R
Walter Zucchini and Iain L. MacDonald (2009)
111 ROC Curves for Continuous Data
Wojtek J. Krzanowski and David J. Hand (2009)
112 Antedependence Models for Longitudinal Data
Dale L. Zimmerman and Vicente A. Núñez-Antón (2009)
113 Mixed Effects Models for Complex Data
Lang Wu (2010)
114 Intoduction to Time Series Modeling
Genshiro Kitagawa (2010)
115 Expansions and Asymptotics for Statistics
Christopher G. Small (2010)
116 Statistical Inference: An Integrated Bayesian/Likelihood Approach
Murray Aitkin (2010)
117 Circular and Linear Regression: Fitting Circles and Lines by Least Squares
Nikolai Chernov (2010)
118 Simultaneous Inference in Regression *Wei Liu* (2010)

Monographs on Statistics and Applied Probability 118

Simultaneous Inference in Regression

Wei Liu
University of Southampton
UK

CRC Press
Taylor & Francis Group
Boca Raton London New York

CRC Press is an imprint of the
Taylor & Francis Group an **informa** business
A CHAPMAN & HALL BOOK

CRC Press
Taylor & Francis Group
6000 Broken Sound Parkway NW, Suite 300
Boca Raton, FL 33487-2742

First issued in paperback 2017

ISBN 13: 978-1-138-11168-4 (pbk)
ISBN 13: 978-1-4398-2809-0 (hbk)

Library of Congress Cataloging-in-Publication Data

Liu, Wei, 1964 Apr. 15-
Simultaneous inference in regression / Wei Liu.
p. cm. -- (Monographs on statistics and applied probability ; 118)
Includes bibliographical references and index.
ISBN 978-1-4398-2809-0 (hardback : alk. paper)
1. Regression analysis. 2. Confidence intervals. I. Title. II. Series.

QA278.2.L58 2010
519.5'36--dc22 2010022103

Visit the Taylor & Francis Web site at
http://www.taylorandfrancis.com

and the CRC Press Web site at
http://www.crcpress.com

Dedication

To My Family

Preface

Simultaneous confidence bands have been used to quantify unknown functions in various statistical problems. The most well-known is probably the confidence band for a cumulative distribution function based on the empirical distribution function and the large sample Kolmogorov distribution (cf. Stuart, 1999); see Frey (2008), Xu *et al.* (2009) and the references therein for recent advances on this. Hall (1993) studied the construction of simultaneous confidence bands for a nonparametric density function. Hall and Wellner (1980) and Lin (1994) investigated the construction of simultaneous confidence bands for the hazard function, whilst Hollander *et al.* (1997) considered simultaneous confidence bands for a survival function. For construction of simultaneous confidence bands for nonparametric regression models, the reader is referred to Genovese and Wasserman (2008), Zhao and Wu (2008) and the references therein. Khorasani (1982) and Cheng (1987) considered simultaneous confidence bands for non-linear parametric regression models. Ma and Hall (1993) and Horvath *et al.* (2008) considered simultaneous confidence bands for receiver operating characteristic (ROC) curves. This list serves the purpose to show that simultaneous confidence bands are useful statistical inferential tools that can be used in many statistical branches. Indeed, using simultaneous confidence bands to bound an unknown function or the differences between unknown functions is a direct generalization of using confidence intervals to bound an unknown parameter or the differences between unknown parameters.

This book, however, focuses on simultaneous confidence bands for linear regression models. Linear regression analysis is a simple but very useful statistical tool about which numerous books have been written. The methodology of simultaneous confidence bands provides additional inferential tools for linear regression analysis that are often more intuitive and informative than the usual approaches of hypotheses testing and confidence interval/set estimations of the regression coefficients. While the methodology was started about eighty years ago by Working and Hotelling (1929) and studied by many researchers, it was reviewed only very briefly, in no more than one chapter, by Seber (1977),

Miller (1980) and Casella and Berger (1990). This book provides a comprehensive overview of the construction methods and applications of simultaneous confidence bands for various inferential purposes. The first seven chapters of the book are on the normal-error linear regression models, for which the methodology is most well developed. The eighth chapter is on the logistic regression model to give the reader a glimpse into how simultaneous confidence bands can be constructed and used for generalized linear regression models by combining the methods for normal-error linear regression models and the large sample normal distribution of the maximum likelihood estimations for a generalized linear regression model. The ideas and approaches presented in Chapter 8 can be used for not only other generalized linear models, but also linear mixed-effects, generalized linear mixed-effects models (cf. McCulloch and Searle, 2001) and indeed general parametric regression models.

The chapters are arranged in the order that I hope they will be read. Chapter 1 provides a very brief review of the normal-error linear regression models. Chapter 2 considers simultaneous confidence bands for a simple regression line, which contains many ideas that are relatively easy to understand and used for a multiple regression model. Chapter 3 then looks at simultaneous confidence bands for a multiple linear regression model. Chapter 4 describes how simultaneous confidence bands can be used to assess whether a part of a multiple linear regression model is significantly different from the zero function and so cannot be deleted from the regression model, or is practically equivalent to the zero function and so can be deleted from the regression model. Chapter 5 discusses how to use simultaneous confidence bands for the comparison of two regression models. Chapter 6 uses simultaneous confidence bands to compare more than two regression models. Chapter 7 shows how simultaneous confidence bands can be used for polynomial regression models, where the covariates have functional relationships among themselves. Chapter 8 demonstrates how simultaneous confidence bands can be used in generalized linear regression models by focusing on logistic regression models for the purpose of illustration. Appendices A, B, C and E collect the details of computation methods necessary in Monte Carlo simulation for approximating the exact critical constants of confidence bands, while Appendix D gives a very brief introduction to the principle of intersection-union tests.

This book is intended to serve the needs of both practitioners and researchers. Several numerical examples are provided to illustrate the methods and the MATLAB® programs for implementing the necessary computation, as well as all the figures in colour, are available at the author's website `http://www.personal.soton.ac.uk/wl/research.html` for downloading. Usually only the first part of a program for entering the dataset needs to be changed to deal with a new dataset. This helps a practitioner to apply the methods provided in this book to solve her or his own problems. The comprehensive review of the published literature and some unsolved problems noted in the book should prove to be valuable to researchers. It is hoped that this book will generate more research interests in simultaneous confidence bands for linear regression models and beyond.

I have assumed that the reader has had a course in mathematical statistics covering the basic concepts of inference and linear model and is familiar with matrix algebra. All the programming is in MATLAB (Version 7.6.0.324(R2008a)), a powerful and user-friendly technical computing environment for computation and data visualization. There are many free resources about MATLAB available on the internet. The book by Hanselman and Littlefield (2005) provides a comprehensive overview of MATLAB, while the book by Martinez and Martinez (2008) demonstrates how to implement many standard statistical procedures in MATLAB.

I am very grateful to Pascal Ah-Kine, Frank Bretz, Jonathan Donnelly, Xiaoliang Han, Tony Hayter, Mori Jamshidian, Bob Jenrich, Shan Lin, Walt Piegorsch, Henry Wynn and Ying Zhang, with whom I have had the good fortune to work on simultaneous confidence band problems. Tony Hayter, Frank Bretz, and two anonymous referees read part or the whole book and provided many useful comments and suggestions. Of course, any errors, omissions or obscurities which remain are entirely my responsibility. I am indebted to Rob Calver, Jessica Vakili, Sarah Morris, David Tumarkin, and Shashi Kumar of the Taylor & Francis Group for their patience and support throughout this project.

Data in Tables 5.3 and 6.1 are from Ruberg and Stegeman (1991) by permission of the International Biometric Society. Data in Table 4.1 are from SAS/STAT User's Guide (1990) by permission of SAS Institute Inc. Data in Tables 7.1 and 7.2 are from Selvin (1998) by permission of Oxford University Press, Inc. Data in Table 8.1 are from Myers *et al.* (2002) by permission of John Wiley & Sons, Ltd. Data in Table 2.1 are from Kleinbaum *et al.* (1998) by permission of Brooks/Cole, a part of Cengage, Inc.

MATLAB® is registered trademark of The MathWorks, Inc. For product information, please contact:

The MathWorks, Inc.
3 Apple Hill Drive
Natick, MA 01760-2098 USA
Tel: 508 647 7000
Fax: 508-647-7000
E-mail: info@mathworks.com
Web: www.mathworks.com

Wei Liu, Southampton, 2010

Contents

List of Figures

List of Tables

1

Introduction to Linear Regression Analysis

1.1 Linear regression models

Linear regression analysis is a statistical tool for evaluating the relationship between a continuous random variable Y and several independent variables $x_1, \cdots, x_p$ according to the model

$$Y = \beta_0 + \beta_1 x_1 + \cdots + \beta_p x_p + e. \tag{1.1}$$

Here $\beta_0, \cdots, \beta_p$ are unknown constants, and e is an unobservable random error which is assumed to have a normal distribution with mean zero and unknown variance $\sigma^2 > 0$ throughout this book. The parameters β_i and σ^2 characterize the model and are estimated from the observed data. The quantity Y is often referred to as the *response* or *dependent variable*, as it depends on the x_i's through the linear relationship $\beta_0 + \beta_1 x_1 + \cdots + \beta_p x_p$. As the value of Y may be predicted from the x_i's by using the relationship (1.1), the x_i's are often called the *predictors* or *predictor variables* or *covariates*. Throughout this book it is assumed that the predictors are not random variables.

Throughout the book letter x_i indicates the ith covariate in the model and the corresponding x_{ji} denote a specific (jth) observed value of the covariate x_i. Similarly, Y defines the response and Y_j is its jth observed value. Let $\mathbf{Y} = (Y_1, \cdots, Y_n)'$ and $\mathbf{x}_i = (x_{1i}, \cdots, x_{ni})'$ denote n observations on Y and x_i, $i = 1, \cdots, p$, respectively. Then, according to (1.1), the jth observed value satisfies the *data model*

$$Y_j = \beta_0 + \beta_1 x_{j1} + \cdots + \beta_p x_{jp} + e_j, \quad j = 1, \cdots, n.$$

Note that the constant coefficients $\beta_0, \cdots, \beta_p$ are the same for all the observations, and the random error e_i's are not observable since the values of $\beta_0, \cdots, \beta_p$ are unknown. It is assumed throughout this book that the errors $e_1, \cdots, e_n$ are independent $N(0, \sigma^2)$ random variables. The n observations can be represented in matrix form as

$$\mathbf{Y} = \mathbf{X}\boldsymbol{\beta} + \mathbf{e} \tag{1.2}$$

where $\mathbf{X} = (\mathbf{1}, \mathbf{x}_1, \cdots, \mathbf{x}_p)$, $\boldsymbol{\beta} = (\beta_0, \beta_1, \cdots, \beta_p)'$, $\mathbf{e} = (e_1, \cdots, e_n)'$, and $\mathbf{1}$ is a column vector of size n with all elements equal to 1. The matrix $\mathbf{X}$ is called the design matrix as its components x_{ji} may be chosen via a suitable design.

Table 1.1: *Infant data*

Observation i	Birthweight in oz (x_1)	Age in days (x_2)	Blood Pressure (mm Hg) (Y)
1	125	3	86
2	101	4	87
3	104	4	87
4	143	5	100
5	92	5	89
6	119	3	86
7	100	4	89
8	149	3	89
9	133	2	83
10	120	4	92
11	118	4	88
12	94	3	79
13	131	5	98
14	93	4	85
15	94	4	87
16	121	5	97
17	96	4	87

There are $p+2$ unknown parameters involved in the linear regression model (1.1), including the $p+1$ coefficients β_i and the variance σ^2 of the random error e. In order to estimate the $p+2$ parameters, at least $p+2$ observations are required (i.e. $n \geq p+2$). Moreover, for the β_i's to be uniquely estimable it is required that the design matrix $\mathbf{X}$ have a full column rank $p+1$. If the design matrix is not of full column rank, then its columns are linearly dependent, and a subset of its columns can be obtained that are linearly independent and span the same column space. In other words, some covariates are linear combinations of other covariates. Some covariates are therefore redundant and can be deleted from model (1.1). The resultant model after deleting the redundant covariates will have its design matrix being full column rank. Thus, without loss of generality, throughout this book it is assumed that $\mathbf{X}$ is of full column rank. The relationship (1.1) and the assumptions made above constitute the *standard* linear regression model, on which the first seven chapters of this book focus.

Example 1.1 Suppose the relationship between systolic blood pressure (Y), birth weight in oz (x_1), and age in days (x_2) of an infant can be modelled by

$$Y = \beta_0 + \beta_1 x_1 + \beta_2 x_2 + e.$$

The measurements on 17 infants are shown in Table 1.1.

We may write this in matrix form as

$$\begin{pmatrix} 86 \\ 87 \\ \vdots \\ 97 \\ 87 \end{pmatrix} = \begin{pmatrix} 1 & 125 & 3 \\ 1 & 101 & 4 \\ \vdots & \vdots & \vdots \\ 1 & 121 & 5 \\ 1 & 96 & 4 \end{pmatrix} \begin{pmatrix} \beta_0 \\ \beta_1 \\ \beta_2 \end{pmatrix} + \begin{pmatrix} e_1 \\ e_2 \\ \vdots \\ e_{16} \\ e_{17} \end{pmatrix}.$$

Assuming that the linear relationship between the response and the predictors holds, then the parameters $\beta_0, \beta_1, \beta_2$ and σ^2 can be estimated based on the $n = 17$ observations.

1.2 Parameter estimation

By far the most popular method for estimating the parameter vector β of the linear regression model (1.2) is the method of least squares. Under the assumption that the errors $e_1, \cdots, e_n$ are independent $N(0, \sigma^2)$ random variables, the least squares estimator of β coincides with its maximum likelihood estimator. The least squares estimator $\hat{\beta} = (\hat{\beta}_0, \cdots, \hat{\beta}_p)'$ of $\beta = (\beta_0, \cdots, \beta_p)'$ is obtained by using the n observations in (1.2) and minimizing the least squares criterion

$$L(\beta) = \|\mathbf{Y} - \mathbf{X}\beta\|^2 = (\mathbf{Y} - \mathbf{X}\beta)'(\mathbf{Y} - \mathbf{X}\beta)$$

over $\beta \in \Re^{p+1}$. Since $\mathbf{Y}'\mathbf{X}\beta$ is a scalar, it is equal to its transpose $\beta'\mathbf{X}'\mathbf{Y}$. Using this fact, the expression $L(\beta)$ above can be expanded to

$$L(\beta) = \mathbf{Y}'\mathbf{Y} - 2\beta'\mathbf{X}'\mathbf{Y} + \beta'\mathbf{X}'\mathbf{X}\beta.$$

Thus, the least squares estimator must satisfy

$$\left.\frac{\partial L(\beta)}{\partial \beta}\right|_{\beta=\hat{\beta}} = -2\mathbf{X}'\mathbf{Y} + 2\mathbf{X}'\mathbf{X}\hat{\beta} = \mathbf{0}$$

which simplifies to the so-called least squares *normal equations*

$$\mathbf{X}'\mathbf{X}\hat{\beta} = \mathbf{X}'\mathbf{Y}.$$

Because $\mathbf{X}$ is assumed to have full column rank, $\mathbf{X}'\mathbf{X}$ is non-singular and the normal equations lead to the unique least squares estimator

$$\hat{\beta} = (\mathbf{X}'\mathbf{X})^{-1}\mathbf{X}'\mathbf{Y}. \tag{1.3}$$

The *fitted regression model* is

$$\hat{Y} = \hat{\beta}_0 + \hat{\beta}_1 x_1 + \cdots + \hat{\beta}_p x_p$$

and can be used to predict the value of Y at a given value of $(x_1,\cdots,x_p)$. In particular, the fitted Y-values corresponding to the n observations $(x_{j1},\cdots,x_{jp})$, $j=1,\cdots,n$, in each row of the design matrix $\mathbf{X}$, are

$$\hat{\mathbf{Y}} = \mathbf{X}\hat{\beta} = \mathbf{X}(\mathbf{X}'\mathbf{X})^{-1}\mathbf{X}'\mathbf{Y} = \mathbf{H}\mathbf{Y}$$

where $\mathbf{H} = \mathbf{X}(\mathbf{X}'\mathbf{X})^{-1}\mathbf{X}'$ is called the *hat matrix* because it transforms the vector of the observed responses $\mathbf{Y}$ into the vector of fitted responses $\hat{\mathbf{Y}}$. The vector of *residuals* is defined by

$$\hat{\mathbf{e}} = (\hat{e}_1,\cdots,\hat{e}_n)' = \mathbf{Y} - \mathbf{X}\hat{\beta} = \mathbf{Y} - \hat{\mathbf{Y}} = (\mathbf{I}-\mathbf{H})\mathbf{Y}.$$

The estimator $\hat{\sigma}^2$ of σ^2 is based on the residual sum of squares

$$SS_E = \sum_{j=1}^{n} \hat{e}_j^2 = \|\hat{\mathbf{e}}\|^2 = \|\mathbf{Y} - \mathbf{X}\hat{\beta}\|^2 = \|(\mathbf{I}-\mathbf{H})\mathbf{Y}\|^2 = L(\hat{\beta}).$$

Specifically, define

$$\hat{\sigma}^2 = SS_E/(n-p-1) = \|\hat{\mathbf{e}}\|^2/(n-p-1). \tag{1.4}$$

The total variation in the observed Y-values is $SS_T = \sum_{j=1}^n (Y_j - \bar{Y})^2$, where $\bar{Y} = \sum_{j=1}^n Y_j/n$. This total variation can be decomposed into two parts according to model (1.2): SS_E which is the variation due to the uncontrollable random errors e in the Y-values, and $SS_R = SS_T - SS_E$ which is the variation in the Y-values due to the systematic component $E(\mathbf{Y}) = \mathbf{X}\beta$ of model (1.2). It can be shown that $SS_R = \sum_{j=1}^n (\hat{Y}_j - \bar{Y})^2$. This leads to the popular exploratory index

$$R^2 = \frac{SS_T - SS_E}{SS_T}$$

that can be interpreted as the proportion of the total variation in the Y-values that is explained by (the systematic component of) the regression model, namely $E(Y) = \mathbf{X}\beta$. The quantity R^2 measures how well model (1.1) fits the observed data. If R^2 is close to one, then all the Y_j's are close to the corresponding fitted $\hat{Y}_j$'s and so model (1.1) fits the data well. On the other hand if R^2 is close to zero, then the Y_j's can be far away from the corresponding fitted $\hat{Y}_j$'s and so model (1.1) does not fit the data well and may not be useful for, e.g., making predictions. Of course, this may well be due to the large error variance σ^2 and so nothing can be done.

Example 1.2 For Example 1.1 it is easy to verify that

$$\mathbf{X}'\mathbf{X} = \begin{pmatrix} 17 & 1933 & 66 \\ 1933 & 225269 & 7466 \\ 66 & 7466 & 268 \end{pmatrix}$$

$$(\mathbf{X}'\mathbf{X})^{-1} = \begin{pmatrix} 4.3288 & -0.0236 & -0.4075 \\ -0.0236 & 0.0002 & 0.0006 \\ -0.4075 & 0.0006 & 0.0870 \end{pmatrix}$$

$$\hat{\beta} = (\mathbf{X}'\mathbf{X})^{-1}\mathbf{X}'\mathbf{Y} = \begin{pmatrix} 47.5828 \\ 0.1808 \\ 5.3129 \end{pmatrix}, \quad \hat{\mathbf{e}} = \mathbf{Y} - \mathbf{X}\hat{\beta} = \begin{pmatrix} -0.1185 \\ -0.0928 \\ -0.6352 \\ 0.0016 \\ -1.7788 \\ 0.9661 \\ 2.0879 \\ -1.4571 \\ 0.7482 \\ 1.4724 \\ -2.1660 \\ -1.5145 \\ 0.1709 \\ -0.6466 \\ 1.1726 \\ 0.9787 \\ 0.8110 \end{pmatrix}$$

$$SS_E = \hat{\mathbf{e}}'\hat{\mathbf{e}} = \|\mathbf{Y} - \mathbf{X}\hat{\beta}\|^2 = 24.1575, \quad \hat{\sigma}^2 = \hat{\mathbf{e}}'\hat{\mathbf{e}}/(17-2-1) = 1.7255.$$

For this example $R^2 = 94.76\%$ and so the linear regression model fits the observations well.

The least squares method can be represented geometrically in two ways; one is the *observation space* representation, and another is the *variable space* representation. In the observation space representation, *points* are the observations $(Y_j; x_{j1}, \cdots, x_{jp})$, $j = 1, \cdots, n$ and β is determined so that the sum of square of the "vertical distances" from Y_j to $\beta_0 + \beta_1 x_{j1} + \cdots + \beta_p x_{jp}$,

$$\sum_{j=1}^{n} (Y_j - \beta_0 - \beta_1 x_{j1} - \cdots - \beta_p x_{jp})^2,$$

is minimized. In the variable space representation, *points* are the variables in $\Re^n$, namely $\mathbf{Y}$, and columns of the design matrix $\mathbf{X} = (\mathbf{1}, \mathbf{x}_1, \cdots, \mathbf{x}_p)$. Let $\mathcal{L}(\mathbf{X})$ be the linear space spanned by the $p+1$ columns of the design matrix $\mathbf{X}$, i.e. $\mathcal{L}(\mathbf{X}) = \{\mathbf{X}\beta : \beta \in \Re^{p+1}\}$. Since $\mathbf{X}$ is full column rank, $\mathcal{L}(\mathbf{X})$ has dimension $p+1$. The least squares criterion to obtain $\hat{\beta}$ requires minimizing $L(\beta) = \|\mathbf{Y} - \mathbf{X}\beta\|^2$ over all $\beta \in \Re^{p+1}$. For each β, the quantity $\mathbf{X}\beta$ is in $\mathcal{L}(\mathbf{X})$, and $L(\beta)$ is the square of the distance between $\mathbf{Y}$ and $\mathbf{X}\beta$. Thus, minimizing over all possible β is equivalent to obtaining the point in $\mathcal{L}(\mathbf{X})$ that is closest to $\mathbf{Y}$. This point is the projection of $\mathbf{Y}$ onto $\mathcal{L}(\mathbf{X})$, which is given by $\hat{\mathbf{Y}} = \mathbf{X}\hat{\beta} = \mathbf{H}\mathbf{Y}$. Note that in this view the hat matrix $\mathbf{H}$ is the projection operator that projects a vector $\mathbf{Y} \in \Re^n$ to $\hat{\mathbf{Y}} \in \mathcal{L}(\mathbf{X})$.

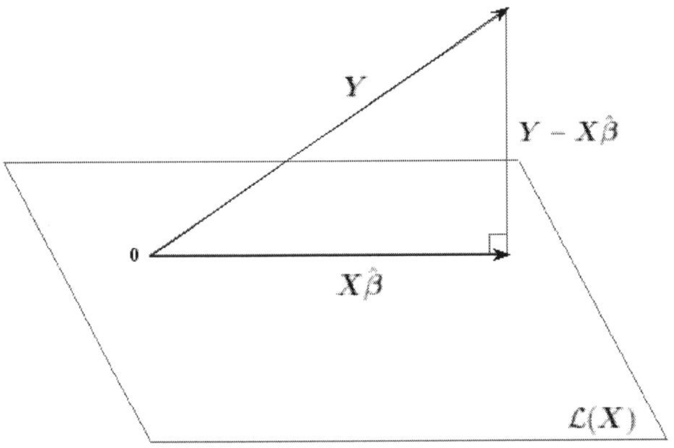

Figure 1.1: *Geometric representation of the least squares method*

This geometric representation of the least squares method is illustrated in Figure 1.1.

The SS_E is simply the square of the distance between $\mathbf{Y}$ and its projection onto the space $\mathcal{L}(\mathbf{X})$, $\mathbf{X}\hat{\beta}$. The independence between $\hat{\beta}$ and $\hat{\mathbf{e}}$ is due to the fact that the two vectors $\mathbf{X}\hat{\beta} = \mathbf{HY}$ and $\hat{\mathbf{e}} = \mathbf{Y} - \mathbf{X}\hat{\beta} = (\mathbf{I} - \mathbf{H})\mathbf{Y}$ are perpendicular to each other; note that $\mathbf{H}(\mathbf{I} - \mathbf{H}) = 0$ since $\mathbf{H}^2 = \mathbf{H}$, i.e., $\mathbf{H}$ is idempotent. The following theorem gives a rigorous proof of this statement and other statistical properties of $\hat{\beta}$, $\hat{\mathbf{e}}$ and $\hat{\sigma}^2$.

Theorem 1.1 Under the standard normality assumptions made in Section 1.1, the following properties hold:

(i) $\hat{\beta} \sim \mathcal{N}_{p+1}(\beta, \sigma^2(\mathbf{X}'\mathbf{X})^{-1})$.

(ii) $\hat{\mathbf{e}} \sim \mathcal{N}_n(\mathbf{0}, \sigma^2(\mathbf{I} - \mathbf{H}))$.

(iii) $\hat{\sigma}^2 \sim \frac{\sigma^2}{n-p-1}\chi^2_{n-p-1}$.

(iv) $\hat{\beta}$ and $\hat{\mathbf{e}}$ are independent.

(v) $\hat{\beta}$ and $\hat{\sigma}^2$ are independent.

Proof 1.1 (i) *Using model (1.2) and the assumption that* $\mathbf{e} \sim \mathcal{N}_n(0, \sigma^2\mathbf{I})$, *it is clear that* $\mathbf{Y} \sim \mathcal{N}_n(\mathbf{X}\beta, \sigma^2\mathbf{I})$. *Since the elements of* $\hat{\beta}$ *are linear combinations of the elements of* $\mathbf{Y}$ *from expression (1.3),* $\hat{\beta}$ *has a normal distribution too. To conclude the proof of* (i), *note that*

$$E(\hat{\beta}) = E\left((\mathbf{X}'\mathbf{X})^{-1}\mathbf{X}'\mathbf{Y}\right) = (\mathbf{X}'\mathbf{X})^{-1}\mathbf{X}'E(\mathbf{Y}) = (\mathbf{X}'\mathbf{X})^{-1}\mathbf{X}'\mathbf{X}\beta = \beta,$$

and

$$\begin{aligned} Cov(\hat{\beta}) &= Cov\left((\mathbf{X}'\mathbf{X})^{-1}\mathbf{X}'\mathbf{Y}\right) \\ &= \left((\mathbf{X}'\mathbf{X})^{-1}\mathbf{X}'\right) Cov(\mathbf{Y}) \left(\mathbf{X}(\mathbf{X}'\mathbf{X})^{-1}\right) \\ &= \sigma^2(\mathbf{X}'\mathbf{X})^{-1}, \quad \text{since } Cov(\mathbf{Y}) = Cov(\mathbf{Y},\mathbf{Y}) = \sigma^2\mathbf{I}. \end{aligned}$$

(ii) *Since* $\hat{\mathbf{e}} = (\mathbf{I}-\mathbf{H})\mathbf{Y}$, *each* $\hat{e}_i$ *is a linear combination of the elements of* $\mathbf{Y}$. *And again, since* $\mathbf{Y}$ *is a normal random vector,* $\hat{\mathbf{e}}$ *is also a normal random vector. To conclude the proof of* (ii), *note that*

$$E(\hat{\mathbf{e}}) = E((\mathbf{I}-\mathbf{H})\mathbf{Y}) = (\mathbf{I}-\mathbf{H})E(\mathbf{Y}) = (\mathbf{I}-\mathbf{H})\mathbf{X}\beta = 0$$

since $(\mathbf{I}-\mathbf{H})\mathbf{X} = \mathbf{0}$, *and*

$$Cov(\hat{\mathbf{e}}) = Cov\left((\mathbf{I}-\mathbf{H})\mathbf{Y}\right) = (\mathbf{I}-\mathbf{H})Cov(\mathbf{Y})(\mathbf{I}-\mathbf{H}) = \sigma^2(\mathbf{I}-\mathbf{H})$$

since the matrix $(\mathbf{I}-\mathbf{H})$ *is symmetric and idempotent, i.e.,* $(\mathbf{I}-\mathbf{H})^2 = (\mathbf{I}-\mathbf{H})$.
(iii) *Let* $\mathbf{Q} = \mathbf{I}-\mathbf{H}$. *Then* $\mathbf{Q}$ *is symmetric and idempotent, and so*

$$\begin{aligned} Rank(\mathbf{Q}) &= Trace(\mathbf{Q}) = Trace(\mathbf{I}) - Trace(\mathbf{H}) = n - Trace(\mathbf{X}(\mathbf{X}'\mathbf{X})^{-1}\mathbf{X}') \\ &= n - Trace((\mathbf{X}'\mathbf{X})^{-1}\mathbf{X}'\mathbf{X}) = n-(p+1) \end{aligned}$$

since $Trace(AB) = Trace(BA)$. *It follows therefore that* $\mathbf{Q}$ *can be expressed as* $\mathbf{Q} = \mathbf{T}'\Lambda\mathbf{T}$ *where* $\mathbf{T}$ *is an orthogonal matrix, and* Λ *is a diagonal matrix with the first* $n-(p+1)$ *diagonal elements equal to one and the remaining diagonal elements equal to zero.*

Note $\mathbf{e} = \mathbf{Y} - \mathbf{X}\beta \sim \mathcal{N}_n(0, \sigma^2\mathbf{I})$. *Let* $\mathbf{z} = \mathbf{Te}$ *then* $\mathbf{z} \sim \mathcal{N}_n(0, \sigma^2\mathbf{I})$ *since* $\mathbf{T}$ *is orthogonal. Also note* $\mathbf{Qe} = (\mathbf{I}-\mathbf{H})(\mathbf{Y}-\mathbf{X}\beta) = (\mathbf{I}-\mathbf{H})\mathbf{Y} = \hat{\mathbf{e}}$ *since* $(\mathbf{I}-\mathbf{H})\mathbf{X} = \mathbf{0}$. *So*

$$\| \hat{\mathbf{e}} \|^2 = \mathbf{e}'\mathbf{Qe} = (\mathbf{Te})'\Lambda(\mathbf{Te}) = z_1^2 + \cdots + z_{n-p-1}^2 \sim \sigma^2\chi^2_{n-p-1}.$$

The proof of (iii) *is thus complete by using (1.4) which defines* $\hat{\sigma}^2 = \|\hat{\mathbf{e}}\|^2/(n-p-1)$.
(iv) *Since* $\hat{\beta}$ *and* $\hat{\mathbf{e}}$ *are normally distributed, to show they are independent, it suffices to show that they have zero covariance. This is easily shown as follows:*

$$\begin{aligned} Cov(\hat{\beta}, \hat{\mathbf{e}}) &= Cov((\mathbf{X}'\mathbf{X})^{-1}\mathbf{X}'\mathbf{Y}, (\mathbf{I}-\mathbf{H})\mathbf{Y}) \\ &= (\mathbf{X}'\mathbf{X})^{-1}\mathbf{X}'Cov(\mathbf{Y},\mathbf{Y})(\mathbf{I}-\mathbf{H}) \\ &= \sigma^2(\mathbf{X}'\mathbf{X})^{-1}\mathbf{X}'(\mathbf{I}-\mathbf{H}) = 0 \end{aligned}$$

since $Cov(\mathbf{Y},\mathbf{Y}) = \sigma^2\mathbf{I}$ *and* $(\mathbf{I}-\mathbf{H})\mathbf{X} = \mathbf{0}$.
(v) *The independence of* $\hat{\boldsymbol{\beta}}$ *and* $\hat{\sigma}^2$ *follows directly from the result of* (iv), *since* $\hat{\sigma}^2 = \|\hat{\mathbf{e}}\|^2/(n-p-1)$. ∎

The residuals $\hat{\mathbf{e}}$ play a very important role in assessing whether (1.2) is a reasonable model that is underlying the observed data. If model (1.2) with the standard normality assumptions hold, then according to Theorem 1.1, $\hat{\mathbf{e}}$ should behave like a random vector from $\mathcal{N}_n(\mathbf{0}, \sigma^2(\mathbf{I}-\mathbf{H}))$. This is usually assessed by various residual plots (see e.g., Draper and Smith, 1998, or Weisberg, 2005). Residual analysis is an integral part of regression analysis and should always be carried out to ensure that the model is a reasonable one before other inferential methods, such as hypotheses tests and simultaneous confidence bands, are deployed since these methods hinge on the validity of the model.

Using the results of Theorem 1.1, one can construct a confidence region for the unknown coefficients $\boldsymbol{\beta}$ which quantifies the plausible values of $\boldsymbol{\beta}$.

Theorem 1.2 An exact $1-\alpha$ confidence region for $\boldsymbol{\beta}$ is given by

$$\left\{ \boldsymbol{\beta} : \frac{(\hat{\boldsymbol{\beta}}-\boldsymbol{\beta})'(\mathbf{X}'\mathbf{X})(\hat{\boldsymbol{\beta}}-\boldsymbol{\beta})}{(p+1)\hat{\sigma}^2} \le f^{\alpha}_{p+1,n-p-1} \right\}$$

where $f^{\alpha}_{p+1,n-p-1}$ is the upper α point of the F distribution with degrees of freedom $p+1$ and $n-p-1$.

Proof 1.2 *Let square matrix* $\mathbf{P}$ *satisfy* $(\mathbf{X}'\mathbf{X})^{-1} = \mathbf{P}'\mathbf{P}$, *and define* $\mathbf{N} = (\mathbf{P}')^{-1}(\hat{\boldsymbol{\beta}}-\boldsymbol{\beta})/\sigma$. *Then*

$$E(\mathbf{N}) = \mathbf{0},\ \ Cov(\mathbf{N}) = (\mathbf{P}')^{-1}Cov(\hat{\boldsymbol{\beta}}-\boldsymbol{\beta})\mathbf{P}^{-1}/\sigma^2 = (\mathbf{P}')^{-1}(\mathbf{X}'\mathbf{X})^{-1}\mathbf{P}^{-1} = \mathbf{I}_{p+1},$$

and so $\mathbf{N} \sim \mathcal{N}_{p+1}(\mathbf{0},\mathbf{I})$ *and*

$$\mathbf{N}'\mathbf{N} = (\hat{\boldsymbol{\beta}}-\boldsymbol{\beta})'\mathbf{P}^{-1}(\mathbf{P}')^{-1}(\hat{\boldsymbol{\beta}}-\boldsymbol{\beta})/\sigma^2 = (\hat{\boldsymbol{\beta}}-\boldsymbol{\beta})'(\mathbf{X}'\mathbf{X})(\hat{\boldsymbol{\beta}}-\boldsymbol{\beta})/\sigma^2$$

has the chi-square distribution with $p+1$ *degrees of freedom. Also note that* $\mathbf{N}'\mathbf{N}$ *is independent of* $\hat{\sigma}^2$. *So*

$$\frac{\mathbf{N}'\mathbf{N}}{(p+1)\hat{\sigma}^2/\sigma^2} = \frac{(\hat{\boldsymbol{\beta}}-\boldsymbol{\beta})'(\mathbf{X}'\mathbf{X})(\hat{\boldsymbol{\beta}}-\boldsymbol{\beta})}{(p+1)\hat{\sigma}^2}$$

has the F distribution with degrees of freedom $p+1$ *and* $n-p-1$, *from which the theorem follows immediately.* ∎

1.3 Testing hypotheses

For regression model (1.1), it is often required to assess whether the regression coefficients β satisfy certain constraints $\mathbf{A}\beta = \mathbf{b}$, where $\mathbf{A}$ is a given $r \times (p+1)$ matrix having full row rank r, where $1 \leq r \leq p+1$ is a given number and $\mathbf{b}$ is a given vector in $\Re^r$. One frequently used approach is to test

$$H_0 : \mathbf{A}\beta = \mathbf{b} \quad \text{against} \quad H_a : \mathbf{A}\beta \neq \mathbf{b}. \tag{1.5}$$

For instance, if one wants to assess whether the lth predictor variable x_l is useful in explaining the variability in the Y value then one wants to know whether $\beta_l = 0$. In this case, $\mathbf{A}$ is set as a $1 \times (p+1)$ matrix with the $(1, l+1)$ element equal to one and all other elements equal to zero, and $\mathbf{b} = 0$. For another example, if one wants to assess whether any of the explanatory variables in model (1.1) is of use in explaining the variability in Y then the question becomes whether $\beta_1 = \cdots = \beta_p = 0$. In this example, $\mathbf{A}$ can be chosen as the $p \times (p+1)$ matrix having the $(i, i+1)$ element equal to one for $i = 1, \cdots, p$ and all the other elements equal to zero, and $\mathbf{b} = \mathbf{0}$.

Before giving the test for (1.5), we consider the least squares estimator $\hat{\beta}_{\mathbf{A}}$ of β for model (1.1) under the constraints $\mathbf{A}\beta = \mathbf{b}$, that is, $\hat{\beta}_{\mathbf{A}}$ minimizes

$$L(\beta) = (\mathbf{Y} - \mathbf{X}\beta)'(\mathbf{Y} - \mathbf{X}\beta) = \|\mathbf{Y} - \mathbf{X}\beta\|^2$$

over all those $\beta \in \Re^{p+1}$ that satisfy the constraints $\mathbf{A}\beta = \mathbf{b}$. Now $\hat{\beta}_{\mathbf{A}}$ can be found using the method of Lagrange multipliers. Define

$$Q(\beta, \mathbf{c}) = (\mathbf{Y} - \mathbf{X}\beta)'(\mathbf{Y} - \mathbf{X}\beta) - \mathbf{c}'(\mathbf{A}\beta - \mathbf{b})$$

then $\hat{\beta}_{\mathbf{A}}$ must satisfy simultaneously

$$\left.\frac{\partial Q(\beta, \mathbf{c})}{\partial \beta}\right|_{\beta = \hat{\beta}_{\mathbf{A}}} = \mathbf{0}, \quad \left.\frac{\partial Q(\beta, \mathbf{c})}{\partial \mathbf{c}}\right|_{\beta = \hat{\beta}_{\mathbf{A}}} = \mathbf{0}.$$

These two equations simplify to

$$\mathbf{X}'\mathbf{X}\hat{\beta}_{\mathbf{A}} - \mathbf{A}'\mathbf{c} = \mathbf{X}'\mathbf{Y}, \quad \mathbf{A}\hat{\beta}_{\mathbf{A}} - \mathbf{b} = \mathbf{0} \tag{1.6}$$

from which $\hat{\beta}_{\mathbf{A}}$ and $\mathbf{c}$ will be solved. The explicit values are given in expressions (1.13) and (1.14) below.

Again $\hat{\beta}_{\mathbf{A}}$ can be interpreted geometrically. Note that

$$\mathcal{L}_{\mathbf{A}}(\mathbf{X}) = \{\mathbf{X}\beta : \beta \in \Re^{p+1}, \mathbf{A}\beta = \mathbf{b}\}$$

is a linear subspace of $\mathcal{L}(\mathbf{X})$. Since $\mathcal{L}_{\mathbf{A}}(\mathbf{X})$ is formed from $\mathcal{L}(\mathbf{X})$ by imposing r constraints $\mathbf{A}\beta = \mathbf{b}$, $\mathcal{L}_{\mathbf{A}}(\mathbf{X})$ is of $p+1-r$ dimensions. The estimator $\hat{\beta}_{\mathbf{A}}$ is simply the vector such that $\mathbf{X}\hat{\beta}_{\mathbf{A}}$ is the projection of $\mathbf{Y}$ on the space $\mathcal{L}_{\mathbf{A}}(\mathbf{X})$ since the smallest distance between $\mathbf{Y}$ and all the vectors in $\mathcal{L}_{\mathbf{A}}(\mathbf{X})$ is attained at the projection of $\mathbf{Y}$ on $\mathcal{L}_{\mathbf{A}}(\mathbf{X})$. This geometric interpretation of $\hat{\beta}_{\mathbf{A}}$ is illustrated in Figure 1.2. From this geometric interpretation it is clear that $\hat{\beta}_{\mathbf{A}}$ must exist.

One consequence of this geometric interpretation is the following.

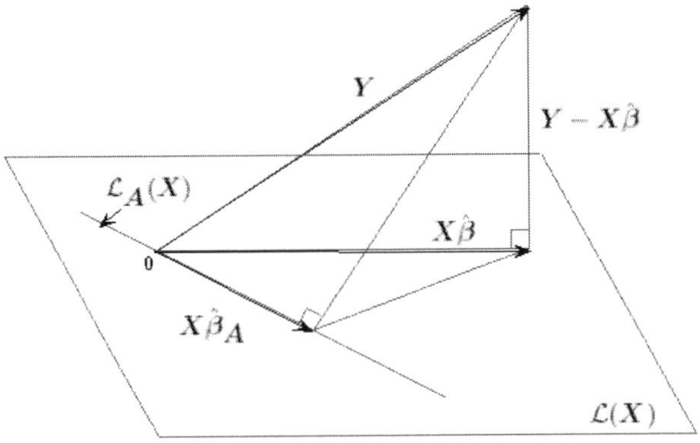

Figure 1.2: *Geometric representation of the estimator* $\hat{\beta}_{\mathbf{A}}$

Theorem 1.3 Let $\hat{\mathbf{Y}} = \mathbf{X}\hat{\beta}$ be the projection of $\mathbf{Y}$ on $\mathcal{L}(\mathbf{X})$ and, analogously, let $\hat{\mathbf{Y}}_{\mathbf{A}} = \mathbf{X}\hat{\beta}_{\mathbf{A}}$ denote the projection of $\mathbf{Y}$ on $\mathcal{L}_{\mathbf{A}}(\mathbf{X})$. Then

$$||\mathbf{Y} - \hat{\mathbf{Y}}_{\mathbf{A}}||^2 - ||\mathbf{Y} - \hat{\mathbf{Y}}||^2 = ||\hat{\mathbf{Y}}_{\mathbf{A}} - \hat{\mathbf{Y}}||^2.$$

Proof 1.3 *We can write*

$$\mathbf{Y} - \hat{\mathbf{Y}}_{\mathbf{A}} = (\hat{\mathbf{Y}} - \hat{\mathbf{Y}}_{\mathbf{A}}) + (\mathbf{Y} - \hat{\mathbf{Y}}).$$

Note that $\hat{\mathbf{Y}} - \hat{\mathbf{Y}}_{\mathbf{A}} \in \mathcal{L}(\mathbf{X})$ *and* $(\mathbf{Y} - \hat{\mathbf{Y}})$ *is orthogonal to* $\mathcal{L}(\mathbf{X})$. *Therefore,* $\mathbf{Y} - \hat{\mathbf{Y}}$ *is orthogonal to* $\hat{\mathbf{Y}} - \hat{\mathbf{Y}}_{\mathbf{A}}$ *which implies*

$$||\mathbf{Y} - \hat{\mathbf{Y}}_{\mathbf{A}}||^2 = ||(\hat{\mathbf{Y}} - \hat{\mathbf{Y}}_{\mathbf{A}})||^2 + ||(\mathbf{Y} - \hat{\mathbf{Y}})||^2$$

and, in turn, the result of the theorem. ▮

Now if the coefficients β of model (1.1) do satisfy the constraints $\mathbf{A}\beta = \mathbf{b}$, i.e., the $\mathbf{Y}$ are generated from model (1.1) with the β satisfying $\mathbf{A}\beta = \mathbf{b}$, then $\hat{\mathbf{Y}}$ should be close to $\hat{\mathbf{Y}}_{\mathbf{A}}$ and, using Theorem 1.3, the distance $||\mathbf{Y} - \mathbf{X}\hat{\beta}_{\mathbf{A}}||$ should not be significantly larger than the distance $||\mathbf{Y} - \mathbf{X}\hat{\beta}||$. On the other hand, if β does not satisfy $\mathbf{A}\beta = \mathbf{b}$ then the distance $||\mathbf{Y} - \mathbf{X}\hat{\beta}_{\mathbf{A}}||$ tends to be substantially larger than the distance $\|\mathbf{Y}-\mathbf{X}\hat{\beta}\|$; see Figure 1.2. This intuition is behind the following size α test of the hypotheses in (1.5):

$$\text{Reject } H_0 \text{ if and only if } \frac{\left\|\mathbf{X}\hat{\beta}_{\mathbf{A}}-\mathbf{X}\hat{\beta}\right\|^2/r}{\|\mathbf{Y}-\mathbf{X}\hat{\beta}\|^2/(n-p-1)} > f^{\alpha}_{r,n-p-1} \qquad (1.7)$$

where $f^{\alpha}_{r,n-p-1}$ is the upper α point of the F distribution with degrees of freedom r and $n-p-1$. It can be shown that this test is actually the likelihood ratio test for the hypotheses in (1.5). The relevant distributional results are given by the following Theorem, whose proof is adapted from Hsu (1941).

Theorem 1.4 Under the standard normality assumptions made in Section 1.1, we have

(i) $\|\mathbf{X}\hat{\beta}-\mathbf{X}\hat{\beta}_{\mathbf{A}}\|^2 \sim \sigma^2\chi^2_r(\delta)$ with the non-centrality parameter

$$\delta = \|\mathbf{X}\beta-\mathbf{X}E(\hat{\beta}_{\mathbf{A}})\|^2/\sigma^2 = (\beta-E(\hat{\beta}_{\mathbf{A}}))'\mathbf{X}'\mathbf{X}(\beta-E(\hat{\beta}_{\mathbf{A}}))/\sigma^2.$$

(ii) $\|\mathbf{Y}-\mathbf{X}\hat{\beta}_{\mathbf{A}}\|^2 \sim \sigma^2\chi^2_{n-(p+1)+r}(\delta)$, with δ defined in (i).

(iii) $\|\mathbf{X}\hat{\beta}-\mathbf{X}\hat{\beta}_{\mathbf{A}}\|^2$ and $\|\mathbf{Y}-\mathbf{X}\hat{\beta}\|^2$ are independent.

Proof 1.4 (i) *Let* $\mathbf{t}_{r+1},\cdots,\mathbf{t}_{p+1}$ *be an orthonormal basis of the* $p+1-r$ *dimensional linear space* $\mathcal{L}_{\mathbf{A}}(\mathbf{X})$. *Since* $\mathcal{L}_{\mathbf{A}}(\mathbf{X})$ *is a linear subspace of the* $p+1$ *dimensional linear space* $\mathcal{L}(\mathbf{X})$, $\mathbf{t}_{r+1},\cdots,\mathbf{t}_{p+1}$ *can be expanded to* $\mathbf{t}_1,\cdots,\mathbf{t}_{p+1}$ *to form an orthonormal basis of* $\mathcal{L}(\mathbf{X})$. *Similarly* $\mathbf{t}_1,\cdots,\mathbf{t}_{p+1}$ *can further be expanded to* $\mathbf{t}_1,\cdots,\mathbf{t}_n$ *to form an orthonormal basis of* $\Re^n$. *The following scheme shows visually the vectors that span each of the given spaces:*

$$\underbrace{\underbrace{\mathbf{t}_1,\cdots,\mathbf{t}_r,\underbrace{\mathbf{t}_{r+1},\cdots,\mathbf{t}_{p+1}}_{\mathcal{L}_A(X)}}_{\mathcal{L}(X)},\mathbf{t}_{p+2},\cdots,\mathbf{t}_n}_{\Re^n}.$$

Define the orthogonal matrix $\mathbf{T}=(\mathbf{T}_1,\mathbf{T}_2,\mathbf{T}_3)$, *where* $\mathbf{T}_1=(\mathbf{t}_1,\cdots,\mathbf{t}_r)$, $\mathbf{T}_2=(\mathbf{t}_{r+1},\cdots,\mathbf{t}_{p+1})$, *and* $\mathbf{T}_3=(\mathbf{t}_{p+2},\cdots,\mathbf{t}_n)$. *Accordingly define* $\mathbf{z}=\mathbf{T}'\mathbf{Y}$ *with* $\mathbf{z}_i=\mathbf{T}_i'\mathbf{Y}$, $i=1,2,3$.

As before, let $\hat{\mathbf{Y}}=\mathbf{X}\hat{\beta}$ *and* $\hat{\mathbf{Y}}_{\mathbf{A}}=\mathbf{X}\hat{\beta}_{\mathbf{A}}$. *Since* $\mathbf{Y}=\mathbf{T}\mathbf{z}=\mathbf{T}_1\mathbf{z}_1+\mathbf{T}_2\mathbf{z}_2+\mathbf{T}_3\mathbf{z}_3$ *and the columns of* $(\mathbf{T}_1,\mathbf{T}_2)$ *are an orthonormal basis for* $\mathcal{L}(\mathbf{X})$ *we have*

$$\hat{\mathbf{Y}}=(\mathbf{T}_1,\mathbf{T}_2)\begin{pmatrix}\mathbf{z}_1\\ \mathbf{z}_2\end{pmatrix}=\mathbf{T}_1\mathbf{z}_1+\mathbf{T}_2\mathbf{z}_2.$$

Similarly, because the columns of $\mathbf{T}_2$ *form an orthonormal basis for* $\mathcal{L}_{\mathbf{A}}(\mathbf{X})$ *we have* $\hat{\mathbf{Y}}_A=\mathbf{T}_2\mathbf{z}_2$. *This and the expression of* $\hat{\mathbf{Y}}$ *above give*

$$\hat{\mathbf{Y}}-\hat{\mathbf{Y}}_{\mathbf{A}}=\mathbf{T}_1\mathbf{z}_1. \qquad (1.8)$$

Furthermore, using (1.8) and the fact that $\mathbf{T}_1$ *has orthonormal columns, we have*

$$\|\hat{\mathbf{Y}} - \hat{\mathbf{Y}}_\mathbf{A}\|^2 = \|\mathbf{T}_1\mathbf{z}_1\|^2 = \|\mathbf{z}_1\|^2. \tag{1.9}$$

Now, by the standard normality assumptions $\mathbf{Y} \sim \mathcal{N}_n(\mathbf{X}\beta, \sigma^2\mathbf{I})$ *and thus* $\mathbf{z}_1 = \mathbf{T}_1'\mathbf{Y}$ *has a normal distribution with covariance matrix* $\sigma^2 I_r$. *This together with (1.9) and the definition of the non-central chi-squared distribution lead to*

$$\frac{\|\mathbf{X}\hat{\beta} - \mathbf{X}\hat{\beta}_\mathbf{A}\|^2}{\sigma^2} = \frac{\|\hat{\mathbf{Y}} - \hat{\mathbf{Y}}_\mathbf{A}\|^2}{\sigma^2} = \frac{\|\mathbf{z}_1\|^2}{\sigma^2} \sim \chi_r^2(\delta),$$

where δ *is the non-centrality parameter. Now that we have established* $\|\mathbf{X}\hat{\beta} - \mathbf{X}\hat{\beta}_\mathbf{A}\|^2/\sigma^2$ *has a* χ^2 *distribution, its non-centrality parameter* δ *can be determined (cf. Plackett, 1960) by replacing* $\hat{\beta}$ *and* $\hat{\beta}_\mathbf{A}$ *with their respective expectations, that is*

$$\delta = \|\mathbf{X}E(\hat{\beta}) - \mathbf{X}E(\hat{\beta}_\mathbf{A})\|^2/\sigma^2 = (\beta - E(\hat{\beta}_\mathbf{A}))'\mathbf{X}'\mathbf{X}(\beta - E(\hat{\beta}_\mathbf{A}))/\sigma^2.$$

This completes the proof of (i).
(ii) *To prove* (ii), *again using the notation* $\hat{\mathbf{Y}}_\mathbf{A} = \mathbf{X}\hat{\beta}_\mathbf{A}$ *we have*

$$\begin{aligned}
\|\mathbf{Y} - \hat{\mathbf{Y}}_\mathbf{A}\|^2 &= \|\mathbf{T}'(\mathbf{Y} - \hat{\mathbf{Y}}_\mathbf{A})\|^2 \\
&= \left\| \begin{pmatrix} \mathbf{T}_1' \\ \mathbf{T}_2' \\ \mathbf{T}_3' \end{pmatrix} (\mathbf{Y} - \hat{\mathbf{Y}}_\mathbf{A}) \right\|^2 \\
&= \left\| \begin{pmatrix} \mathbf{T}_1' \\ \mathbf{T}_3' \end{pmatrix} (\mathbf{Y} - \hat{\mathbf{Y}}_\mathbf{A}) \right\|^2 \\
&= \left\| \begin{pmatrix} \mathbf{T}_1' \\ \mathbf{T}_3' \end{pmatrix} \mathbf{Y} \right\| = \left\| \begin{pmatrix} \mathbf{z}_1 \\ \mathbf{z}_3 \end{pmatrix} \right\|^2.
\end{aligned} \tag{1.10}$$

The first equality holds since $\mathbf{T}'$ *is an orthogonal matrix, the third equality holds since* $\mathbf{Y} - \hat{\mathbf{Y}}_\mathbf{A}$ *is orthogonal to* $\mathcal{L}_A(\mathbf{X})$ *and thus to the columns of* $\mathbf{T}_2$, *and the fourth equality holds since columns of* $(\mathbf{T}_1, \mathbf{T}_3)$ *span the linear space perpendicular to* $\mathcal{L}_\mathbf{A}(\mathbf{X})$, $\mathcal{L}_\mathbf{A}(\mathbf{X})^\perp$, *and so* $\hat{\mathbf{Y}}_\mathbf{A}$ *is orthogonal to columns of* $(\mathbf{T}_1, \mathbf{T}_3)$.

Now using similar arguments as in the proof of assertion (i), *we have* $\mathbf{z}_1 \sim \mathcal{N}_r(T_1'\mathbf{X}\beta, \sigma^2\mathbf{I})$ *and* $\mathbf{z}_3 \sim \mathcal{N}_{n-p-1}(T_3'\mathbf{X}\beta, \sigma^2\mathbf{I})$. *Also, it is clear that* $\mathbf{z}_1$ *and* $\mathbf{z}_3$ *are independent since the columns of* $\mathbf{T}_1$ *are orthogonal to the columns of* $\mathbf{T}_3$. *Using these facts, (1.10), and the definition of the non-central chi-square we have*

$$\frac{\|\mathbf{Y} - \mathbf{X}\hat{\beta}_\mathbf{A}\|^2}{\sigma^2} = \frac{\|\mathbf{Y} - \hat{\mathbf{Y}}_\mathbf{A}\|^2}{\sigma^2} = \left\| \begin{pmatrix} \mathbf{z}_1 \\ \mathbf{z}_3 \end{pmatrix} \right\|^2 / \sigma^2 \sim \chi^2_{n+r-p-1}(\delta),$$

where again δ *is the non-centrality parameter and can be obtained by replacing* $\mathbf{Y}$ *and* $\hat{\beta}_\mathbf{A}$ *in the above expression by their expectations. This completes the proof of* (ii).

(iii) *We have*

$$||\mathbf{Y}-\hat{\mathbf{Y}}||^2 = ||\mathbf{Y}-\hat{\mathbf{Y}}_\mathbf{A}||^2 - ||\hat{\mathbf{Y}}-\hat{\mathbf{Y}}_\mathbf{A}||^2 = (||\mathbf{z}_1||^2+||\mathbf{z}_3||^2) - ||\mathbf{z}_1||^2 = ||\mathbf{z}_3||^2,$$

where the first equality follows from Theorem 1.3, and the second equality follows from (1.9) and (1.10). Since $\mathbf{z}_1$ *and* $\mathbf{z}_3$ *are independent, the above equality and (1.9) complete the proof of assertion* (iii). ∎

Note that Theorem 1.4 holds whether or not β satisfies the constraints $\mathbf{A}\beta = \mathbf{b}$. If the null hypothesis $\mathbf{A}\beta = \mathbf{b}$ is true, then one can show that $E(\mathbf{z}_1) = 0$ and so $||\mathbf{X}\hat{\beta}-\mathbf{X}\hat{\beta}_\mathbf{A}||^2/\sigma^2 = ||\mathbf{z}_1||^2/\sigma^2$ has a central χ^2_r distribution. Therefore, under the null hypothesis, the test statistic in (1.7) has a central $F_{r,n-p-1}$ distribution, which is the reason for using the critical value $f^\alpha_{r,n-p-1}$.

A test of the hypotheses in (1.5) can also be constructed in the following way. Note that if the null hypothesis $H_0 : \mathbf{A}\beta = \mathbf{b}$ is true then $\mathbf{A}\hat{\beta} - \mathbf{b}$ should not be excessively different from the zero vector since

$$\mathbf{A}\hat{\beta}-\mathbf{b} \sim \mathcal{N}(\mathbf{A}\beta-\mathbf{b},\ \sigma^2\mathbf{A}(\mathbf{X}'\mathbf{X})^{-1}\mathbf{A}').$$

Hence we

reject H_0 if and only if

$$\frac{(\mathbf{A}\hat{\beta}-\mathbf{b})'(\mathbf{A}(\mathbf{X}'\mathbf{X})^{-1}\mathbf{A}')^{-1}(\mathbf{A}\hat{\beta}-\mathbf{b})/r}{||\mathbf{Y}-\mathbf{X}\hat{\beta}||^2/(n-p-1)} > f^\alpha_{r,n-p-1}. \tag{1.11}$$

The next theorem shows that test (1.11) is actually the same as test (1.7).

Theorem 1.5 We have

$$||\mathbf{X}\hat{\beta}-\mathbf{X}\hat{\beta}_\mathbf{A}||^2 = (\mathbf{A}\hat{\beta}-\mathbf{b})'(\mathbf{A}(\mathbf{X}'\mathbf{X})^{-1}\mathbf{A}')^{-1}(\mathbf{A}\hat{\beta}-\mathbf{b}).$$

Proof 1.5 *From the least squares normal equation and the first equation of (1.6) we have*

$$\begin{aligned} ||\mathbf{X}\hat{\beta}-\mathbf{X}\hat{\beta}_\mathbf{A}||^2 &= (\hat{\beta}_\mathbf{A}-\hat{\beta})'(\mathbf{X}'\mathbf{X})(\hat{\beta}_\mathbf{A}-\hat{\beta}) \\ &= (\hat{\beta}_\mathbf{A}-\hat{\beta})'\mathbf{A}'\mathbf{c} \\ &= (\mathbf{b}-\mathbf{A}\hat{\beta})'\mathbf{c} \end{aligned} \tag{1.12}$$

where the last equality follows from the second equation of (1.6). Next we solve $\mathbf{c}$ *from (1.6). From the first equation of (1.6) one gets*

$$\hat{\beta}_\mathbf{A} = (\mathbf{X}'\mathbf{X})^{-1}(\mathbf{X}'\mathbf{Y}+\mathbf{A}'\mathbf{c}) = \hat{\beta}+(\mathbf{X}'\mathbf{X})^{-1}\mathbf{A}'\mathbf{c}. \tag{1.13}$$

Substituting this into the second equation of (1.6), one solves

$$\mathbf{c} = (\mathbf{A}(\mathbf{X}'\mathbf{X})^{-1}\mathbf{A}')^{-1}(\mathbf{b}-\mathbf{A}\hat{\beta}). \tag{1.14}$$

Substituting this into (1.12) gives the result of the theorem. ∎

This proof provides an explicit formula of the constrained least squares estimator $\hat{\beta}_{\mathbf{A}}$ in (1.13) with $\mathbf{c}$ given in (1.14). Tests (1.7) or (1.11) are often called the (partial) F-test for (1.5). From this theorem, we have an alternative expression for the non-central parameter δ:

$$\delta = (\mathbf{A}\beta - \mathbf{b})'(\mathbf{A}(\mathbf{X}'\mathbf{X})^{-1}\mathbf{A}')^{-1}(\mathbf{A}\beta - \mathbf{b})/\sigma^2.$$

An exact $1-\alpha$ confidence region for $\mathbf{A}\beta$ is given by

$$\left\{ \mathbf{b} : \frac{(\mathbf{A}\hat{\beta} - \mathbf{b})'(\mathbf{A}(\mathbf{X}'\mathbf{X})^{-1}\mathbf{A}')^{-1}(\mathbf{A}\hat{\beta} - \mathbf{b})/r}{\|\mathbf{Y} - \mathbf{X}\hat{\beta}\|^2/(n-p-1)} \le f^{\alpha}_{r,n-p-1} \right\}.$$

Any $\mathbf{b}$ that belongs to this region is a plausible value of $\mathbf{A}\beta$. In particular, test (1.11) is implied by this confidence region: the null hypothesis $H_0 : \mathbf{A}\beta = \mathbf{b}$ is rejected if and only if $\mathbf{b}$ is judged by this confidence region not to be a plausible value, that is, $\mathbf{b}$ is not in this confidence region.

Example 1.3 For Example 1.1, if one wants to test

$$H_0 : \beta_2 = 0 \;\text{ against }\; H_a : \beta_2 \neq 0 \tag{1.15}$$

then $\mathbf{A} = (0,0,1)$ and $\mathbf{b} = (0)$. If one wants to test

$$H_0 : \beta_1 = \beta_2 = 0 \;\text{ against }\; H_a : \text{not } H_0$$

then

$$\mathbf{A} = \begin{pmatrix} 0 & 1 & 0 \\ 0 & 0 & 1 \end{pmatrix} \quad \mathbf{b} = \begin{pmatrix} 0 \\ 0 \end{pmatrix}.$$

For the hypotheses in (1.15), if the null hypothesis H_0 is true then the original model which has dependent variables x_1 and x_2 is reduced to the simpler model

$$Y = \beta_0 + \beta_1 x_1 + e, \tag{1.16}$$

which has just one dependent variable x_1. So $\|\mathbf{Y} - \mathbf{X}\hat{\beta}_{\mathbf{A}}\|^2$ is simply the residual sum of squares of the reduced model (1.16), and is calculated to be 348.5678. Hence the test statistic in (1.7) is given by

$$\frac{(348.5678 - 24.1575)/1}{24.1575/14} = 188.0056$$

which is much larger than the corresponding critical value $f^{0.05}_{1,14} = 4.60$. The p-value is actually given by 1.66×10^{-9}. So H_0 is rejected at $\alpha = 5\%$. From the alternative form (1.11) of this test, one can see that the test statistic is in fact the square of the usual t-test statistic:

$$\frac{(0,0,1)\hat{\beta}/\sqrt{(0,0,1)(\mathbf{X}'\mathbf{X})^{-1}(0,0,1)'}}{\sqrt{\|\mathbf{Y} - \mathbf{X}\hat{\beta}\|^2/(17-2-1)}} = 13.7115.$$

The constrained least squares estimator in this case is given by $\hat{\beta}_{\mathbf{A}} = (72.4669, 0.1433, 0)$ where $(72.4669, 0.1433)$ is the least squares estimator of (β_0, β_1) of the reduced model (1.16).

1.4 Confidence and prediction intervals

A confidence interval may be constructed for the mean response of model (1.1) at a specific $\mathbf{x} = (1, x_1, \cdots, x_p)'$, i.e. for $\mathbf{x}'\beta$. Clearly $\mathbf{x}'\beta$ can be estimated by $\mathbf{x}'\hat{\beta}$. Since

$$\mathbf{x}'(\hat{\beta} - \beta) \sim \mathcal{N}(0, \sigma^2 \mathbf{x}'(\mathbf{X}'\mathbf{X})^{-1}\mathbf{x})$$

and $\hat{\beta}$ is independent of $\hat{\sigma}$ from Theorem 1.1,

$$\frac{\mathbf{x}'(\hat{\beta} - \beta)}{\hat{\sigma}\sqrt{\mathbf{x}'(\mathbf{X}'\mathbf{X})^{-1}\mathbf{x}}}$$

has a t distribution with $n - p - 1$ degrees of freedom. Hence a $1 - \alpha$ confidence interval for $\mathbf{x}'\beta$ is given by

$$P\left\{ \mathbf{x}'\beta \in \mathbf{x}'\hat{\beta} \pm t_{n-p-1}^{\alpha/2} \hat{\sigma} \sqrt{\mathbf{x}'(\mathbf{X}'\mathbf{X})^{-1}\mathbf{x}} \right\} = 1 - \alpha \tag{1.17}$$

where $t_{n-p-1}^{\alpha/2}$ is the upper $\alpha/2$ point of the t distribution with $n - p - 1$ degrees of freedom.

One may predict the magnitude of a future observation Y_f at a specific $\mathbf{x}$ by constructing a prediction interval for Y_f . Since $Y_f = \mathbf{x}'\beta + e_f$ according to model (1.1) where e_f is the random error associated with Y_f, a prediction for Y_f is still given by $\mathbf{x}'\hat{\beta}$. However

$$Y_f - \mathbf{x}'\hat{\beta} = e_f - \mathbf{x}'(\hat{\beta} - \beta) \sim \mathcal{N}(0, \sigma^2 + \sigma^2 \mathbf{x}'(\mathbf{X}'\mathbf{X})^{-1}\mathbf{x})$$

since $e_f \sim \mathcal{N}(0, \sigma^2)$ and is independent of $\hat{\beta}$. So

$$\frac{Y_f - \mathbf{x}'\hat{\beta}}{\hat{\sigma}\sqrt{1 + \mathbf{x}'(\mathbf{X}'\mathbf{X})^{-1}\mathbf{x}}}$$

has a t distribution with $n - p - 1$ degrees of freedom. Hence a $1 - \alpha$ prediction interval for Y_f is given by

$$P\left\{ Y_f \in \mathbf{x}'\hat{\beta} \pm t_{n-p-1}^{\alpha/2} \hat{\sigma} \sqrt{1 + \mathbf{x}'(\mathbf{X}'\mathbf{X})^{-1}\mathbf{x}} \right\} = 1 - \alpha. \tag{1.18}$$

It is clear that the prediction interval (1.18) is wider than the confidence interval (1.17). This is due to the extra source of random error e_f associated with prediction.

Example 1.4 For Example 1.1, suppose one is interested in predicting the response Y at $x_1 = 110$ and $x_2 = 3$. Then the 95% confidence interval for the mean response $\mathrm{E}(Y) = \beta_0 + 110\beta_1 + 3\beta_2$ is calculated from (1.17) to be $(82.38, 84.43)$, and the 95% prediction interval for a new observation Y corresponding to $x_1 = 110$ and $x_2 = 3$ is calculated from (1.18) to be $(80.41, 86.41)$.

All the numerical results in this chapter were computed using MATLAB® program example01.m.

2

Confidence Bands for One Simple Regression Model

One important inferential task in regression analysis is to assess where lies the true model $\mathbf{x}'\beta$ from which the observed data have been generated.

From Chapter 1, the unknown coefficients β can be estimated by $\hat{\beta}$ given in (1.3). So the true regression model $\mathbf{x}'\beta$ is somewhere about $\mathbf{x}'\hat{\beta}$. Furthermore, a $1-\alpha$ confidence region for β is given in Theorem 1.2 by

$$\left\{ \beta : \frac{(\beta-\hat{\beta})'(\mathbf{X}'\mathbf{X})(\beta-\hat{\beta})}{(p+1)\parallel Y-\mathbf{X}\hat{\beta}\parallel^2/(n-p-1)} \le f^{\alpha}_{p+1,n-p-1} \right\}. \tag{2.1}$$

This provides information on where the true model $\mathbf{x}'\beta$ can be: $\mathbf{x}'\beta$ is a plausible model if and only if β is contained in the confidence region (2.1). Equivalently one can test

$$H_0 : \beta = \beta_0 \quad \text{against} \quad H_a : \beta \neq \beta_0; \tag{2.2}$$

$$H_0 \text{ is rejected if and only if } \frac{(\beta_0-\hat{\beta})'(\mathbf{X}'\mathbf{X})(\beta_0-\hat{\beta})}{(p+1)\parallel Y-\mathbf{X}\hat{\beta}\parallel^2/(n-p-1)} > f^{\alpha}_{p+1,n-p-1}. \tag{2.3}$$

This test can also be deduced from the (partial) F test (1.11) with $\mathbf{A} = \mathbf{I}_{p+1}$ and $\mathbf{b} = \beta_0$. However these methods of inference do not provide information directly on the possible range of the true regression model function $\mathbf{x}'\beta$.

A simultaneous confidence band quantifies the plausible range of the true model $\mathbf{x}'\beta$ directly. Any regression model $\mathbf{x}'\beta_0$ that lies inside the simultaneous confidence band is deemed by this band as a plausible candidate of the true model $\mathbf{x}'\beta$. On the other hand, any regression model $\mathbf{x}'\beta_0$ that does not lie completely inside the simultaneous confidence band is not a plausible candidate of the true model $\mathbf{x}'\beta$. As a by-product, a confidence band can also be used to test the hypotheses in (2.2): H_0 is rejected if and only if $\mathbf{x}'\beta_0$ does not lie completely inside the simultaneous confidence band. Since the confidence band contains the true regression model with probability $1-\alpha$, the size of this test is α.

In this chapter we focus on a simple linear regression model, which has only one predictor variable x_1. For notation simplicity, in the rest of this chapter, the subscript '1' of x_1 is suppressed and let $n-p-1 = n-2$ be denoted by ν. We look at the construction of simultaneous confidence bands for a simple linear regression model. Due to the simplicity of this special case, exact simultaneous confidence bands can be constructed. The expositions in Sections 2.1–2.4 follow closely Liu, Lin and Piegorsch (2008). We also look at the two criteria in the statistical literature for the comparison of simultaneous confidence bands. Many of the ideas introduced in this chapter are applicable to a multiple regression model considered in Chapter 3. First, some preparations are necessary before presenting the derivations.

2.1 Preliminaries

Let $\mathbf{P}$ be the unique square root matrix of $(\mathbf{X}'\mathbf{X})^{-1}$ and so $(\mathbf{X}'\mathbf{X})^{-1} = \mathbf{P}^2$. The matrix $\mathbf{P}$ is used only in the derivations but not the final formulas of the simultaneous confidence level. Let

$$\begin{aligned} v(c,d) &:= \mathrm{Var}\{(c,d)\hat{\beta}\}/\sigma^2 = (c,d)(\mathbf{X}'\mathbf{X})^{-1}\begin{pmatrix} c \\ d \end{pmatrix} \\ &= \left\{\mathbf{P}\begin{pmatrix} c \\ d \end{pmatrix}\right\}'\left\{\mathbf{P}\begin{pmatrix} c \\ d \end{pmatrix}\right\} = \left\|\mathbf{P}\begin{pmatrix} c \\ d \end{pmatrix}\right\|^2. \end{aligned} \tag{2.4}$$

Since $\hat{\beta} \sim N_2(\beta, \sigma^2(\mathbf{X}'\mathbf{X})^{-1})$, it is clear that

$$\mathbf{N} := \mathbf{P}^{-1}(\hat{\beta} - \beta)/\sigma \sim \mathcal{N}_2(\mathbf{0}, I).$$

Furthermore, since $\hat{\sigma}/\sigma \sim \sqrt{\chi^2_\nu/\nu}$ and $\hat{\beta}$ and $\hat{\sigma}$ are independent random variables it follows immediately that

$$\mathbf{T} := \mathbf{N}/(\hat{\sigma}/\sigma) = \mathbf{P}^{-1}(\hat{\beta} - \beta)/\hat{\sigma} \tag{2.5}$$

has a standard bivariate $\mathcal{T}_{2,\nu}$ distribution with ν degrees of freedom, which was introduced by Dunnett and Sobel (1954a) (also see Tong, 1990, Kotz and Nadarajah, 2004, and Genz and Bretz, 2009) and has the probability density function (pdf)

$$f_{\mathbf{T}}(t_1,t_2) = \frac{1}{2\pi}\left(1 + \frac{1}{\nu}(t_1^2 + t_2^2)\right)^{-(2+\nu)/2}, \quad (t_1,t_2) \in \Re^2.$$

Define the polar coordinates of $\mathbf{N} = (N_1, N_2)'$, $(R_{\mathbf{N}}, \theta_{\mathbf{N}})$, by

$$N_1 = R_{\mathbf{N}} \cos\theta_{\mathbf{N}}, \quad N_2 = R_{\mathbf{N}} \sin\theta_{\mathbf{N}} \quad \text{for } R_{\mathbf{N}} \geq 0 \text{ and } \theta_{\mathbf{N}} \in [0, 2\pi).$$

It is well known (see e.g., Ross, 1988) that $R_{\mathbf{N}}$ has the distribution $\sqrt{\chi^2_2}$, $\theta_{\mathbf{N}}$ has a uniform distribution on the interval $[0, 2\pi)$, and $R_{\mathbf{N}}$ and $\theta_{\mathbf{N}}$ are independent random variables. The polar coordinates of $\mathbf{T}$, $(R_{\mathbf{T}}, \theta_{\mathbf{T}})$, can be represented in terms of $(R_{\mathbf{N}}, \theta_{\mathbf{N}})$ by

$$R_{\mathbf{T}} = \|\mathbf{T}\| = R_{\mathbf{N}}/(\hat{\sigma}/\sigma), \quad \theta_{\mathbf{T}} = \theta_{\mathbf{N}}.$$

So $R_{\mathbf{T}}$ has the distribution $\sqrt{2F_{2,\nu}}$ where F_{k_1,k_2} denotes an F random variable with degrees of freedom k_1 and k_2, $\theta_{\mathbf{T}}$ has a uniform distribution on the interval $[0, 2\pi)$, and $R_{\mathbf{T}}$ and $\theta_{\mathbf{T}}$ are independent random variables. From $R_{\mathbf{T}} = \sqrt{2F_{2,\nu}}$, direct calculation shows that the cumulative distribution function (cdf) of $R_{\mathbf{T}}$ is given by

$$F_{R_{\mathbf{T}}}(x) = 1 - (1 + x^2/\nu)^{-\nu/2}, \quad x > 0. \tag{2.6}$$

For a given vector $\mathbf{v} \in \Re^2$ and a number $r > 0$, the set

$$L(\mathbf{v}, r) := \{\mathbf{T}: \ \mathbf{v}'\mathbf{T}/\|\mathbf{v}\| = r\} \subset \Re^2$$

is given by the straight line that is perpendicular to the vector $\mathbf{v}$ and r distance away, in the direction of $\mathbf{v}$, from the origin. This straight line divides the whole $\Re^2$ plane into two half planes. The half plane that contains the origin is given by

$$\{\mathbf{T}: \ \mathbf{v}'\mathbf{T}/\|\mathbf{v}\| < r\} \subset \Re^2. \tag{2.7}$$

The set

$$\{\mathbf{T}: \ |\mathbf{v}'\mathbf{T}|/\|\mathbf{v}\| < r\} \subset \Re^2 \tag{2.8}$$

can be expressed as $\{\mathbf{T}: \ \mathbf{v}'\mathbf{T}/\|\mathbf{v}\| < r\} \cap \{\mathbf{T}: \ (-\mathbf{v})'\mathbf{T}/\| - \mathbf{v}\| < r\}$ and so is simply the stripe bounded by the parallel straight lines $L(\mathbf{v}, r)$ and $L(-\mathbf{v}, r)$.

2.2 Hyperbolic bands

2.2.1 *Two-sided band*

A two-sided hyperbolic confidence band over an interval $x \in (a, b)$ has the form

$$\beta_0 + \beta_1 x \in \hat{\beta}_0 + \hat{\beta}_1 x \pm c\hat{\sigma}\sqrt{v(1,x)} \quad \forall\, x \in (a, b) \tag{2.9}$$

where the critical constant c is chosen so that the simultaneous confidence level of this band is equal to $1 - \alpha$. Working and Hotelling (1929) considered this band for the special case of $(a, b) = (-\infty, \infty)$.

Figure 2.1 illustrates the shape of this confidence band. The center of the band at x is given by $\hat{\beta}_0 + \hat{\beta}_1 x$ whilst the width of the band is given by $2c\hat{\sigma}\sqrt{v(1,x)}$, which is proportional to $\sqrt{Var(\hat{\beta}_0 + \hat{\beta}_1 x)}$. The width of the band is the smallest at $x = \bar{x}$, the mean of the observed covariate values, and increases as x moves away from $\bar{x}$ on either sides. Furthermore, the upper part of the band, $\hat{\beta}_0 + \hat{\beta}_1 x + c\hat{\sigma}\sqrt{v(1,x)}$ for $x \in (a, b)$, is convex, whilst the lower part of the band is concave.

In order to determine the critical value c, it is necessary to find an expression of the simultaneous confidence level of the band that is amenable to computation. The simultaneous confidence level is given by

$$\mathrm{P}\left\{\beta_0 + \beta_1 x \in \hat{\beta}_0 + \hat{\beta}_1 x \pm c\hat{\sigma}\sqrt{v(1,x)} \ \text{ for all } x \in (a, b)\right\}$$

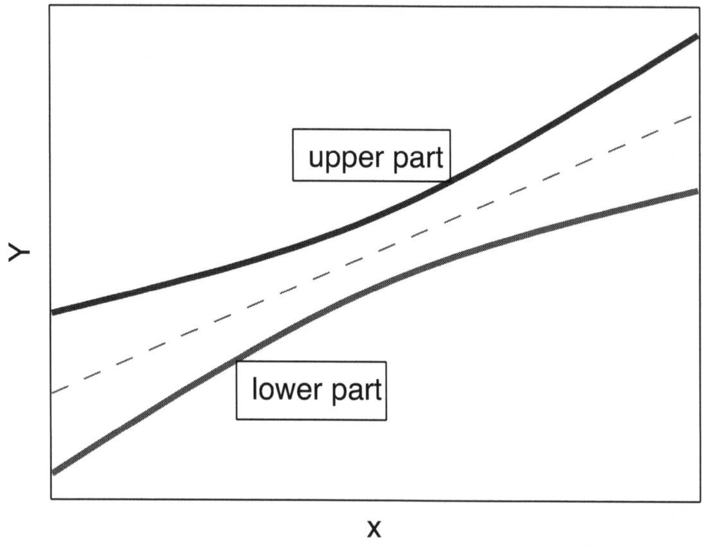

Figure 2.1: *The shape of a two-sided hyperbolic band*

$$= \mathrm{P}\left\{ \sup_{x\in(a,b)} |(1,x)(\hat{\beta}-\beta)/\hat{\sigma}|/\sqrt{v(1,x)} < c \right\}$$

$$= \mathrm{P}\left\{ \sup_{x\in(a,b)} \left| \left\{ \mathbf{P}\begin{pmatrix} 1 \\ x \end{pmatrix} \right\}' \mathbf{T} \right| \Big/ \left\| \mathbf{P}\begin{pmatrix} 1 \\ x \end{pmatrix} \right\| < c \right\} \quad (2.10)$$

$$= \mathrm{P}\{\mathbf{T} \in R_{h,2}\} \quad (2.11)$$

where the equality in (2.10) follows from the definition of **T** in (2.5) and expression (2.4), and the set $R_{h,2} \subset \Re^2$ in (2.11) is given by

$$R_{h,2} = \cap_{x\in(a,b)} R_{h,2}(x) \quad (2.12)$$

where

$$R_{h,2}(x) = \left\{ \mathbf{T} : \left| \left\{ \mathbf{P}\begin{pmatrix} 1 \\ x \end{pmatrix} \right\}' \mathbf{T} \right| \Big/ \left\| \mathbf{P}\begin{pmatrix} 1 \\ x \end{pmatrix} \right\| < c \right\}.$$

Note that set $R_{h,2}(x)$ is of the form (2.8) and so given by the stripe bounded by the two parallel lines that are perpendicular to the vector $\mathbf{P}\begin{pmatrix} 1 \\ x \end{pmatrix}$ and c distance from the origin. Hence $R_{h,2}$ in (2.12) is the intersection of all such stripes indexed by $x \in (a,b)$ and given by the spindle region depicted in Figure 2.2. In particular the angle ϕ depicted in the picture is formed by the vectors $\mathbf{P}\begin{pmatrix}1\\a\end{pmatrix}$ and $\mathbf{P}\begin{pmatrix}1\\b\end{pmatrix}$, and can be calculated from

$$\begin{aligned}\cos\phi &= \left\{\mathbf{P}\begin{pmatrix}1\\a\end{pmatrix}\right\}'\left\{\mathbf{P}\begin{pmatrix}1\\b\end{pmatrix}\right\}\Big/\left\|\mathbf{P}\begin{pmatrix}1\\a\end{pmatrix}\right\|\left\|\mathbf{P}\begin{pmatrix}1\\b\end{pmatrix}\right\| \\ &= (1,a)(\mathbf{X}'\mathbf{X})^{-1}\begin{pmatrix}1\\b\end{pmatrix}\Big/\sqrt{v(1,a)v(1,b)}. \qquad (2.13)\end{aligned}$$

It is noteworthy that $\cos\phi$ is simply the correlation coefficient between $\hat{\beta}_0+\hat{\beta}_1 a$ and $\hat{\beta}_0+\hat{\beta}_1 b$, and the matrix $\mathbf{P}$ is not required in the calculation of $\cos\phi$ from (2.13).

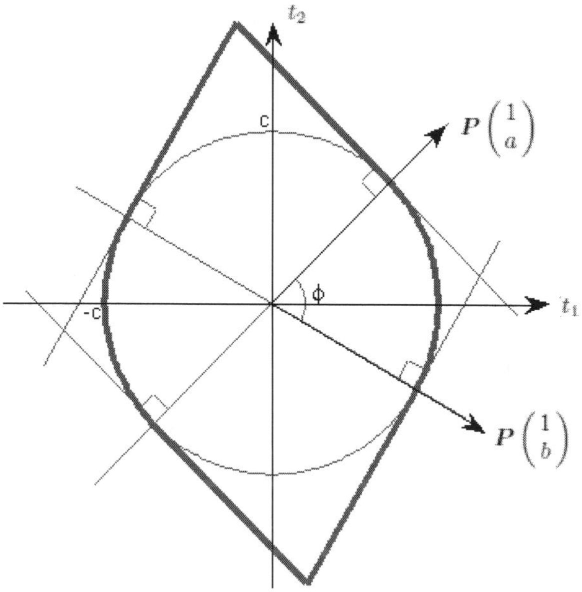

Figure 2.2: *The region $R_{h,2}$*

Note that the probability of $\mathbf{T}$ in any region that results from rotating $R_{h,2}$ around the origin is equal to the probability of $\mathbf{T}$ in $R_{h,2}$ since the pdf of $\mathbf{T}$ is rotational invariant. In particular, let $R^*_{h,2}$ be the region that is resulted from rotating $R_{h,2}$ around the origin to the position so that the angle ϕ is divided into two equal halves by the t_2-axis, as depicted in Figure 2.3. This region $R^*_{h,2}$ has the expression

$$R^*_{h,2} = \{\mathbf{T}:\ |\mathbf{v}'\mathbf{T}|/\|\mathbf{v}\| < c \ \text{ for all } \mathbf{v}\in\mathcal{V}(\phi)\}$$

where

$$\mathcal{V}(\phi) = \{\mathbf{v} = (v_1, v_2)' : \; v_2 > \|\mathbf{v}\| \cos(\phi/2)\} \tag{2.14}$$

is a cone and illustrated in Figure 2.3. Hence the simultaneous confidence level is equal to

$$P\{\mathbf{T} \in R_{h,2}\} = P\{\mathbf{T} \in R^*_{h,2}\} = P\left\{ \sup_{\mathbf{v}\in\mathcal{V}(\phi)} |\mathbf{v}'\mathbf{T}|/\|\mathbf{v}\| < c \right\}. \tag{2.15}$$

Next three different methods are used to derive expressions of the confidence level from expression (2.15).

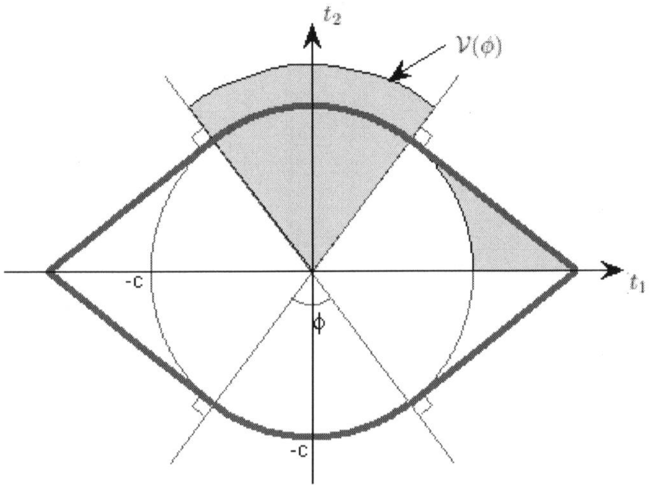

Figure 2.3: *The region $R^*_{h,2}$*

2.2.1.1 *The method of Wynn and Bloomfield (1971)*

This method was given by Wynn and Bloomfield (1971) and calculates the probability $P\{\mathbf{T} \in R^*_{h,2}\}$ directly. From Figure 2.3, $R^*_{h,2}$ can be partitioned into the whole disc of radius c and the remaining region. The probability of $\mathbf{T}$ in the disc is given by

$$\mathrm{P}\{\|\mathbf{T}\| < c\} = \mathrm{P}\{R_{\mathbf{T}} < c\} = 1 - \left(1 + c^2/\nu\right)^{-\nu/2} \tag{2.16}$$

where the last equality follows immediately from the cdf of $R_{\mathbf{T}}$ in (2.6). Due to symmetry, the probability of $\mathbf{T}$ in the remaining region is equal to four times the probability of $\mathbf{T}$ in the shaded region in Figure 2.3 that has the expression:

$$\{\mathbf{T}:\ \theta_{\mathbf{T}} \in [0, (\pi-\phi)/2],\ \|\mathbf{T}\| > c,\ (\cos((\pi-\phi)/2), \sin((\pi-\phi)/2))\mathbf{T} < c\}$$

and so the probability is further equal to

$$\begin{aligned}
& 4\mathrm{P}\{\ \theta_{\mathbf{T}} \in [0, (\pi-\phi)/2],\ \|\mathbf{T}\| > c,\ (\cos((\pi-\phi)/2), \sin((\pi-\phi)/2))\mathbf{T} < c\ \} \\
= \;& 4\mathrm{P}\{\ 0 < \theta_{\mathbf{T}} < (\pi-\phi)/2,\ c < R_{\mathbf{T}} < c/\cos((\pi-\phi)/2 - \theta_{\mathbf{T}})\ \} && (2.17) \\
= \;& 4\int_0^{(\pi-\phi)/2} \frac{1}{2\pi}\mathrm{P}\{\ c < R_{\mathbf{T}} < c/\cos((\pi-\phi)/2-\theta)\ \}\, d\theta && (2.18) \\
= \;& \frac{2}{\pi}\int_0^{(\pi-\phi)/2} \left(\left(1+\frac{c^2}{\nu}\right)^{-\nu/2} - \left(1+\frac{c^2}{\nu\sin^2(\theta+\phi/2)}\right)^{-\nu/2}\right) d\theta && (2.19)
\end{aligned}$$

where equality (2.17) follows directly by representing $\mathbf{T}$ in its polar coordinates $\mathbf{T} = R_{\mathbf{T}}(\cos\theta_{\mathbf{T}}, \sin\theta_{\mathbf{T}})'$, equality (2.18) follows immediately from the uniform distribution of $\theta_{\mathbf{T}}$ on $[0, 2\pi)$ and the independence of $\theta_{\mathbf{T}}$ and $R_{\mathbf{T}}$, and equality (2.19) follows directly from the cdf of $R_{\mathbf{T}}$ in (2.6). Combining the expressions (2.16) and (2.19) gives the simultaneous confidence level equal to

$$1 - \frac{\phi}{\pi}\left(1+\frac{c^2}{\nu}\right)^{-\nu/2} - \frac{2}{\pi}\int_0^{(\pi-\phi)/2}\left(1+\frac{c^2}{\nu\sin^2(\theta+\phi/2)}\right)^{-\nu/2} d\theta. \tag{2.20}$$

2.2.1.2 An algebraic method

This method evaluates the confidence level

$$\mathrm{P}\left\{\sup_{\mathbf{v}\in\mathcal{V}(\phi)} |\mathbf{v}'\mathbf{T}|/\|\mathbf{v}\| < c\right\} = \mathrm{P}\left\{\|\mathbf{T}\|\sup_{\mathbf{v}\in\mathcal{V}(\phi)} |\mathbf{v}'\mathbf{T}|/(\|\mathbf{v}\|\|\mathbf{T}\|) < c\right\} \tag{2.21}$$

by finding an explicit expression for the supremum.

Note that $\mathbf{v}'\mathbf{T}/(\|\mathbf{v}\|\|\mathbf{T}\|)$ is equal to the cosine of the angle between $\mathbf{T}$ and $\mathbf{v}$, and $\cos\theta$ is monotone decreasing in $\theta \in [0, \pi]$. So the supremum is attained at the $\mathbf{v} \in \mathcal{V}(\phi) \cup -\mathcal{V}(\phi)$ that forms the smallest angle with $\mathbf{T}$; throughout this book set $-A$ is defined to be $\{\mathbf{a}:\ -\mathbf{a} \in A\}$. From this geometry it is clear that

$$\sup_{\mathbf{v}\in\mathcal{V}(\phi)} \frac{|\mathbf{v}'\mathbf{T}|}{\|\mathbf{v}\|} = \begin{cases} \|\mathbf{T}\| & \text{if } \theta_{\mathbf{T}} \in [(\pi-\phi)/2, (\pi+\phi)/2] \\ & \quad \cup[(3\pi-\phi)/2, (3\pi+\phi)/2], \\ \|\mathbf{T}\||\sin(\theta_{\mathbf{T}}+\phi/2)| & \text{if } \theta_{\mathbf{T}} \in [0, (\pi-\phi)/2] \cup [\pi, (3\pi-\phi)/2], \\ \|\mathbf{T}\||\sin(\theta_{\mathbf{T}}-\phi/2)| & \text{if } \theta_{\mathbf{T}} \in [(\pi+\phi)/2, \pi] \cup [(3\pi+\phi)/2, 2\pi). \end{cases}$$

The simultaneous confidence level in (2.21) is therefore equal to

$$\begin{aligned}
& \mathrm{P}\{\ \theta_{\mathbf{T}} \in [(\pi-\phi)/2, (\pi+\phi)/2] \cup [(3\pi-\phi)/2, (3\pi+\phi)/2],\ \ \|\mathbf{T}\| < c\ \} \\
+ \ & \mathrm{P}\{\ \theta_{\mathbf{T}} \in [0, (\pi-\phi)/2] \cup [\pi, (3\pi-\phi)/2],\ \ \|\mathbf{T}\| |\sin(\theta_{\mathbf{T}}+\phi/2)| < c\ \} \\
+ \ & \mathrm{P}\{\ \theta_{\mathbf{T}} \in [(\pi+\phi)/2, \pi] \cup [(3\pi+\phi)/2, 2\pi),\ \ \|\mathbf{T}\| |\sin(\theta_{\mathbf{T}}-\phi/2)| < c\ \} \\
= \ & \frac{2\phi}{2\pi}\mathrm{P}\{\ \|\mathbf{T}\| < c\ \} \\
+ \ & 2\int_0^{(\pi-\phi)/2} \frac{1}{2\pi}\mathrm{P}\{\ \|\mathbf{T}\| |\sin(\theta+\phi/2)| < c\ \} d\theta \qquad (2.22) \\
+ \ & 2\int_{(\pi+\phi)/2}^{\pi} \frac{1}{2\pi}\mathrm{P}\{\ \|\mathbf{T}\| |\sin(\theta-\phi/2)| < c\ \} d\theta \qquad (2.23) \\
= \ & \frac{\phi}{\pi}\mathrm{P}\{\ R_{\mathbf{T}} < c\ \} \qquad (2.24) \\
+ \ & \frac{2}{\pi}\int_0^{(\pi-\phi)/2} \mathrm{P}\{\ R_{\mathbf{T}} < c/|\sin(\theta+\phi/2)|\ \} d\theta \\
= \ & \frac{\phi}{\pi}\left(1-\left(1+\frac{c^2}{\nu}\right)^{-\nu/2}\right) \\
+ \ & \frac{2}{\pi}\int_0^{(\pi-\phi)/2} \left(1-\left(1+\frac{c^2}{\nu\sin^2(\theta+\phi/2)}\right)^{-\nu/2}\right) d\theta \qquad (2.25)
\end{aligned}$$

where equality (2.24) follows by observing that the two integrals in (2.22) and (2.23) are equal, and equality (2.25) follows directly from the cdf of $R_{\mathbf{T}}$. It is clear that expression (2.25) is equal to expression (2.20).

2.2.1.3 *The method of Uusipaikka(1983)*

This method was given by Uusipaikka (1983) and hinges on the volume of tubular neighbourhoods of $\mathcal{V}(\phi)$ and $-\mathcal{V}(\phi)$. Due to the simplicity of the cones $\mathcal{V}(\phi)$ and $-\mathcal{V}(\phi)$, the exact volume of tubular neighborhoods of $\mathcal{V}(\phi) \cup -\mathcal{V}(\phi)$ can be easily calculated. From (2.15), the simultaneous confidence level is given by

$$\begin{aligned}
& 1-\mathrm{P}\left\{ \sup_{\mathbf{v}\in\mathcal{V}(\phi)} \frac{|\mathbf{v}'\mathbf{T}|}{\|\mathbf{v}\|\,\|\mathbf{T}\|} \ge \frac{c}{\|\mathbf{T}\|} \right\} \\
= \ & 1-\int_0^{\infty} \mathrm{P}\left\{ \sup_{\mathbf{v}\in\mathcal{V}(\phi)} \frac{|\mathbf{v}'\mathbf{T}|}{\|\mathbf{v}\|\,\|\mathbf{T}\|} \ge \frac{c}{w} \right\} dF_{R_{\mathbf{T}}}(w) \qquad (2.26) \\
= \ & 1-\int_c^{\infty} \mathrm{P}\left\{ \sup_{\mathbf{v}\in\mathcal{V}(\phi)} \frac{|\mathbf{v}'\mathbf{T}|}{\|\mathbf{v}\|\,\|\mathbf{T}\|} \ge \frac{c}{w} \right\} dF_{R_{\mathbf{T}}}(w) \qquad (2.27)
\end{aligned}$$

where (2.26) is due to the independence of $R_{\mathbf{T}}$ and $\theta_{\mathbf{T}}$ and that the supremum depends on $\mathbf{T}$ only through $\theta_{\mathbf{T}}$ (see (2.28) below), and (2.27) follows directly from the fact that the supremum is no larger than one. The probability in (2.27) can be written as

$$P\{\theta_{\mathbf{T}} \in |\mathcal{V}|(\phi, c/w)\}$$

where, for $0 < r < 1$,

$$\begin{aligned} |\mathcal{V}|(\phi,r) &= \left\{ \mathbf{T}: \sup_{\mathbf{v}\in\mathcal{V}(\phi)} \frac{|\mathbf{v}'\mathbf{T}|}{\|\mathbf{v}\|\|\mathbf{T}\|} \geq r \right\} \\ &= \left\{ \theta_{\mathbf{T}}: \sup_{\mathbf{v}\in\mathcal{V}(\phi)} |\cos(\theta_{\mathbf{T}} - \theta_{\mathbf{v}})| \geq \cos(\arccos r) \right\}. \quad (2.28) \end{aligned}$$

$|\mathcal{V}|(\phi,r)$ is called the tubular neighborhoods of $\mathcal{V}(\phi) \cup -\mathcal{V}(\phi)$ of angular radius $\arccos r$, and depicted in Figure 2.4.

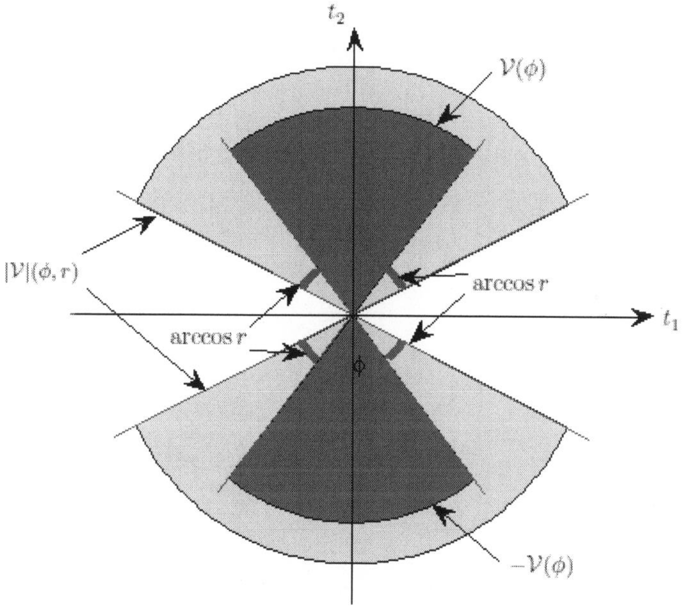

Figure 2.4: *The tubular neighborhoods* $|\mathcal{V}|(\phi,r)$

It is clear from its definition and Figure 2.4 that $|\mathcal{V}|(\phi,r)$ can further be expressed as

$$|\mathcal{V}|(\phi,r) = \left\{ \theta_{\mathbf{T}}: \ \theta_{\mathbf{T}} \in \quad \left[\frac{\pi-\phi}{2} - \arccos r, \ \frac{\pi+\phi}{2} + \arccos r \right] \right.$$

$$\cup \left[\frac{3\pi - \phi}{2} - \arccos r, \ \frac{3\pi + \phi}{2} + \arccos r \right] \Bigg\}$$

when $\phi/2 + \arccos r < \pi/2$ and

$$|\mathcal{V}|(\phi, r) = \{\theta_{\mathbf{T}} : \ \theta_{\mathbf{T}} \in [0, 2\pi]\}$$

when $\phi/2 + \arccos r \geq \pi/2$. Note further that

$$\phi/2 + \arccos(c/w) < \pi/2 \Longleftrightarrow w < c/\sin(\phi/2)$$

and $\theta_{\mathbf{T}}$ has a uniform distribution on $[0, 2\pi)$. We therefore have, for $c \leq w < c/\sin(\phi/2)$,

$$\mathrm{P}\{\theta_{\mathbf{T}} \in |\mathcal{V}|(\phi, c/w)\} = (2\arccos(c/w) + \phi)/\pi \tag{2.29}$$

and, for $w \geq c/\sin(\phi/2)$,

$$\mathrm{P}\{\theta_{\mathbf{T}} \in |\mathcal{V}|(\phi, c/w)\} = 1. \tag{2.30}$$

Substituting (2.29) and (2.30) into (2.27) gives the confidence level equal to

$$\begin{aligned} & 1 - \int_{c}^{c/\sin(\phi/2)} \frac{2\arccos(c/w) + \phi}{\pi} dF_{R_{\mathbf{T}}}(w) - \int_{c/\sin(\phi/2)}^{\infty} 1 dF_{R_{\mathbf{T}}}(w) \\ = \ & 1 - \left(1 + \frac{c^2}{\nu \sin^2(\phi/2)}\right)^{-\nu/2} \\ & - \int_{c}^{c/\sin(\phi/2)} \frac{2\arccos(c/w) + \phi}{\pi} dF_{R_{\mathbf{T}}}(w). \end{aligned} \tag{2.31}$$

It can be shown that expressions (2.31) and (2.25) are equal, as expected, after some calculus manipulations. Both expressions involve only one-dimensional integration and can be used to compute the simultaneous confidence level via numerical integrations.

For the special case of $(a, b) = (-\infty, \infty)$, the angle $\phi = \pi$. From expression (2.15) or (2.25) or (2.31), the simultaneous confidence level becomes

$$\mathrm{P}\{R_{\mathbf{T}} < c\} = \mathrm{P}\{F_{2,\nu} < c^2/2\}$$

and so $c = \sqrt{2 f_{2,\nu}^{\alpha}}$, where $f_{2,\nu}^{\alpha}$ is the upper α-point of the $F_{2,\nu}$ distribution. The critical value of a hyperbolic confidence band over the whole covariate region is provided by Scheffé (1953) and Hoel (1951); see Section 3.1.1 below.

Example 2.1 Kleinbaum *et al.* (1998, pp.192) provided a data set on how systolic blood pressure (Y) changes with age (x) for a group of forty males. The data set is

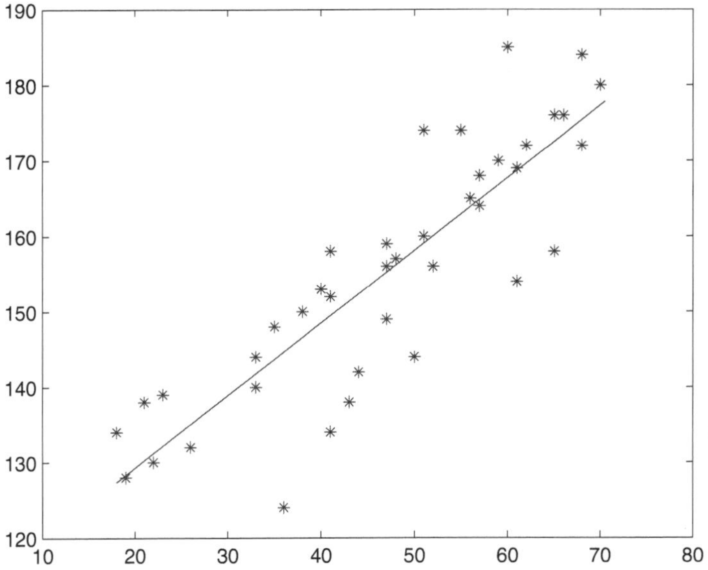

Figure 2.5: *The data points and fitted regression line*

given in Table 2.1. The data points and the fitted regression line, $Y = 110.034 + 0.961\,x$, are plotted in Figure 2.5. Also $\bar{x} = 46.92$, the minimum age $\min(x_i) = 18$, the maximum age $\max(x_i) = 70$, $S_{xx} = \sum_{i=1}^{n}(x_i - \bar{x})^2 = 8623.5591$, and $\hat{\sigma} = 8.479$ with 38 degrees of freedom.

A two-sided hyperbolic simultaneous confidence band can be used to quantify the plausible range of the true regression line. Since $\min(x_i) = 18$ and $\max(x_i) = 70$, it is sensible to construct a simultaneous confidence band over $x \in (18, 70) = (a, b)$. In this case the angle ϕ is calculated from (2.13) to be $\phi = 2.105 = 0.67\pi$. For $\alpha = 0.05$, one can calculate the critical value c of the hyperbolic band (2.9) over the interval $(18, 70)$, by using the expressions (2.20) or (2.25) or (2.31), to get $c = 2.514$. For the pointwise two-sided t confidence interval in (1.17), the critical value $t_{38}^{0.025} = 2.024$ is used instead of $c = 2.514$. The simultaneous confidence band over $x \in (-\infty, \infty)$ uses $\sqrt{2 f_{2,38}^{0.05}} = 2.548$. It should be emphasized that use of pointwsie t confidence interval in (1.17) for the purpose of simultaneously banding the regression line is incorrect; its simultaneous confidence level is less than the nominal $1 - \alpha = 95\%$ level.

All the three bands are plotted in Figure 2.6. The pointwise band is given by the two inner curves, the simultaneous band over $(18, 70)$ and the simultaneous band over $(-\infty, \infty)$ almost overlap each other, shown by the two outer curves.

Table 2.1: *Data from Kleinbaum et al. (1998, page 192)*

Person i	Age in years (x)	Blood Pressure (mm Hg) (Y)	Person i	Age in years (x)	Blood Pressure (mm Hg) (Y)
1	41	158	21	38	150
2	60	185	22	52	156
3	41	152	23	41	134
4	47	159	24	18	134
5	66	176	25	51	174
6	47	156	26	55	174
7	68	184	27	65	158
8	43	138	28	33	144
9	68	172	29	23	139
10	57	168	30	70	180
11	65	176	31	56	165
12	57	164	32	62	172
13	61	154	33	51	160
14	36	124	34	48	157
15	44	142	35	59	170
16	50	144	36	40	153
17	47	149	37	35	148
18	19	128	38	33	140
19	22	130	39	26	132
20	21	138	40	61	169

As expected, the pointwise band is the narrowest, the simultaneous band over $(-\infty,\infty)$ is the widest, whilst the simultaneous band over $(18,70)$ is in the middle. For this particular example, the simultaneous band over $(-\infty,\infty)$ is only $(2.548-2.514)/2.514 \approx 0.16\%$ wider than the simultaneous band over $(18,70)$. This is because the interval $(18,70)$ is so wide that the angle ϕ is quite close to the upper limit π. This ratio can be close to $(2.548-2.105)/2.105 \approx 21\%$ when the interval (a,b) is narrow enough for ϕ to be close to the lower limit 0. The computations of this example are done using the MATLAB® program example0201.m.

2.2.2 *One-sided band*

A one-sided lower hyperbolic band has the form

$$\beta_0+\beta_1 x > \hat{\beta}_0+\hat{\beta}_1 x - c\hat{\sigma}\sqrt{v(1,x)} \text{ for all } x \in (a,b) \tag{2.32}$$

while a one-sided upper hyperbolic band has the form

$$\beta_0+\beta_1 x < \hat{\beta}_0+\hat{\beta}_1 x + c\hat{\sigma}\sqrt{v(1,x)} \text{ for all } x \in (a,b).$$

Note that the $1-\alpha$ level upper band uses the same critical constant as the $1-\alpha$ level lower band, and so we will focus on the the lower confidence band below.

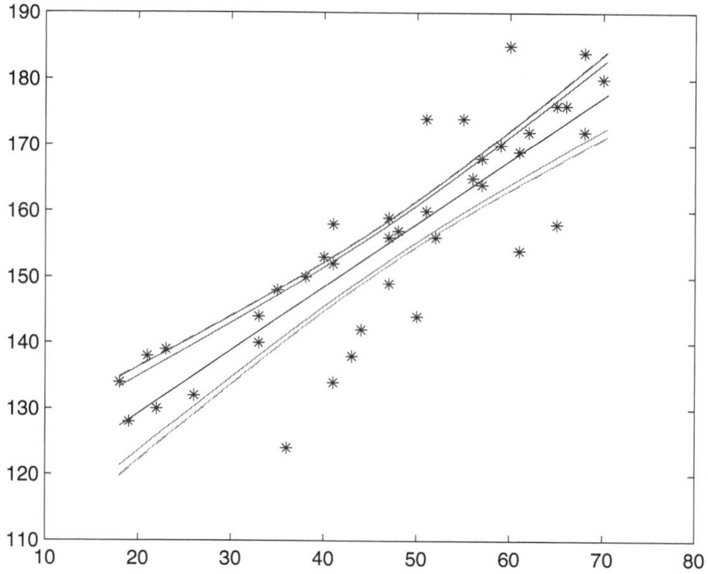

Figure 2.6: *The pointwise band and simultaneous bands over* $(18,70)$ *and* $(-\infty,\infty)$

The lower hyperbolic band (2.32) has a similar shape as the lower part of the two-sided hyperbolic band (2.9); see Figure 2.1.

The simultaneous confidence level of the lower one-sided hyperbolic band (2.32) is given by

$$\begin{aligned} & \mathrm{P}\left\{\beta_0+\beta_1 x>\hat{\beta}_0+\hat{\beta}_1 x-c\hat{\sigma}\sqrt{v(1,x)} \text{ for all } x\in(a,b)\right\} \\ = & \mathrm{P}\left\{\sup_{x\in(a,b)}\left((1,x)(\hat{\boldsymbol{\beta}}-\boldsymbol{\beta})/\hat{\sigma}\right)/\sqrt{v(1,x)}<c\right\} \\ = & \mathrm{P}\{\mathbf{T}\in R_{h,1}\} \end{aligned} \tag{2.33}$$

where the set $R_{h,1}\subset\Re^2$ in (2.33) is given by

$$R_{h,1}=\cap_{x\in(a,b)}R_{h,1}(x)$$

where

$$R_{h,1}(x)=\left\{\mathbf{T}:\ \left\{\mathbf{P}\begin{pmatrix}1\\x\end{pmatrix}\right\}'\mathbf{T}\Big/\left\|\mathbf{P}\begin{pmatrix}1\\x\end{pmatrix}\right\|<c\right\}.$$

Note that set $R_{h,1}(x)$ is of the form (2.7), and so $R_{h,1}(x)$ is made up of all the points on the same side as the origin of the straight line that is perpendicular to the vector $\mathbf{P}\begin{pmatrix}1\\x\end{pmatrix}$ and c distance, in the direction of $\mathbf{P}\begin{pmatrix}1\\x\end{pmatrix}$, from the origin. Hence $R_{h,1}$ is given by the region depicted in Figure 2.7; it is bounded by the two straight lines and part of the circle with angle ϕ. In particular the angle ϕ is the same as in the two-sided case and given by (2.13).

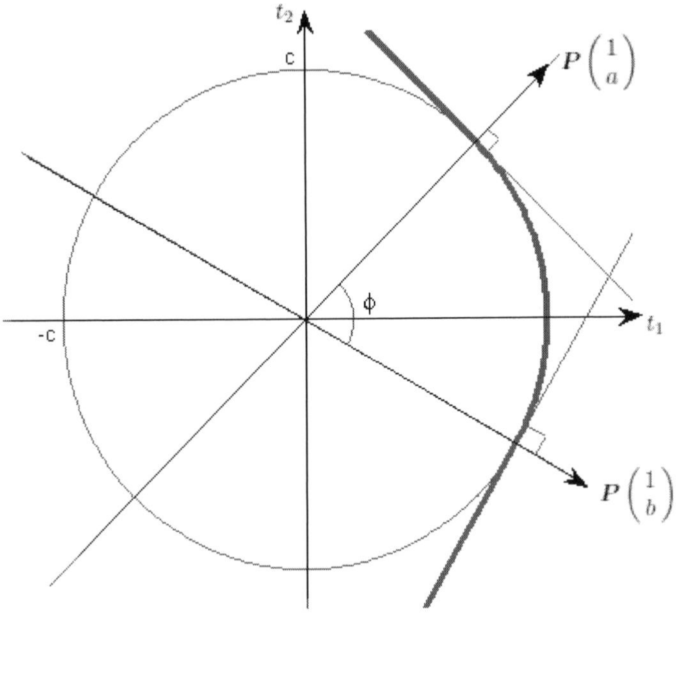

Figure 2.7: *The region* $R_{h,1}$

Let $R^*_{h,1}$ be the region that is resulted from rotating $R_{h,1}$ around the origin to the position so that the angle ϕ is divided into two equal halves by the t_2-axis, as depicted in Figure 2.8. This region $R^*_{h,1}$ has the expression

$$R^*_{h,1} = \{\mathbf{T}: \ \mathbf{v}'\mathbf{T}/\|\mathbf{v}\| < c \ \text{ for all } \mathbf{v} \in \mathcal{V}(\phi)\}$$

where $\mathcal{V}(\phi)$ is the same as in the two-sided case and given by (2.14). Due to the rotational invariance of the $\mathbf{T}$ probability distribution, the simultaneous confidence level is further equal to

$$\mathrm{P}\{\mathbf{T} \in R^*_{h,1}\} = \mathrm{P}\left\{\sup_{\mathbf{v}\in\mathcal{V}(\phi)} \mathbf{v}'\mathbf{T}/\|\mathbf{v}\| < c\right\}. \tag{2.34}$$

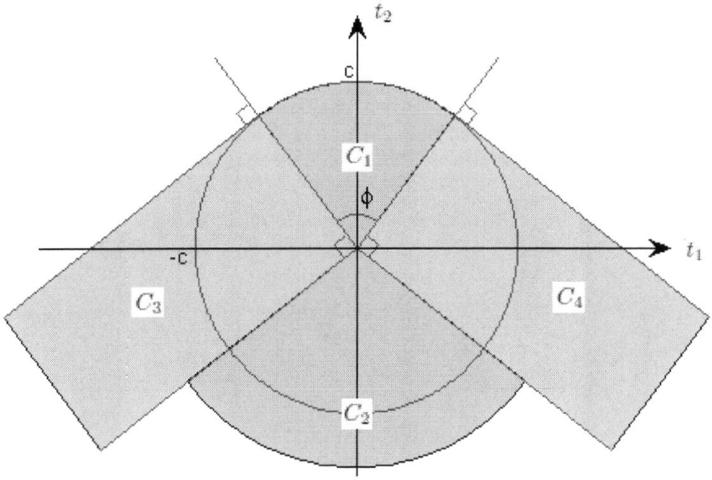

Figure 2.8: *The region $R^*_{h,1}$*

We next derive three expressions for this simultaneous confidence level as in the two-sided case.

2.2.2.1 *The method of Bohrer and Francis (1972a)*

This method was given by Bohrer and Francis (1972a) and calculates the probability $P\{\mathbf{T} \in R^*_{h,1}\}$ directly. Note that $R^*_{h,1}$ can be partitioned into four parts as depicted in Figure 2.8: the fan C_1 which is the intersection of the cone $\mathcal{V}(\phi)$ and the disc of radius c, the cone C_2 which is the dual cone of $\mathcal{V}(\phi)$, and the two half-stripes C_3 and C_4. The probability of $\mathbf{T}$ in the fan C_1 is equal to

$$\begin{aligned} & P\{\ \theta_{\mathbf{T}} \in [(\pi-\phi)/2, (\pi+\phi)/2],\ \|\mathbf{T}\| \le c\ \} \\ = & \frac{\phi}{2\pi} P\{\ \|\mathbf{T}\| \le c\ \} = \frac{\phi}{2\pi} P\left\{\ \frac{\|\mathbf{T}\|^2}{2} \le \frac{c^2}{2}\ \right\} = \frac{\phi}{2\pi} F_{2,\nu}\left(\frac{c^2}{2}\right) \quad (2.35) \end{aligned}$$

since $\|\mathbf{T}\|^2/2$ has an F distribution with degrees of freedom 2 and ν. The probability of $\mathbf{T}$ in the cone C_2 is given by

$$P\{\ \theta_{\mathbf{T}} \in [(3\pi-(\pi-\phi))/2, (3\pi+(\pi-\phi))/2]\ \} = \frac{\pi-\phi}{2\pi} \quad (2.36)$$

since $\theta_{\mathbf{T}}$ has a uniform distribution on $[0, 2\pi)$. Again due to the rotational invariance of the $\mathbf{T}$ probability distribution, the probability of $\mathbf{T}$ in $C_3 \cup C_4$ is equal to the probability of $\mathbf{T}$ in the stripe which is the union of C_4 and the half-stripe that is resulted from rotating C_3 clockwise ϕ angle, and further equal to the probability of $\mathbf{T}$ in the stripe that results from rotating the last stripe clockwise $(\pi - \phi)/2$ angle:

$$\mathrm{P}\{\, 0 < T_1 < c \,\} = \frac{1}{2}\mathrm{P}\{\, T_1^2 < c^2 \,\} = \frac{1}{2}F_{1,\nu}(c^2) \tag{2.37}$$

since T_1^2 has an F distribution with degrees of freedom 1 and ν. Putting (2.35), (2.36) and (2.37) together gives the simultaneous confidence level equal to

$$\frac{\pi - \phi}{2\pi} + \frac{1}{2}F_{1,\nu}(c^2) + \frac{\phi}{2\pi}F_{2,\nu}\left(\frac{c^2}{2}\right). \tag{2.38}$$

2.2.2.2 An algebraic method

This method is similar to that for the two-sided case in Section 2.2.1.2; the key is to find an explicit expression for the supremum in (2.34). Similar to the two-sided case, we have

$$\sup_{\mathbf{v} \in \mathcal{V}(\phi)} \frac{\mathbf{v}'\mathbf{T}}{\|\mathbf{v}\|} = \begin{cases} \|\mathbf{T}\| & \text{if } \theta_{\mathbf{T}} \in [(\pi - \phi)/2, (\pi + \phi)/2], \\ \|\mathbf{T}\| \cos((\pi - \phi)/2 - \theta_{\mathbf{T}}) & \text{if } \theta_{\mathbf{T}} \in [-\pi/2, (\pi - \phi)/2], \\ \|\mathbf{T}\| \cos(\theta_{\mathbf{T}} - (\pi + \phi)/2) & \text{if } \theta_{\mathbf{T}} \in [(\pi + \phi)/2, 3\pi/2]. \end{cases}$$

So, under the assumption $c > 0$, the simultaneous confidence level (2.34) is equal to

$$\mathrm{P}\{\, \theta_{\mathbf{T}} \in [(\pi - \phi)/2, (\pi + \phi)/2],\ \|\mathbf{T}\| < c \,\} \tag{2.39}$$
$$+\ \mathrm{P}\{\, \theta_{\mathbf{T}} \in [-\pi/2, (\pi - \phi)/2],\ \|\mathbf{T}\| \cos((\pi - \phi)/2 - \theta_{\mathbf{T}}) < c \,\} \tag{2.40}$$
$$+\ \mathrm{P}\{\, \theta_{\mathbf{T}} \in [(\pi + \phi)/2, 3\pi/2],\ \|\mathbf{T}\| \cos(\theta_{\mathbf{T}} - (\pi + \phi)/2) < c \,\}. \tag{2.41}$$

Now the probability in (2.39) is equal to $\frac{\phi}{2\pi}\mathrm{P}\{\|\mathbf{T}\| < c\}$ as in the two-sided case. The probability in (2.40) is equal to

$$\left\{ \int_{(\pi-\phi)/2-\pi/2}^{(\pi-\phi)/2} + \int_{-\pi/2}^{(\pi-\phi)/2-\pi/2} \right\} \frac{1}{2\pi}\mathrm{P}\{\, \|\mathbf{T}\| \cos((\pi - \phi)/2 - \theta) < c \,\} d\theta$$
$$= \int_{(\pi-\phi)/2-\pi/2}^{(\pi-\phi)/2} \frac{1}{2\pi}\mathrm{P}\{\, \|\mathbf{T}\| \cos[(\pi - \phi)/2 - \theta] < c \,\} d\theta + \frac{(\pi - \phi)/2}{2\pi}$$

by noting that if $\theta \in (-\pi/2, (\pi - \phi)/2 - \pi/2)$ then $\|\mathbf{T}\| \cos((\pi - \phi)/2 - \theta)$ is negative and hence less than $c > 0$ with probability one. Similarly, the probability in (2.41) is equal to

$$\left\{ \int_{(\pi+\phi)/2}^{(\pi+\phi)/2+\pi/2} + \int_{(\pi+\phi)/2+\pi/2}^{3\pi/2} \right\} \frac{1}{2\pi}\mathrm{P}\{\, \|\mathbf{T}\| \cos(\theta - (\pi + \phi)/2) < c \,\} d\theta$$
$$= \int_{(\pi+\phi)/2}^{(\pi+\phi)/2+\pi/2} \frac{1}{2\pi}\mathrm{P}\{\, \|\mathbf{T}\| \cos(\theta - (\pi + \phi)/2) < c \,\} d\theta + \frac{(\pi - \phi)/2}{2\pi}$$

by noting that if $\theta \in ((\pi+\phi)/2+\pi/2, 3\pi/2)$ then $\|\mathbf{T}\|\cos[\theta-(\pi+\phi)/2]$ is negative and so less than c with probability one. Substituting these into (2.39), (2.40) and (2.41), respectively, the simultaneous confidence level is further equal to

$$\begin{aligned}
& \frac{\phi}{2\pi}F_{R_{\mathbf{T}}}(c) \\
+ & \int_{(\pi-\phi)/2-\pi/2}^{(\pi-\phi)/2} \frac{1}{2\pi}F_{R_{\mathbf{T}}}\left(\frac{c}{\cos((\pi-\phi)/2-\theta)}\right)d\theta + \frac{(\pi-\phi)/2}{2\pi} \\
+ & \int_{(\pi+\phi)/2}^{(\pi+\phi)/2+\pi/2} \frac{1}{2\pi}F_{R_{\mathbf{T}}}\left(\frac{c}{\cos(\theta-(\pi+\phi)/2)}\right)d\theta + \frac{(\pi-\phi)/2}{2\pi} \\
= & \frac{\phi}{2\pi}F_{R_{\mathbf{T}}}(c) + \frac{\pi-\phi}{2\pi} + \frac{1}{\pi}\int_0^{\pi/2} F_{R_{\mathbf{T}}}\left(\frac{c}{\cos(\theta)}\right)d\theta, \qquad (2.42)
\end{aligned}$$

where equality (2.42) follows from some simple change of variables.

2.2.2.3 *The method of Pan, Piegorsch and West (2003)*

Uusipaikka's (1983) method was manipulated by Pan, Piegorsch and West (2003) to calculate the confidence level of the one-sided hyperbolic band. From (2.34), the simultaneous confidence level is given by

$$\begin{aligned}
& 1-\mathrm{P}\left\{\sup_{\mathbf{v}\in\mathcal{V}(\phi)} \frac{\mathbf{v}'\mathbf{T}}{\|\mathbf{v}\|\|\mathbf{T}\|} \geq \frac{c}{\|\mathbf{T}\|}\right\} \\
= & 1-\int_0^\infty \mathrm{P}\left\{\sup_{\mathbf{v}\in\mathcal{V}(\phi)} \frac{\mathbf{v}'\mathbf{T}}{\|\mathbf{v}\|\|\mathbf{T}\|} \geq \frac{c}{w}\right\} dF_{R_{\mathbf{T}}}(w) \\
= & 1-\int_c^\infty \mathrm{P}\left\{\sup_{\mathbf{v}\in\mathcal{V}(\phi)} \frac{\mathbf{v}'\mathbf{T}}{\|\mathbf{v}\|\|\mathbf{T}\|} \geq \frac{c}{w}\right\} dF_{R_{\mathbf{T}}}(w). \qquad (2.43)
\end{aligned}$$

The probability in the last expression can be written as

$$\mathrm{P}\{\theta_{\mathbf{T}} \in \mathcal{V}(\phi, c/w)\}$$

where, for $0<r<1$,

$$\begin{aligned}
\mathcal{V}(\phi,r) &= \left\{\mathbf{T}: \sup_{\mathbf{v}\in\mathcal{V}(\phi)} \frac{\mathbf{v}'\mathbf{T}}{\|\mathbf{v}\|\|\mathbf{T}\|} \geq r\right\} \\
&= \left\{\theta_{\mathbf{T}}: \sup_{\mathbf{v}\in\mathcal{V}(\phi)} \cos(\theta_{\mathbf{T}}-\theta_{\mathbf{v}}) \geq \cos(\arccos r)\right\}.
\end{aligned}$$

$\mathcal{V}(\phi,r)$ is the tubular neighborhoods of $\mathcal{V}(\phi)$ of angular radius $\arccos r$, and depicted in Figure 2.9. It is clear from its definition and Figure 2.9 that $\mathcal{V}(\phi,r)$ can

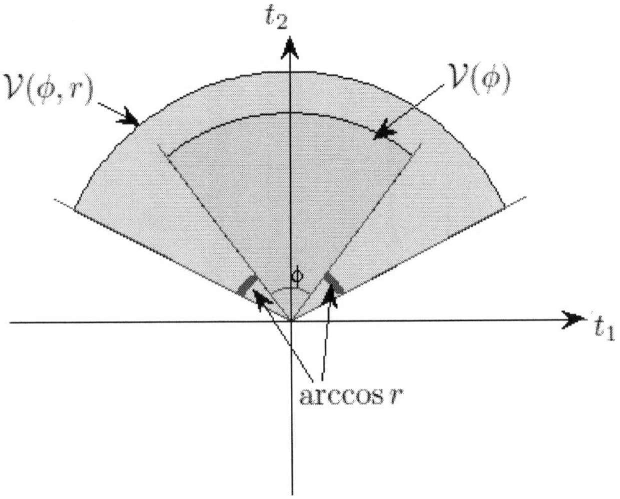

Figure 2.9: *The tubular neighborhoods* $\mathcal{V}(\phi,r)$

further be expressed as

$$\mathcal{V}(\phi,r)=\left\{\theta_{\mathbf{T}}\in\left[\frac{\pi-\phi}{2}-\arccos r,\frac{\pi+\phi}{2}+\arccos r\right]\right\}.$$

It follows therefore from the uniform distribution of $\theta_{\mathbf{T}}$ that

$$\mathrm{P}\left\{\theta_{\mathbf{T}}\in\mathcal{V}\left(\phi,\frac{c}{w}\right)\right\}=\frac{2\arccos(c/w)+\phi}{2\pi}.$$

Substituting this into (2.43) gives the confidence level equal to

$$1-\int_c^\infty\frac{2\arccos(c/w)+\phi}{2\pi}dF_{R_{\mathbf{T}}}(w). \tag{2.44}$$

Expressions (2.44), (2.42) and (2.38) can be shown to be equal, as expected, by some calculus manipulations. But of course expression (2.38) is the simplest to use as it involves only the cdf's of $F_{1,\nu}$ and $F_{2,\nu}$.

When $(a,b)=(-\infty,\infty)$, the angle $\phi=\pi$. In this case expression (2.38) simplifies to

$$\frac{1}{2}F_{1,\nu}(c^2)+\frac{1}{2}F_{2,\nu}\left(\frac{c^2}{2}\right),$$

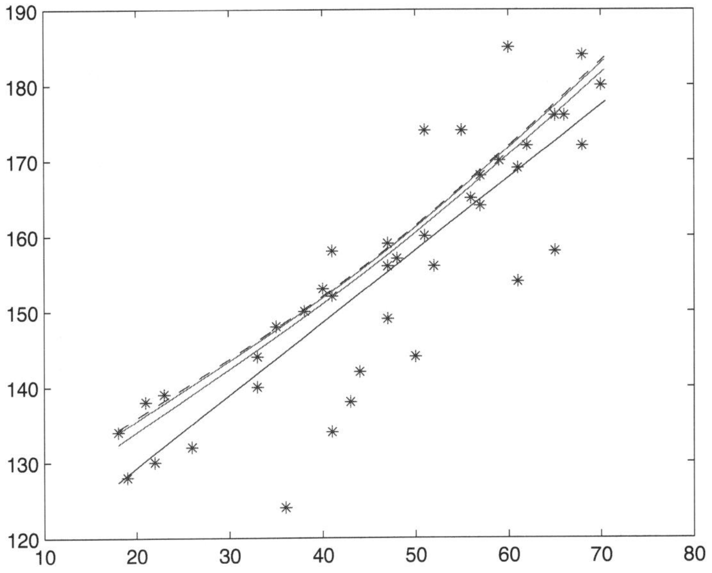

Figure 2.10: *The one-sided pointwise and simultaneous confidence bands*

which agrees with the result of Hochberg and Quade (1975; see Theorem 3.3 in Chapter 3).

Example 2.2 Consider the data set in Example 2.1. An upper simultaneous confidence band for the regression line provides information on plausible upper limits on the mean SBP over a chosen age range. This information is of interest since unduly high SBP is a health risk. To construct a 95% one-sided hyperbolic simultaneous confidence band over the *age* range $(18,70)$, expression (2.38) can be used to calculate the critical value c. First, the angle ϕ is already calculated in the two-sided case in Example 2.1. Next, c is calculated to be 2.213. For the one-sided simultaneous band over $x \in (-\infty,\infty)$, the critical constant is given by $c = 2.350$. The one-sided pointwise t interval uses $c = t_{38}^{0.05} = 1.686$. It should be re-emphasized that use of the one-sided pointwise t interval for the purpose of simultaneously bounding the regression line is incorrect since its simultaneous confidence level is less than the nominal level $1-\alpha = 95\%$.

All three bands are plotted in Figure 2.10, with the simultaneous band over $(-\infty,\infty)$ being given by the highest curve, the simultaneous band over $(18,70)$ given by the middle curve, and the pointwise band given by the lowest curve. All three upper bands are above the fitted regression line as expected. The two simultaneous confidence bands are quite close to each other for the same reason as pointed out in the two-sided case: the interval $(18,70)$ is so wide that ϕ is quite close to the upper limit π. But the simultaneous band over $(-\infty,\infty)$ can be as much as nearly $(2.350-1.686)/1.686 \approx 39\%$ wider than a simultaneous band over (a,b) when the interval (a,b) is narrow enough for ϕ to be close to the lower limit zero.

Also, the upper one-sided hyperbolic band over $(18,70)$ is below the upper part of the two-sided hyperbolic band over $(18,70)$ which uses critical value $c = 2.514$. So the one-sided upper band is more precise than the upper part of the two-sided band for the purpose of bounding the regression line from above. The computations of this example are done using the MATLAB program example0202.m.

2.3 Three-segment bands

2.3.1 Two-sided band

A two-sided three-segment band has the form

$$\beta_0+\beta_1 x \in \hat{\beta}_0+\hat{\beta}_1 x \pm \hat{\sigma} H_3(x) \text{ for all } x \in (a,b) \tag{2.45}$$

where

$$H_3(x) = \frac{1}{b-a}\left((x-a)c_1\sqrt{v(1,b)}+(b-x)c_2\sqrt{v(1,a)}\right), \quad x \in (a,b). \tag{2.46}$$

The two critical constants c_1 and c_2 are chosen so that the confidence level is equal to $1-\alpha$. This band was proposed by Bowden and Graybill (1966). Its shape is illustrated in Figure 2.11. The center of the band at x is given by $\hat{\beta}_0+\hat{\beta}_1 x$. Both the upper and lower parts of the band are straight line segments over $x \in (a,b)$. The width of the band is $2c_2\sqrt{v(1,a)}$ at $x=a$ and $2c_1\sqrt{v(1,b)}$ at $x=b$. When $c_2\sqrt{v(1,a)} = c_1\sqrt{v(1,b)}$, the three-segment band becomes a constant width band over $x \in (a,b)$ which was considered by Gafarian (1964). See Section 2.6.2 and Figure 2.26 below for why this band is called a three-segment band.

Note that the two-sided three-segment band (2.45) is constructed by putting the following two-sided bounds on the regression line at both $x=a$ and $x=b$

$$\beta_0+\beta_1 a \in \hat{\beta}_0+\hat{\beta}_1 a \pm \hat{\sigma} H_3(a), \quad \beta_0+\beta_1 b \in \hat{\beta}_0+\hat{\beta}_1 b \pm \hat{\sigma} H_3(b). \tag{2.47}$$

For $x \in (a,b)$, the upper and lower parts of the band are formed of the two line segments that connect the two upper limits and the two lower limits of the two bounds in (2.47) respectively. Therefore the simultaneous confidence level of the band is given by

$$\begin{aligned} & \mathrm{P}\left\{\beta_0+\beta_1 a \in \hat{\beta}_0+\hat{\beta}_1 a \pm \hat{\sigma} H_3(a),\ \beta_0+\beta_1 b \in \hat{\beta}_0+\hat{\beta}_1 b \pm \hat{\sigma} H_3(b)\right\} \\ = \ & \mathrm{P}\left\{|(1,a)(\hat{\beta}-\beta)/\hat{\sigma}|/H_3(a)<1,\ |(1,b)(\hat{\beta}-\beta)/\hat{\sigma}|/H_3(b)<1\right\}. \end{aligned} \tag{2.48}$$

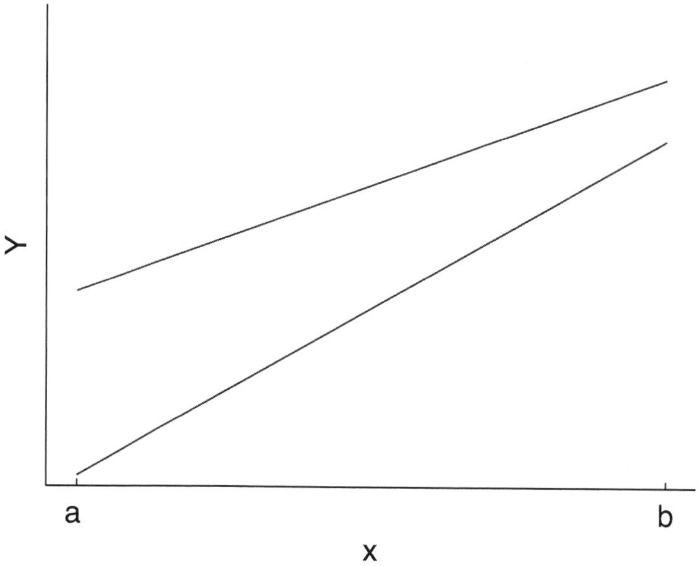

Figure 2.11: *The shape of a two-sided three-segment band*

Alternatively, the simultaneous confidence level of the band is given by

$$\begin{aligned} & P\left\{\beta_0+\beta_1 x \in \hat{\beta}_0+\hat{\beta}_1 x \pm \hat{\sigma} H_3(x) \text{ for all } x \in (a,b)\right\} \\ = & P\left\{\sup_{x\in(a,b)} |(1,x)(\hat{\beta}-\beta)/\hat{\sigma}|/H_3(x) < 1\right\}. \end{aligned} \quad (2.49)$$

Note that

$$\frac{\partial}{\partial x}\left(\left((1,x)(\hat{\beta}-\beta)/\hat{\sigma}\right)/H_3(x)\right)$$

has a fixed sign, either positive or negative, over $x \in (-\infty, \infty)$. The supremum in (2.49) is therefore attained at either $x = a$ or $x = b$, and so the confidence level can further be expressed as

$$P\left\{\max_{x=a \text{ or } b} |(1,x)(\hat{\beta}-\beta)/\hat{\sigma}|/H_3(x) < 1\right\} = P\{\mathbf{T} \in R_{3,2}\} \quad (2.50)$$

where

$$R_{3,2} = R_{3,2}(a) \cap R_{3,2}(b) \quad (2.51)$$

with

$$R_{3,2}(a) = \{\mathbf{T}:\ |(1,a)(\hat{\beta}-\beta)/\hat{\sigma}|/H_3(a) < 1\}$$

$$= \left\{ \mathbf{T} : \left| \left\{ \mathbf{P} \begin{pmatrix} 1 \\ a \end{pmatrix} \right\}' \mathbf{T} \right| \Big/ \left(c_2 \sqrt{v(1,a)} \right) < 1 \right\}$$

$$= \left\{ \mathbf{T} : \left| \left\{ \mathbf{P} \begin{pmatrix} 1 \\ a \end{pmatrix} \right\}' \mathbf{T} \right| \Big/ \left\| \mathbf{P} \begin{pmatrix} 1 \\ a \end{pmatrix} \right\| < c_2 \right\}$$

and, in a similar way,

$$R_{3,2}(b) = \{ \mathbf{T} : |(1,b)(\hat{\beta} - \beta)/\hat{\sigma}| / H_3(b) < 1 \}$$

$$= \left\{ \mathbf{T} : \left| \left\{ \mathbf{P} \begin{pmatrix} 1 \\ b \end{pmatrix} \right\}' \mathbf{T} \right| \Big/ \left\| \mathbf{P} \begin{pmatrix} 1 \\ b \end{pmatrix} \right\| < c_1 \right\}.$$

It is clear that expressions (2.48) and (2.50) are equal.

Note that both $R_{3,2}(a)$ and $R_{3,2}(b)$ are of the form (2.8). Hence $R_{3,2}$ is given by the parallelogram depicted in Figure 2.12. In particular the angle ϕ depicted in the picture is formed by the vectors $\mathbf{P}\begin{pmatrix} 1 \\ a \end{pmatrix}$ and $\mathbf{P}\begin{pmatrix} 1 \\ b \end{pmatrix}$ as in the hyperbolic bands and so given by (2.13).

Let $R_{3,2}^*$ be the region that is resulted from rotating $R_{3,2}$ around the origin to the position so that $\mathbf{P}\begin{pmatrix} 1 \\ b \end{pmatrix}$ is in the direction of the t_1-axis, as depicted in Figure 2.13. Due to the rotational invariance of the probability distribution of $\mathbf{T}$, the confidence level is equal to the probability of $\mathbf{T}$ in $R_{3,2}^*$ and further equal to twice of the probability of $\mathbf{T}$ in the top-right half of $R_{3,2}^*$ which has the expression

$$\begin{aligned} & \{ \mathbf{T} : \theta_{\mathbf{T}} \in [-(\pi - \eta_1), \xi_1], R_{\mathbf{T}} \cos \theta_{\mathbf{T}} \leq c_1 \} \\ \cup \; & \{ \mathbf{T} : \theta_{\mathbf{T}} \in [\xi_1, \eta_1], R_{\mathbf{T}} \cos(\theta_{\mathbf{T}} - \phi) \leq c_2 \} \end{aligned}$$

where the angles ξ_1 and η_1 are depicted in Figure 2.13 and given by

$$\xi_1 = \arcsin \left(\frac{c_2 - c_1 \cos \phi}{\sqrt{c_2^2 + c_1^2 - 2c_2 c_1 \cos \phi}} \right), \tag{2.52}$$

$$\eta_1 = \arccos \left(\frac{-c_1 \sin \phi}{\sqrt{c_2^2 + c_1^2 + 2c_2 c_1 \cos \phi}} \right). \tag{2.53}$$

These two expressions can be found by first writing down the (t_1, t_2)-equations of the four straight lines that form the parallelogram $R_{3,2}^*$, then solving the (t_1, t_2)-coordinates of the four vertices of $R_{3,2}^*$ and finally converting the (t_1, t_2)-coordinates into polar coordinates.

Hence the simultaneous confidence level is equal to

$$2\mathrm{P}\{ \theta_{\mathbf{T}} \in [-(\pi - \eta_1), \xi_1], R_{\mathbf{T}} \cos \theta_{\mathbf{T}} \leq c_1 \}$$

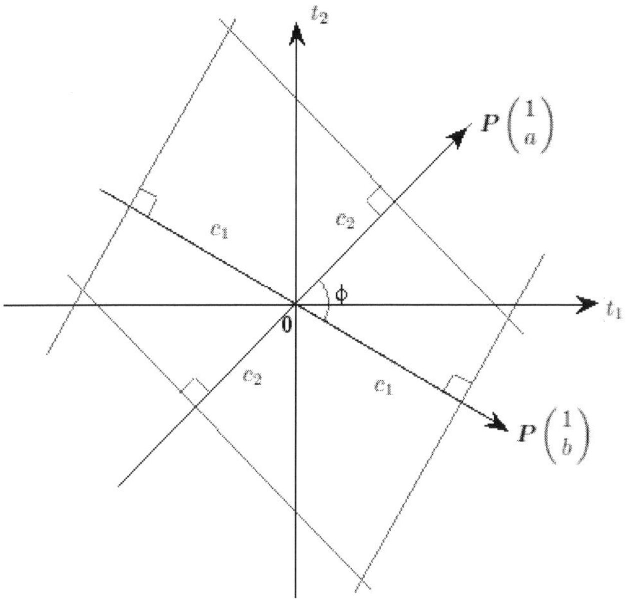

Figure 2.12: *The region* $R_{3,2}$

$$
\begin{aligned}
& +2\mathrm{P}\{\theta_{\mathbf{T}} \in [\xi_1, \eta_1], R_{\mathbf{T}} \cos(\theta_{\mathbf{T}} - \phi) \le c_2\} \\
= \; & 2\int_{-(\pi-\eta_1)}^{\xi_1} \frac{1}{2\pi} \mathrm{P}\{R_{\mathbf{T}} \cos\theta \le c_1\} d\theta + 2\int_{\xi_1}^{\eta_1} \frac{1}{2\pi} \mathrm{P}\{R_{\mathbf{T}} \cos(\theta - \phi) \le c_2\} d\theta \\
= \; & \frac{1}{\pi} \int_{-(\pi-\eta_1)}^{\xi_1} F_{R_{\mathbf{T}}}\left(\frac{c_1}{\cos\theta}\right) d\theta + \frac{1}{\pi} \int_{\xi_1-\phi}^{\eta_1-\phi} F_{R_{\mathbf{T}}}\left(\frac{c_2}{\cos\theta}\right) d\theta. \qquad (2.54)
\end{aligned}
$$

Expression (2.54) involves only one-dimensional integration since $F_{R_{\mathbf{T}}}(\cdot)$ is given explicitly in (2.6). Bowden and Graybill (1966) expressed the simultaneous confidence level as a two-dimensional integral of a bivariate t density function. In the special case of constant width band, Gafarian (1964) expressed the simultaneous confidence level as a one-dimensional integral similar to (2.54) by using the polar coordinates.

Example 2.3 For the data set considered in Example 2.1, it is interesting to construct a 95% simultaneous two-sided three-segment band for the regression line over $x \in (18, 70)$ and compare it with the two-sided hyperbolic band. Using expression (2.54), one can calculate the critical constants c_1 and c_2. Note, however,

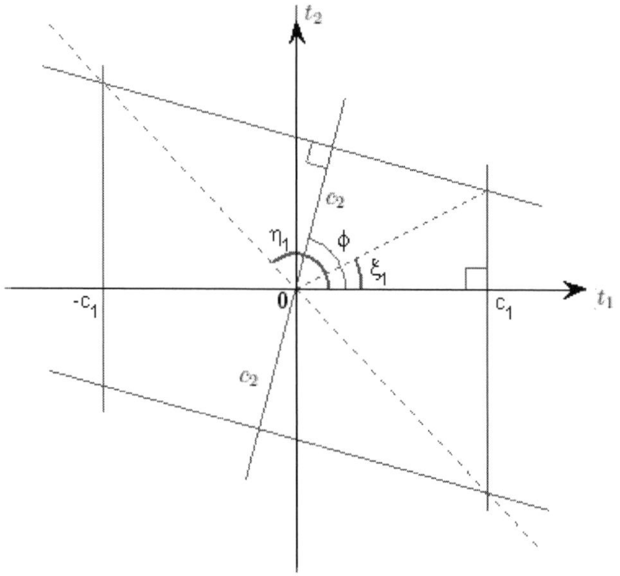

Figure 2.13: *The region* $R^*_{3,2}$

there are infinitely many pairs of (c_1, c_2) that satisfy the 95% simultaneous confidence level constraint. If we impose one more constraint on (c_1, c_2), then (c_1, c_2) can be determined uniquely. We impose specifically the constraint $c_1 = c_2$ so that the resultant three-segment band is optimal among all the three-segment band under the minimum area confidence set optimality criterion (see Section 2.7.2 below). In this case, we have $c_1 = c_2 = 2.296$. This three-segment band and the two-sided hyperbolic band are plotted in Figure 2.14.

It is clear from Figure 2.14 that the three-segment band tends to be narrower than the hyperbolic band for x near the two ends $x = a$ and $x = b$, but much wider than the hyperbolic band for x near the center of the interval $[a, b]$. This example is calculated using MATLAB program example0203.m.

2.3.2 *One-sided band*

The lower one-sided three-segment band has the form

$$\beta_0 + \beta_1 x > \hat{\beta}_0 + \hat{\beta}_1 x - \hat{\sigma} H_3(x) \quad \text{for all } x \in (a, b) \tag{2.55}$$

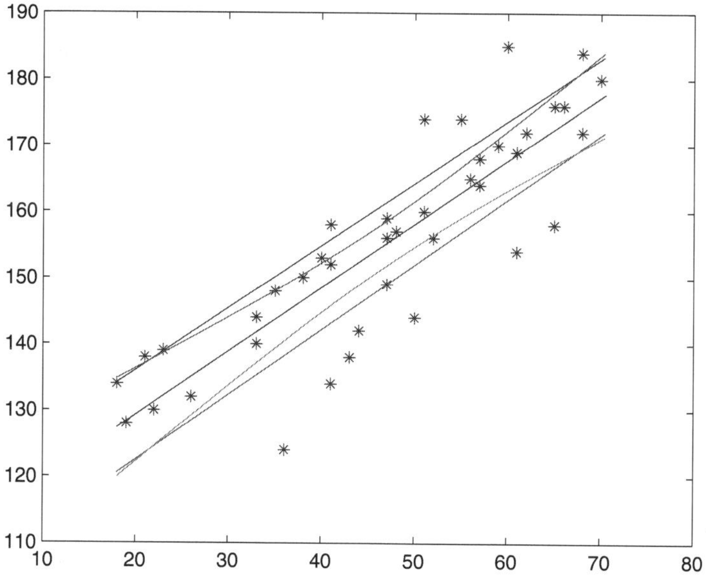

Figure 2.14: *The two-sided three-segment and hyperbolic bands*

and the upper one-sided three-segment band has the form

$$\beta_0 + \beta_1 x < \hat{\beta}_0 + \hat{\beta}_1 x + \hat{\sigma} H_3(x) \text{ for all } x \in (a,b), \tag{2.56}$$

where $H_3(x)$ is given in (2.46). It is clear that a one-sided three-segment band is simply a line segment over $x \in (a,b)$. The upper and lower confidence bands have equal simultaneous confidence levels, and so we evaluate only the simultaneous confidence level of the lower confidence band below.

Similar to the two-sided case, the lower three-segment band (2.55) is constructed by putting the following lower bounds on the regression line at both $x = a$ and $x = b$

$$\beta_0 + \beta_1 a > \hat{\beta}_0 + \hat{\beta}_1 a - \hat{\sigma} H_3(a), \ \beta_0 + \beta_1 b > \hat{\beta}_0 + \hat{\beta}_1 b - \hat{\sigma} H_3(b). \tag{2.57}$$

The confidence level of the band is therefore given by

$$\begin{aligned} & P\left\{\beta_0 + \beta_1 a > \hat{\beta}_0 + \hat{\beta}_1 a - \hat{\sigma} H_3(a), \ \beta_0 + \beta_1 b > \hat{\beta}_0 + \hat{\beta}_1 b - \hat{\sigma} H_3(b)\right\} \\ = \ & P\{(1,a)(\hat{\beta} - \beta)/\hat{\sigma}/H_3(a) < 1, \ (1,b)(\hat{\beta} - \beta)/\hat{\sigma}/H_3(b) < 1\}. \end{aligned} \tag{2.58}$$

Alternatively, the simultaneous confidence level is given by

$$P\left\{\beta_0 + \beta_1 x > \hat{\beta}_0 + \hat{\beta}_1 x - \hat{\sigma} H_3(x) \text{ for all } x \in (a,b)\right\}$$

$$= \mathrm{P}\left\{ \sup_{x\in(a,b)} \left((1,x)(\hat{\beta}-\beta)/\hat{\sigma}\right)/H_3(x) < 1 \right\}$$

$$= \mathrm{P}\left\{ \max_{x=a \text{ or } b} \left((1,x)(\hat{\beta}-\beta)/\hat{\sigma}\right)/H_3(x) < 1 \right\} \tag{2.59}$$

$$= \mathrm{P}\{\mathbf{T} \in R_{3,1}\} \tag{2.60}$$

where the equality in (2.59) follows from a similar argument as in the two-sided case, and

$$R_{3,1} = R_{3,1}(a) \cap R_{3,1}(b)$$

with

$$\begin{aligned} R_{3,1}(a) &= \{\mathbf{T}:\ \left((1,a)(\hat{\beta}-\beta)/\hat{\sigma}\right)/H_3(a) < 1\ \} \\ &= \left\{\mathbf{T}:\ \left\{\mathbf{P}\begin{pmatrix}1\\a\end{pmatrix}\right\}'\mathbf{T}\Big/\left\|\mathbf{P}\begin{pmatrix}1\\a\end{pmatrix}\right\| < c_2\right\} \end{aligned}$$

and

$$\begin{aligned} R_{3,1}(b) &= \{\mathbf{T}:\ \left((1,b)(\hat{\beta}-\beta)/\hat{\sigma}\right)/H_3(b) < 1\ \} \\ &= \left\{\mathbf{T}:\ \left\{\mathbf{P}\begin{pmatrix}1\\b\end{pmatrix}\right\}'\mathbf{T}\Big/\left\|\mathbf{P}\begin{pmatrix}1\\b\end{pmatrix}\right\| < c_1\right\}. \end{aligned}$$

It is clear that the expressions (2.60) and (2.58) are the same.

Note that both $R_{3,1}(a)$ and $R_{3,1}(b)$ are of the form (2.7). Hence $R_{3,1}$ is given by the region depicted in Figure 2.15. In particular the angle ϕ depicted in the picture is the same as in the two-sided case.

Let $R^*_{3,1}$ be the region that results from rotating $R_{3,1}$ around the origin to the position so that $\mathbf{P}\begin{pmatrix}1\\b\end{pmatrix}$ is in the direction of the t_1-axis, as depicted in Figure 2.16. It has the expression

$$\begin{aligned} &\{\mathbf{T}:\ \theta_{\mathbf{T}} \in [\xi_2, \phi+\pi/2], R_{\mathbf{T}}\cos(\theta_{\mathbf{T}}-\phi) \le c_2\} \\ \cup\ &\{\mathbf{T}:\ \theta_{\mathbf{T}} \in [-\pi/2, \xi_2], R_{\mathbf{T}}\cos\theta_{\mathbf{T}} \le c_1\} \\ \cup\ &\{\mathbf{T}:\ \theta_{\mathbf{T}} \in [\phi+\pi/2, 3\pi/2]\} \end{aligned}$$

where the angle ξ_2 corresponds to the angle ξ_1 in the two-sided case and is given by

$$\xi_2 = \arcsin\left(\frac{c_2 - c_1\cos\phi}{\sqrt{c_2^2+c_1^2-2c_2c_1\cos\phi}}\right).$$

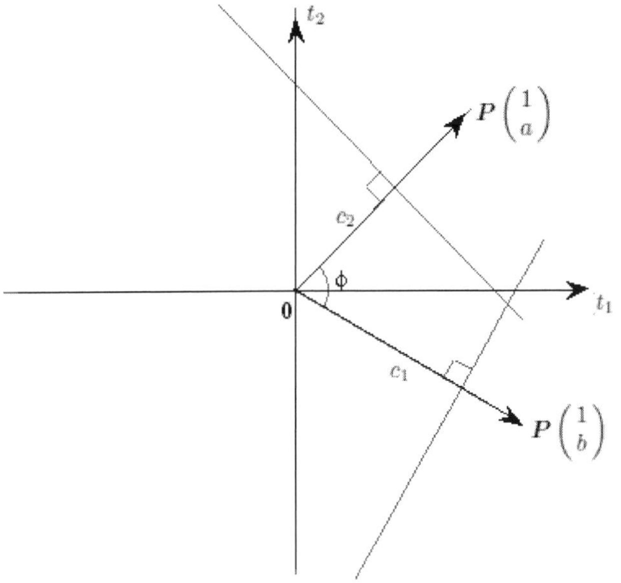

Figure 2.15: *The region* $R_{3,1}$

Due to the rotational invariance of the probability distribution of $\mathbf{T}$, the simultaneous confidence level is equal to

$$\begin{aligned}
& \mathrm{P}\{\mathbf{T} \in R^*_{3,1}\} \\
= \;& \mathrm{P}\{\theta_{\mathbf{T}} \in [\xi_2, \phi + \pi/2],\ R_{\mathbf{T}} \cos(\theta_{\mathbf{T}} - \phi) \le c_2\} \\
& + \mathrm{P}\{\theta_{\mathbf{T}} \in [-\pi/2, \xi_2],\ R_{\mathbf{T}} \cos\theta_{\mathbf{T}} \le c_1\} \\
& + \mathrm{P}\{\theta_{\mathbf{T}} \in [\phi + \pi/2, 3\pi/2]\} \\
= \;& \int_{\xi_2}^{\phi+\pi/2} \frac{1}{2\pi} \mathrm{P}\{R_{\mathbf{T}} \cos(\theta - \phi) \le c_2\}\, d\theta \\
& + \int_{-\pi/2}^{\xi_2} \frac{1}{2\pi} \mathrm{P}\{R_{\mathbf{T}} \cos\theta \le c_1\}\, d\theta + \frac{\pi - \phi}{2\pi} \\
= \;& \frac{1}{2\pi} \int_{\xi_2 - \phi}^{\pi/2} F_{R_{\mathbf{T}}}\left(\frac{c_2}{\cos\theta}\right) d\theta + \frac{1}{2\pi} \int_{-\pi/2}^{\xi_2} F_{R_{\mathbf{T}}}\left(\frac{c_1}{\cos\theta}\right) d\theta + \frac{\pi - \phi}{2\pi}. \quad (2.61)
\end{aligned}$$

If one chooses a and b so that $\hat{\beta}_0 + \hat{\beta}_1 a$ and $\hat{\beta}_0 + \hat{\beta}_1 b$ are independent random variables following the idea of Hayter *et al.* (2007), then $\phi = \pi/2$ from (2.13). In

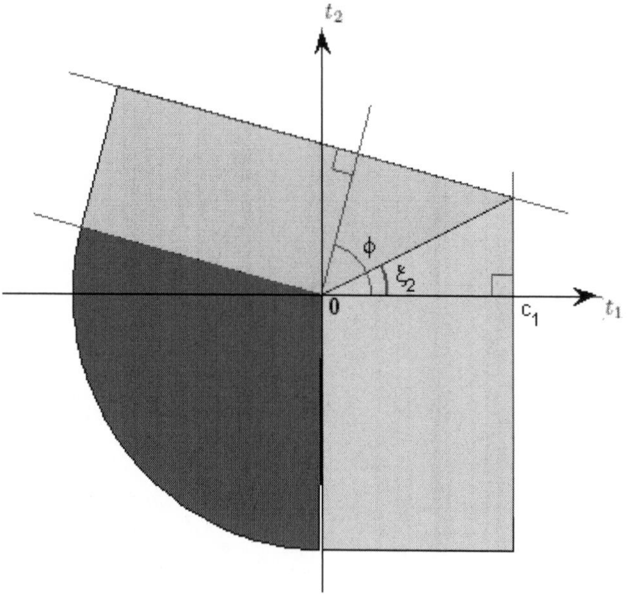

Figure 2.16: *The region* $R^*_{3,1}$

this case the expressions of simultaneous confidence level in Sections 2.2 and 2.3 may be simplified to a certain degree.

Example 2.4 We construct a 95% simultaneous one-sided three-segment band for the regression model over $x \in (18, 70)$ using the data set in Example 2.1. Again we need to impose one more constraint on (c_1, c_2) so that (c_1, c_2) can be determined uniquely. Under the restriction $c_1 = c_2$, we have calculated $c_1 = c_2 = 2.0243$ whilst the two-sided three-segment band uses $c_1 = c_2 = 2.296$. The upper three-segment band and the two-sided three segment band are plotted in Figure 2.17. It is clear that the upper three-segment band is lower than the upper part of the two-sided three-segment band. The curve in the picture is the upper hyperbolic band. All the computations are done using example0204.m.

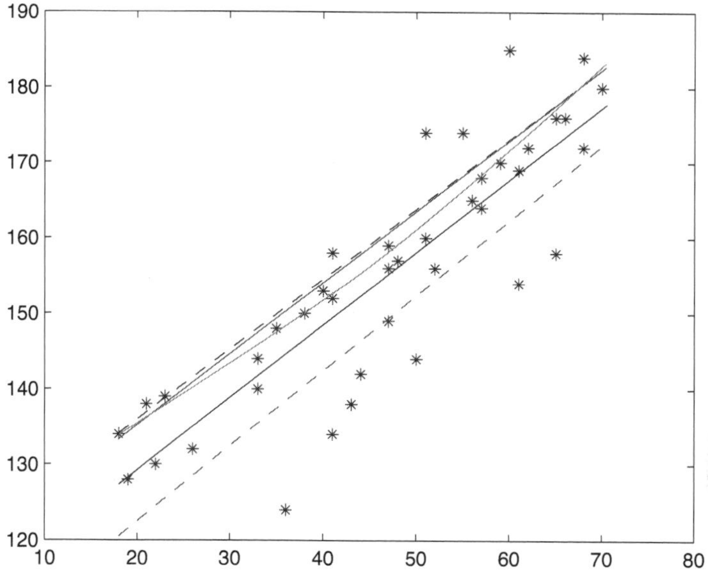

Figure 2.17: *Several simultaneous confidence bands*

2.4 Two-segment bands

2.4.1 Two-sided band

A two-segment band is defined on the whole range $x \in (-\infty, \infty)$, which is intrinsically different from the hyperbolic and three-segment bands. A two-sided two-segment band has the form

$$\beta_0 + \beta_1 x \in \hat{\beta}_0 + \hat{\beta}_1 x \pm \hat{\sigma} H_2(x) \text{ for all } x \in (-\infty, \infty) \tag{2.62}$$

where

$$H_2(x) = c_1 \sqrt{v(1, \bar{x})} + c_2 |x - \bar{x}| \sqrt{v(0, 1)}, \quad x \in (-\infty, \infty) \tag{2.63}$$

where the critical constants c_1 and c_2 are chosen so that the simultaneous confidence level is equal to $1 - \alpha$. The special case $c_1 = c_2$ of this band was first considered by Graybill and Bowden (1967).

Note that this two-sided two-segment band is constructed by putting the following two-sided bounds on the regression line at $x = \bar{x}$ and on the gradient β_1 of the regression line

$$\beta_0 + \beta_1 \bar{x} \in \hat{\beta}_0 + \hat{\beta}_1 \bar{x} \pm \hat{\sigma} c_1 \sqrt{v(1, \bar{x})}, \quad \beta_1 \in \hat{\beta}_1 \pm \hat{\sigma} c_2 \sqrt{v(0, 1)}. \tag{2.64}$$

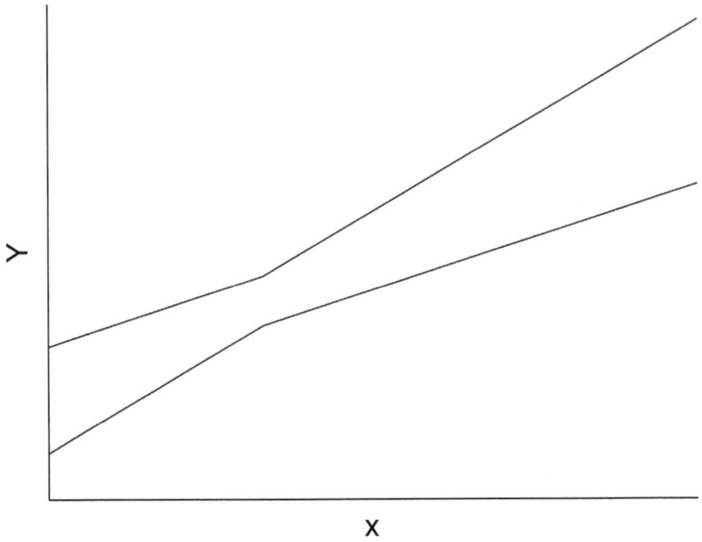

Figure 2.18: *The shape of a two-segment confidence band*

The shape of a two-sided two-segment band is illustrated in Figure 2.18. The center of the band at x is $\hat{\beta}_0 + \hat{\beta}_1 x$. Both the upper and lower parts of the band consist of two line segments. The band is the narrowest at $x = \bar{x}$ at which point the band is given by the two bounds on $\beta_0 + \beta_1 \bar{x}$ in (2.64). The gradients of the four line segments are either $\hat{\beta}_1 - \hat{\sigma} c_2 \sqrt{v(0,1)}$ or $\hat{\beta}_1 + \hat{\sigma} c_2 \sqrt{v(0,1)}$, which are the two bounds on β_1 in (2.64).

From (2.64), the simultaneous confidence level is given by

$$\begin{aligned}
&P\left\{\beta_0 + \beta_1 \bar{x} \in \hat{\beta}_0 + \hat{\beta}_1 \bar{x} \pm \hat{\sigma} c_1 \sqrt{v(1,\bar{x})},\ \ \beta_1 \in \hat{\beta}_1 \pm \hat{\sigma} c_2 \sqrt{v(0,1)}\right\} \\
&= P\left\{|(1,\bar{x})(\hat{\beta} - \beta)/\hat{\sigma}|/H_2(\bar{x}) < 1,\ |(1,\infty)(\hat{\beta} - \beta)/\hat{\sigma}|/H_2(\infty) < 1\right\}. \quad (2.65)
\end{aligned}$$

Alternatively, the simultaneous confidence level is given by

$$\begin{aligned}
&P\{\beta_0 + \beta_1 x \in \hat{\beta}_0 + \hat{\beta}_1 x \pm \hat{\sigma} H_2(x) \ \text{ for all } x \in (-\infty, \infty)\} \\
&= P\left\{\sup_{x \in (-\infty,\infty)} |(1,x)(\hat{\beta} - \beta)/\hat{\sigma}|/H_2(x) < 1\right\}. \quad (2.66)
\end{aligned}$$

Note that

$$\frac{\partial}{\partial x}\left(\left((1,x)(\hat{\beta} - \beta)/\hat{\sigma}\right)/H_2(x)\right)$$

has a fixed sign, either positive or negative, over $x > \bar{x}$ and over $x < \bar{x}$. The supreme in (2.66) is therefore attained at either $x = \bar{x}$ or limits $x \to -\infty$ or $x \to \infty$. So the confidence level can further be expressed as

$$P\left\{ \sup_{x=-\infty \text{ or } \bar{x} \text{ or } \infty} |(1,x)(\hat{\beta}-\beta)/\hat{\sigma}|/H_2(x) < 1 \right\} = P\{\mathbf{T} \in R_{2,2}\} \quad (2.67)$$

where

$$R_{2,2} = R_{2,2}(-\infty) \cap R_{2,2}(\bar{x}) \cap R_{2,2}(\infty) \quad (2.68)$$

with

$$\begin{aligned} R_{2,2}(\bar{x}) &= \{\mathbf{T}: |(1,\bar{x})(\hat{\beta}-\beta)/\hat{\sigma}|/H_2(\bar{x}) < 1\} \\ &= \left\{\mathbf{T}: \left|\left\{\mathbf{P}\begin{pmatrix}1\\ \bar{x}\end{pmatrix}\right\}'\mathbf{T}\right| \Big/ \left[c_1\sqrt{v(1,\bar{x})}\right] < 1\right\} \\ &= \left\{\mathbf{T}: \left|\left\{\mathbf{P}\begin{pmatrix}1\\ \bar{x}\end{pmatrix}\right\}'\mathbf{T}\right| \Big/ \left\|\mathbf{P}\begin{pmatrix}1\\ \bar{x}\end{pmatrix}\right\| < c_1\right\}, \end{aligned}$$

and

$$\begin{aligned} R_{2,2}(\infty) &= \left\{\mathbf{T}: \lim_{x\to\infty} |(1,x)(\hat{\beta}-\beta)/\hat{\sigma}|/H_2(x) < 1\right\} \\ &= \left\{\mathbf{T}: \left|\left\{\mathbf{P}\begin{pmatrix}0\\ 1\end{pmatrix}\right\}'\mathbf{T}\right| \Big/ \left\|\mathbf{P}\begin{pmatrix}0\\ 1\end{pmatrix}\right\| < c_2\right\} = R_{2,2}(-\infty). \end{aligned}$$

It is clear that expressions (2.67) and (2.65) are equal.

Note that $R_{2,2}(\bar{x})$ and $R_{2,2}(\pm\infty)$ are of the form (2.8). Hence $R_{2,2}$ is given by the parallelogram depicted in Figure 2.19. In particular the angle $\phi^* \in (0,\pi)$ depicted in the picture is formed by the vectors $\mathbf{P}\begin{pmatrix}0\\ 1\end{pmatrix}$ and $\mathbf{P}\begin{pmatrix}1\\ \bar{x}\end{pmatrix}$ and can be calculated from

$$\cos\phi^* = (0,1)(\mathbf{X}'\mathbf{X})^{-1}\begin{pmatrix}1\\ \bar{x}\end{pmatrix} \Big/ \sqrt{v(0,1)v(1,\bar{x})} = 0, \quad (2.69)$$

that is, $\phi^* = \pi/2$ and so $R_{2,2}$ is a rectangle.

By comparing the 'parallelogram' $R_{2,2}$ in Figure 2.19 with the parallelogram $R_{3,2}$ for the two-sided three-segment band in Figure 2.12, the confidence level $P\{\mathbf{T} \in R_{2,2}\}$ is given by expression (2.54) but with ϕ replaced with $\phi^* = \pi/2$, i.e.,

$$P\{\mathbf{T} \in R_{2,2}\} = \frac{1}{\pi}\int_{-(\pi-\eta_1^*)}^{\xi_1^*} F_{R_\mathbf{T}}\left(\frac{c_1}{\cos\theta}\right)d\theta + \frac{1}{\pi}\int_{\xi_1^*-\phi^*}^{\eta_1^*-\phi^*} F_{R_\mathbf{T}}\left(\frac{c_2}{\cos\theta}\right)d\theta \quad (2.70)$$

where

$$\xi_1^* = \arcsin\left(c_2/\sqrt{c_2^2+c_1^2}\right) \text{ and } \eta_1^* = \arccos\left(-c_1/\sqrt{c_2^2+c_1^2}\right).$$

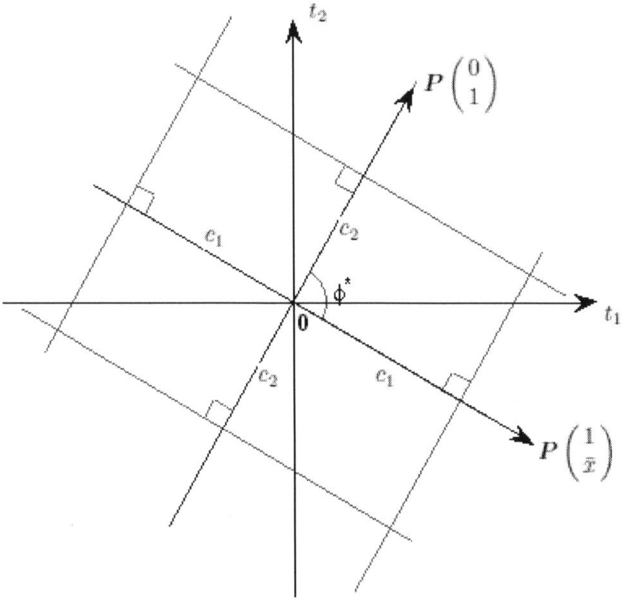

Figure 2.19: *The region* $R_{2,2}$

Note that the angles ξ_1^* and η_1^* correspond to the angles ξ_1 in (2.52) and η_1 in (2.53) respectively.

For the special case of $c_2 = c_1$, Graybill and Bowden (1967) expressed the simultaneous confidence level as a two-dimensional integral of a bivariate t density function. Expression (2.70) involves only one-dimensional integration however.

Example 2.5 We construct the 95% simultaneous two-sided two-segment band for the regression model using the data set in Example 2.1. Again we need to impose one more constraint on (c_1, c_2) so that (c_1, c_2) can be determined uniquely. The two-segment band with $c_1 = c_2$ is optimal among all the two-segment bands under certain criterion (see Section 2.7.2 below). Under the restriction $c_1 = c_2$, we have calculated $c_1 = c_2 = 2.326$. The two-sided hyperbolic, three-segment and two-segment bands are plotted in Figure 2.20. It is clear that the two-segment band is the narrowest at $x = \bar{x}$ compared with the other two bands. All the computations are done using example0205.m.

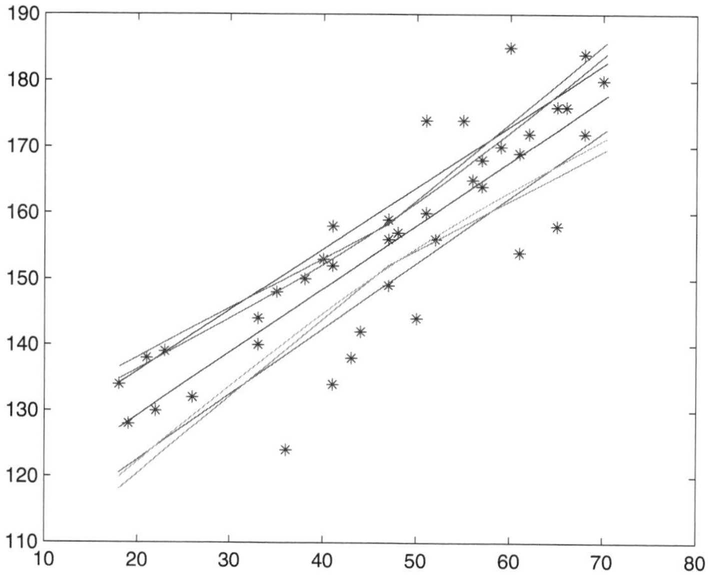

Figure 2.20: *Several simultaneous confidence bands*

2.4.2 *One-sided band*

A lower two-segment band has the form

$$\beta_0 + \beta_1 x > \hat{\beta}_0 + \hat{\beta}_1 x - \hat{\sigma} H_2(x) \ \text{ for all } x \in (-\infty, \infty) \tag{2.71}$$

and an upper two-segment band has the form

$$\beta_0 + \beta_1 x < \hat{\beta}_0 + \hat{\beta}_1 x + \hat{\sigma} H_2(x) \ \text{ for all } x \in (-\infty, \infty), \tag{2.72}$$

where $H_2(x)$ is given in (2.63). The lower and upper confidence bands have the same simultaneous confidence level, and so we evaluate the simultaneous confidence level of the lower band below.

Note that the lower two-segment band is constructed by putting the following lower bound on the regression line at $x = \bar{x}$ and two-sided bounds on the gradient β_1 of the regression line

$$\beta_0 + \beta_1 \bar{x} > \hat{\beta}_0 + \hat{\beta}_1 \bar{x} - \hat{\sigma} c_1 \sqrt{v(1, \bar{x})}, \ \ \beta_1 \in \hat{\beta}_1 \pm \hat{\sigma} c_2 \sqrt{v(0,1)}. \tag{2.73}$$

The shape of the lower two-segment band is the same as the lower part of the two-sided two-segment band. The gradients of the two line segments are $\hat{\beta}_1 - \hat{\sigma} c_2 \sqrt{v(0,1)}$ and $\hat{\beta}_1 + \hat{\sigma} c_2 \sqrt{v(0,1)}$, respectively.

From (2.73), the simultaneous confidence level is given by

$$\begin{aligned} & \mathrm{P}\left\{\beta_0+\beta_1\bar{x} > \hat{\beta}_0+\hat{\beta}_1\bar{x} - \hat{\sigma}c_1\sqrt{v(1,\bar{x})},\ \beta_1 \in \hat{\beta}_1 \pm \hat{\sigma}c_2\sqrt{v(0,1)}\right\} \\ = \ & \mathrm{P}\left\{(1,\bar{x})(\hat{\boldsymbol{\beta}}-\boldsymbol{\beta})/\hat{\sigma}/H_2(\bar{x}) < 1,\ |(1,\infty)(\hat{\boldsymbol{\beta}}-\boldsymbol{\beta})/\hat{\sigma}|/H_2(\infty) < 1\right\}. \quad (2.74)\end{aligned}$$

Alternatively, the simultaneous confidence level is given by

$$\begin{aligned} & \mathrm{P}\{\beta_0+\beta_1 x > \hat{\beta}_0+\hat{\beta}_1 x - \hat{\sigma}H_2(x) \ \text{ for all } x \in (-\infty,\infty)\} \\ = \ & \mathrm{P}\left\{\sup_{x\in(-\infty,\infty)} \left((1,x)(\hat{\boldsymbol{\beta}}-\boldsymbol{\beta})/\hat{\sigma}\right)/H_2(x) < 1\right\} \\ = \ & \mathrm{P}\left\{\sup_{x=-\infty \text{ or } \bar{x} \text{ or } \infty} \left((1,x)(\hat{\boldsymbol{\beta}}-\boldsymbol{\beta})/\hat{\sigma}\right)/H_2(x) < 1\right\} \quad (2.75) \\ = \ & \mathrm{P}\{\mathbf{T} \in R_{2,1}\}\end{aligned}$$

where the equality in (2.75) follows from a similar argument as in the two-sided case, and

$$R_{2,1} = R_{2,1}(-\infty) \cap R_{2,1}(\bar{x}) \cap R_{2,1}(\infty)$$

with

$$\begin{aligned} R_{2,1}(\bar{x}) &= \left\{\mathbf{T}:\ \left((1,\bar{x})(\hat{\boldsymbol{\beta}}-\boldsymbol{\beta})/\hat{\sigma}\right)/H_2(\bar{x}) < 1\right\} \\ &= \left\{\mathbf{T}:\ \left\{\mathbf{P}\begin{pmatrix}1\\ \bar{x}\end{pmatrix}\right\}'\mathbf{T} \Big/ \left\|\mathbf{P}\begin{pmatrix}1\\ \bar{x}\end{pmatrix}\right\| < c_1\right\}, \end{aligned}$$

$$\begin{aligned} R_{2,1}(\infty) &= \left\{\mathbf{T}:\ \lim_{x\to\infty} \left((1,x)(\hat{\boldsymbol{\beta}}-\boldsymbol{\beta})/\hat{\sigma}\right)/H_2(x) < 1\right\} \\ &= \left\{\mathbf{T}:\ \left\{\mathbf{P}\begin{pmatrix}0\\ 1\end{pmatrix}\right\}'\mathbf{T} \Big/ \left\|\mathbf{P}\begin{pmatrix}0\\ 1\end{pmatrix}\right\| < c_2\right\}, \end{aligned}$$

and

$$R_{2,1}(-\infty) = \left\{\mathbf{T}:\ \left\{\mathbf{P}\begin{pmatrix}0\\ 1\end{pmatrix}\right\}'\mathbf{T} \Big/ \left\|\mathbf{P}\begin{pmatrix}0\\ 1\end{pmatrix}\right\| > -c_2\right\}.$$

Hence expressions (2.75) and (2.74) are the same.

Note that $R_{2,1}(\bar{x})$ is of the form (2.7) and $R_{2,1}(\infty) \cap R_{2,1}(-\infty)$ is of the form (2.8). Hence $R_{2,1}$ is given by the region depicted in Figure 2.21 in which the angle $\phi^* = \pi/2$ as in the two-sided case.

Let $R^*_{2,1}$ be the region that results from rotating $R_{2,1}$ around the origin to the position so that $\mathbf{P}\begin{pmatrix}1\\ \bar{x}\end{pmatrix}$ is in the direction of the t_1-axis, as depicted in Figure

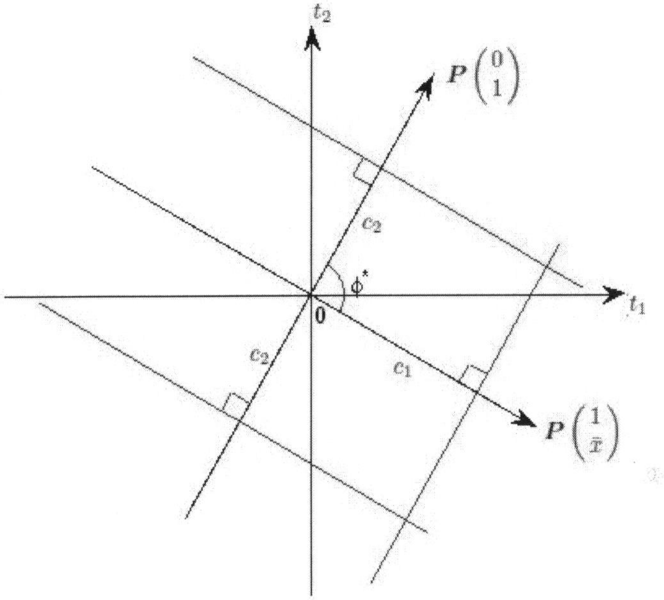

Figure 2.21: *The region $R_{2,1}$*

2.22. Due to the rotational invariance of the probability distribution of $\mathbf{T}$, the confidence level is equal to the probability of $\mathbf{T}$ in $R^*_{2,1}$, and $R^*_{2,1}$ has the expression

$$\begin{aligned} & \{\mathbf{T}:\ \theta_{\mathbf{T}} \in [-(\pi-\eta_2^*), \xi_2^*], R_{\mathbf{T}} \cos\theta_{\mathbf{T}} \le c_1\} \\ \cup\ & \{\mathbf{T}:\ \theta_{\mathbf{T}} \in [\xi_2^*, \pi], R_{\mathbf{T}} \cos(\theta_{\mathbf{T}} - \pi/2) \le c_2\} \\ \cup\ & \{\mathbf{T}:\ \theta_{\mathbf{T}} \in [\pi, \eta_2^* + \pi], R_{\mathbf{T}} \cos(\theta_{\mathbf{T}} - \pi/2) \ge -c_2\} \end{aligned}$$

where the angles ξ_2^* and η_2^*, depicted in Figure 22, are given by

$$\xi_2^* = \arcsin\left(c_2/\sqrt{c_2^2+c_1^2}\right) \text{ and } \eta_2^* = \arccos\left(-c_1/\sqrt{c_2^2+c_1^2}\right) = \pi - \xi_2^*.$$

Hence the required simultaneous confidence level is equal to

$$\begin{aligned} & \mathrm{P}\{\theta_{\mathbf{T}} \in [-(\pi-\eta_2^*), \xi_2^*], R_{\mathbf{T}} \cos\theta_{\mathbf{T}} \le c_1\} \\ & +\mathrm{P}\{\theta_{\mathbf{T}} \in [\xi_2^*, \pi], R_{\mathbf{T}} \cos(\theta_{\mathbf{T}} - \pi/2) \le c_2\} \\ & +\mathrm{P}\{\theta_{\mathbf{T}} \in [\pi, \eta_2^* + \pi], R_{\mathbf{T}} \cos(\theta_{\mathbf{T}} - \pi/2) \ge -c_2\} \\ = & \int_{-(\pi-\eta_2^*)}^{\xi_2^*} \frac{1}{2\pi} \mathrm{P}\{R_{\mathbf{T}} \cos\theta \le c_1\}\, d\theta \end{aligned}$$

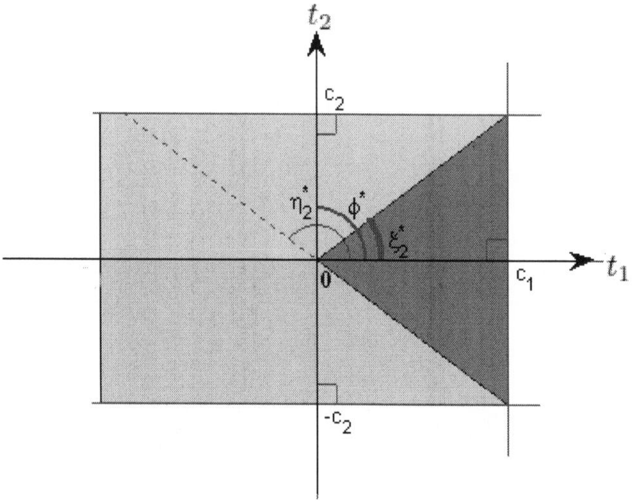

Figure 2.22: *The region* $R^*_{2,1}$

$$+2\int_{\xi_2^*}^{\pi} \frac{1}{2\pi} \mathrm{P}\{R_{\mathbf{T}}\cos(\theta-\pi/2) \le c_2\}\,d\theta$$

$$= \frac{1}{2\pi}\int_{-(\pi-\eta_2^*)}^{\xi_2^*} F_{R_{\mathrm{T}}}\left(\frac{c_1}{\cos\theta}\right)d\theta + \frac{1}{\pi}\int_{\xi_2^*-\pi/2}^{\pi/2} F_{R_{\mathrm{T}}}\left(\frac{c_2}{\cos\theta}\right)d\theta. \quad (2.76)$$

This last expression involves only one-dimensional integration and can be used to compute the critical constants.

Example 2.6 We construct the 95% simultaneous upper two-segment band for the regression model using the data set in Example 2.1. Again we need to impose one more constraint on (c_1, c_2) so that (c_1, c_2) can be determined uniquely. Under the restriction $c_1 = c_2$, we have calculated $c_1 = c_2 = 2.201$. The upper hyperbolic, upper three-segment and upper two-segment bands are all plotted in Figure 2.23. It is clear that the two-segment band is the lowest at $x = \bar{x}$ compared with the other two bands. All the computations are done using example0206.m.

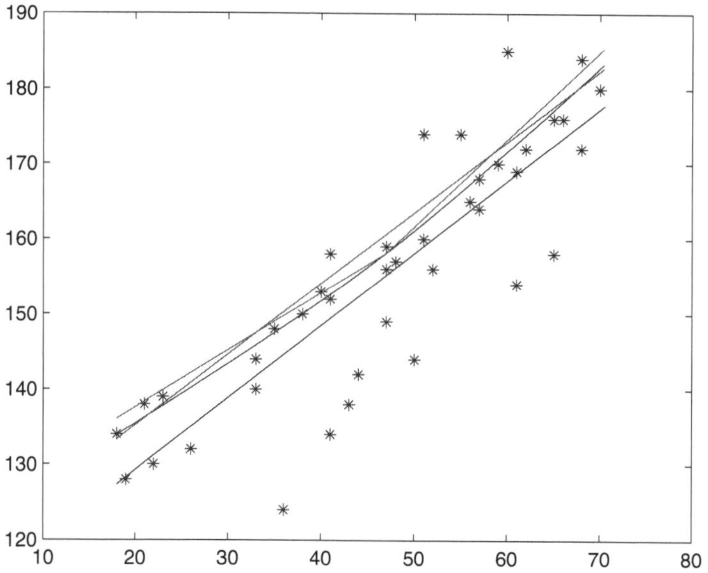

Figure 2.23: *Several one-sided simultaneous confidence bands*

2.5 Other confidence bands

Bowden (1970) showed how to use Hölder's inequality to derive the hyperbolic band over the whole range, three-segment bands, two-segment bands, and a general p-family of confidence bands. Assuming that the explanatory variable x is centered at the mean of the observed values of the explanatory variable, i.e., $\bar{x} = 0$, the p-family of confidence bands has the form

$$\beta_0 + \beta_1 x \in \hat{\beta}_0 + \hat{\beta}_1 x \pm c\hat{\sigma}(1 + \tau|x|^p)^{1/p} \text{ for all } x \in (a, b) \tag{2.77}$$

where $\tau > 0$ and $p \geq 1$ are given constants, and c is a critical constant. Hyperbolic band corresponds to $p = 2$. Piegorsch *et al.* (2000) used both polar coordinates as in Sections 2.2–2.4 of this chapter and double integrations of a bivariate t distribution to compute the simultaneous confidence level of the p-family confidence bands. Dunn (1968) constructed a confidence band for a regression straight line by bounding it at more than two covariate values, which generalizes the idea of the three-segment band. Dalal (1990) provided conservative and approximate confidence bands when the error variances at different x-values may be different. Other published confidence bands include those given by Naiman (1984a, 1987b) and Piegorsch (1985b) which are optimal under average width related criteria. Hayter *et al.* (2006a) considered how to modify a confidence band if it is used only when a preliminary test of the slope of the regression line is significant.

From Sections 2.2–2.4, it is clear that the simultaneous confidence level of any confidence band can be expressed as the probability of $\mathbf{T}$ in a region $R \subset \Re^2$. Note further that the region R can always be represented using the polar coordinates $(R_{\mathbf{T}}, \theta_{\mathbf{T}})$ in the form of

$$R = \{\mathbf{T}:\ \theta_{\mathbf{T}} \in (\theta_1, \theta_2), R_{\mathbf{T}} \leq G(\theta_{\mathbf{T}})\}$$

for some given constants $0 \leq \theta_1 \leq \theta_2 \leq 2\pi$ and given function G. So the simultaneous confidence level is equal to

$$\int_{\theta_1}^{\theta_2} \frac{1}{2\pi} F_{R_{\mathbf{T}}}[G(\theta)] d\theta$$

since $\theta_{\mathbf{T}}$ is uniformly distributed on the interval $[0, 2\pi)$ and independent of $R_{\mathbf{T}}$ whose cdf $F_{R_{\mathbf{T}}}(\cdot)$ is given in (2.6). Therefore the simultaneous confidence level can always be expressed as a one-dimensional integral, even though this one-dimensional integral may further be simplified in some special cases as for the one-sided hyperbolic band. Hayter *et al.* (2008) used this idea to compute the simultaneous confidence intervals for pairwise comparison of three normal means.

The key distributional assumption to the construction of simultaneous confidence bands is that $(\hat{\beta} - \beta)/\hat{\sigma}$ has a bivariate t distribution. This distributional assumption holds asymptotically for models other than the standard linear regression models considered here and, in particular, the linear mixed-effects models, generalized linear models and generalized linear mixed-effects models (see e.g., Pinheiro and Bates, 2000, and McCulloch and Searle, 2001). Construction of simultaneous confidence bands for linear mixed-effects models has been considered by Rao (1959), Elston and Grizzle (1962), Dawson *et al.* (1980), Stewart (1991a) and Sun *et al.* (1999) among others.

2.6 Extensions and restrictions of a confidence band

2.6.1 *Confidence band for $\mathbf{x}'\beta$ and confidence set for β*

Before we start the discussion, recall the following conventional notations. For a given vector (or point) $\mathbf{a} \in \Re^2$ and a given set (or region) $S \subset \Re^2$, let

$$\mathbf{a} + S := \{\mathbf{a} + \mathbf{s}:\ \mathbf{s} \in S\},$$

which is the set resulting from shifting S by an amount of $\mathbf{a}$. For a given 2×2 non-singular matrix M and a given region $S \subset \Re^2$, let

$$MS := \{M\mathbf{s}:\ \mathbf{s} \in S\},$$

which is a set resulting from linearly transforming S. It is clear that $M^{-1}MS = S$.

Every simultaneous confidence band for $\mathbf{x}'\beta$ induces a confidence set for the regression coefficients β. For the two-sided hyperbolic band (2.9), the induced confidence set is given by

$$C_{h,2}(\hat{\beta},\hat{\sigma}) := \{\beta : \mathbf{P}^{-1}(\beta-\hat{\beta})/\hat{\sigma} \in R_{h,2}\} \tag{2.78}$$

where $R_{h,2} \subset \Re^2$ is given in (2.12). This confidence set can further be expressed as

$$C_{h,2}(\hat{\beta},\hat{\sigma}) = \hat{\beta} + \hat{\sigma}\mathbf{P}R_{h,2}.$$

For the special case of $(a,b) = (-\infty,\infty)$, $R_{h,2}$ is a disc centered at the origin. Region $\hat{\sigma}\mathbf{P}R_{h,2}$ is a linear transformation of the disc and given by an ellipse centered at the origin. Region $\hat{\beta} + \hat{\sigma}\mathbf{P}R_{h,2}$ is the ellipse of same size but centered at $\hat{\beta}$. This is the usual ellipsoidal confidence region for β given in (2.1).

For a general (a,b), $R_{h,2}$ is a spindle centered at the origin as depicted in Figure 2.2. The region $\hat{\sigma}\mathbf{P}R_{h,2}$ is a linear transformation of this spindle and given by another spindle which is centered at the origin and may not be reflectionally symmetric. Finally the confidence region $\hat{\beta} + \hat{\sigma}\mathbf{P}R_{h,2}$ is generated by shifting the latter spindle to be centered at $\hat{\beta}$.

For the two-sided three-segment band (2.45), the induced confidence set is given by

$$C_{3,2}(\hat{\beta},\hat{\sigma}) := \{\beta : \mathbf{P}^{-1}(\beta-\hat{\beta})/\hat{\sigma} \in R_{3,2}\} \tag{2.79}$$

where $R_{3,2} \subset \Re^2$ is given in (2.51). This confidence set can further be expressed as

$$C_{3,2}(\hat{\beta},\hat{\sigma}) = \hat{\beta} + \hat{\sigma}\mathbf{P}R_{3,2}.$$

Region $R_{3,2}$ is a parallelogram centered at the origin as depicted in Figure 2.12. The region $\hat{\sigma}\mathbf{P}R_{h,2}$ is a linear transformation of this parallelogram and given by another parallelogram which is also centered at the origin. Finally the confidence region $\hat{\beta} + \hat{\sigma}\mathbf{P}R_{3,2}$ is generated by shifting the latter parallelogram to be centered at $\hat{\beta}$.

For the two-sided two-segment band (2.62), the induced confidence set is given by

$$C_{2,2}(\hat{\beta},\hat{\sigma}) := \{\beta : \mathbf{P}^{-1}(\beta-\hat{\beta})/\hat{\sigma} \in R_{2,2}\} \tag{2.80}$$

where $R_{2,2} \subset \Re^2$ is given in (2.68). This confidence set can further be expressed as

$$C_{2,2}(\hat{\beta},\hat{\sigma}) = \hat{\beta} + \hat{\sigma}\mathbf{P}R_{2,2}.$$

Region $R_{2,2}$ is a rectangle centered at the origin as depicted in Figure 2.19. The region $\hat{\sigma}\mathbf{P}R_{h,2}$ is a linear transformation of this rectangle and given by a parallelogram which is also centered at the origin. Finally the confidence region $\hat{\beta} + \hat{\sigma}\mathbf{P}R_{2,2}$ is generated by shifting the parallelogram to be centered at $\hat{\beta}$.

In a similar way, one can write the confidence sets that are induced by the one-sided hyperbolic, three-segment and two-segment bands. Indeed any simultaneous confidence band for the regression model $\mathbf{x}'\beta$ induces a confidence set for β.

One interesting question is whether any confidence set for β induces a simultaneous confidence band for the regression model $\mathbf{x}'\beta$. The answer is positive if a confidence set is convex. The correspondence between confidence bands for $\mathbf{x}'\beta$ and confidence sets for β has been studied by Folks and Antle (1967), Haplerin (1963a, 1963b), Khorasani and Milliken (1979), Kanoh and Kusunoki (1984), and Piegorsch (1985a, 1987a), to which the reader is referred.

One general method for constructing a simultaneous confidence band for $\mathbf{x}'\beta$ is from a given confidence set for β in the following way. Let $C(\hat{\beta}, \hat{\sigma})$ be a confidence set for β. Then for each given x, the possible range of the regression model $\beta_0 + \beta_1 x$ at x, when β varies in the confidence set $\beta \in C(\hat{\beta}, \hat{\sigma})$, is given by $(l(x), u(x))$ where

$$l(x) = inf_{\beta \in C(\hat{\beta}, \hat{\sigma})} \beta_0 + \beta_1 x, \text{ and } u(x) = sup_{\beta \in C(\hat{\beta}, \hat{\sigma})} \beta_0 + \beta_1 x.$$

Note that

$$\beta_0 + \beta_1 x = \left\{ \beta^T \begin{pmatrix} 1 \\ x \end{pmatrix} \Big/ \left\| \begin{pmatrix} 1 \\ x \end{pmatrix} \right\| \right\} \left\| \begin{pmatrix} 1 \\ x \end{pmatrix} \right\|$$

and the quantity inside the braces is just the projection of β on the vector $(1, x)'$. The maximum of the projection of β when β varies over $C(\hat{\beta}, \hat{\sigma})$ is clearly attained at β^u where β^u is depicted in Figure 2.24 and so $u(x) = (1, x)\beta^u$. Similarly the minimum of the projection of β when β varies over $C(\hat{\beta}, \hat{\sigma})$ is attained at β^l where β^l is depicted in Figure 2.24 and $l(x) = (1, x)\beta^l$. Consequently, the confidence band is given by

$$\beta_0 + \beta_1 x \in (l(x),\ u(x)) \text{ for all } x \in (-\infty,\ \infty).$$

This is indeed the method used by Working and Hotelling (1929) to derive the hyperbolic confidence band over $x \in (-\infty, \infty)$ from the confidence ellipse for β. This idea has been used by Hoel (1951, 1954) and Gafarian (1978) to derive conservative confidence bands for multiple linear regression models. It has also been used for non-linear regressions by Haplerin (1963a, 1963b), Khorasani (1982) and Cheng (1987) among others. Note, however, a confidence band derived in this way is often conservative. A different method of construction of simultaneous confidence bands for a general parametric regression model is given in Gsteiger *et al.* (2010).

2.6.2 Extension and restriction of a confidence band

Note that the hyperbolic and three-segment bands over a finite interval $x \in (a, b)$ can be extended to any interval (a^*, b^*) that contains interval (a, b). Specifically, the extension of the hyperbolic band is illustrated in Figure 2.25, in which the

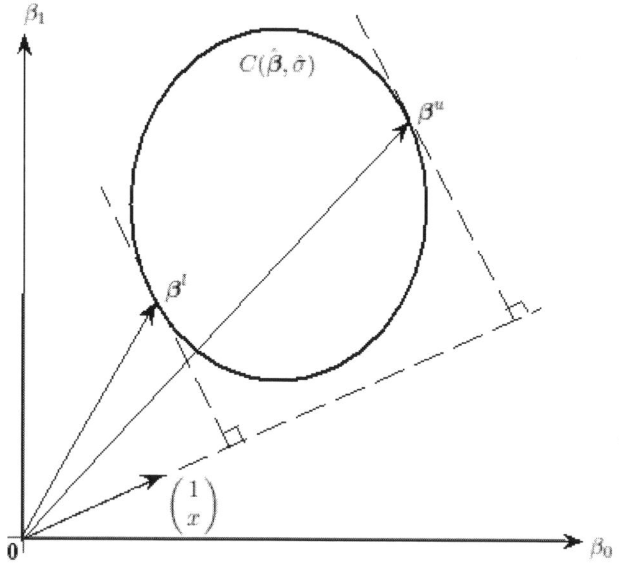

Figure 2.24: *Construction of confidence band from confidence set*

extension overlaps with the original hyperbolic band over $x \in (a,b)$, and over $x \in (a^*,a) \cup (b,b^*)$ the extension is made up of line segments of the straight line that passes through the lower bound at $x = a$ and upper bound at $x = b$ and of the straight line that passes through the upper bound at $x = a$ and lower bound at $x = b$. The extension of the three-segment band can be constructed in a similar way and is illustrated in Figure 2.26. Note that both the upper and lower parts of this extension are made up of three line segments, hence the name three-segment band.

The construction of an extension utilizes the shape of the regression model, which is a straight line in this case. The original confidence band on $x \in (a,b)$ and the straight line shape of the regression model stipulate that the regression straight line cannot go beyond the extended band over any interval $x \in (a^*,b^*) \supset (a,b)$. So the constraints of the extended band on the regression model over $x \in (a^*,a) \cup (b,b^*)$ are implicitly implied by the original confidence band. The simultaneous confidence level of the extended band is therefore equal to that of the original band. Indeed the confidence set induced by the original band is the same as that induced by the extended band.

On the other hand, note that the two-segment band is defined over $x \in$

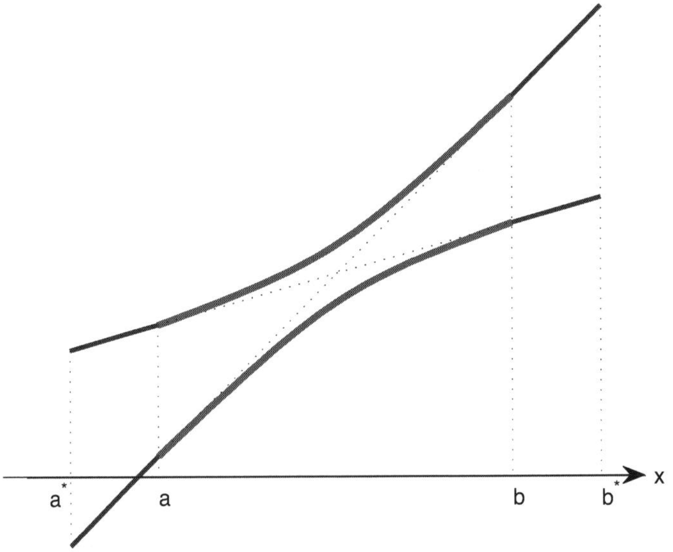

Figure 2.25: *Extension of the hyperbolic band*

$(-\infty,\infty)$ intrinsically. If we restrict the two-segment band to a finite interval $x \in (a,b)$, by which we mean that only the constraints of the two-segment band over $x \in (a,b)$ are used while the constraints of the two-segment outside the interval (a,b) are deleted, then this restriction to the interval $x \in (a,b)$ of the original band has a simultaneous confidence level larger than the simultaneous confidence level of the original two-segment band. In fact, the restriction of the two-segment band to a finite interval $x \in (a,b)$ becomes a four-segment band whose exact simultaneous confidence level can be computed involving one-dimensional integration by using the formula in Section 2.5.

2.7 Comparison of confidence bands

By using the expressions of simultaneous confidence level derived in Sections 2.2–2.4, one can calculate the critical constants to achieve a required confidence level. Note, however, there are two critical constants in either the three-segment band or the two-segment band. So a constraint on the two critical constants in addition to the $1-\alpha$ confidence level requirement is necessary in order to determine the two critical constants uniquely. Gafarian (1964) imposed

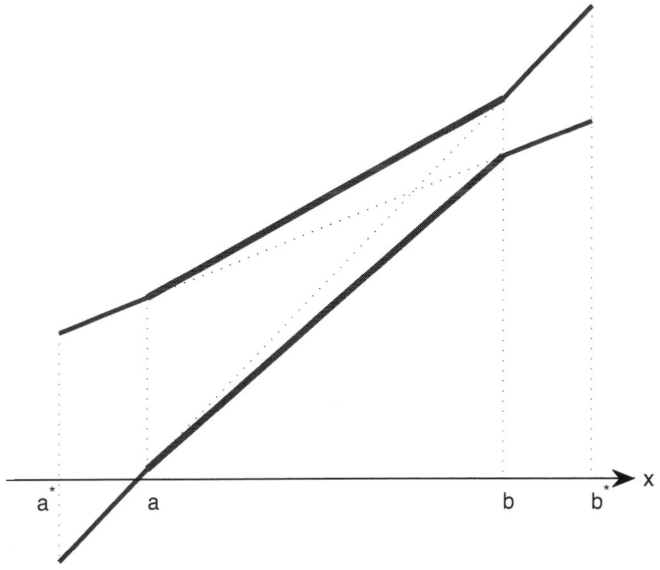

Figure 2.26: *Extension of the three-segment band*

$c_1\sqrt{v(1,b)} = c_2\sqrt{v(1,a)}$ and Graybill and Bowden (1967) imposed $c_1 = c_2$. There is also the question 'which confidence band among the hyperbolic, three-segment and two-segment bands is the best?' In this section we look at the average width and minimum area confidence set criteria for comparing confidence bands. Naiman (1987b) considered minimax regret simultaneous confidence bands.

2.7.1 Average width criterion

The average width of a given confidence band can be written straightforwardly. For example, the average width of the hyperbolic band in (2.9) is given by

$$\int_a^b 2c\hat{\sigma}\sqrt{v(1,x)}dx/(b-a) \tag{2.81}$$

since the width of the band at x is $2c\hat{\sigma}\sqrt{v(1,x)}$, while the average width of the three-segment band in (2.45) is given by

$$\int_a^b 2\hat{\sigma}H_3(x)dx/(b-a). \tag{2.82}$$

The average width of a confidence band can be generalized to weighted average width

$$\int_a^b w(x)du(x) \tag{2.83}$$

where $w(x)$ denotes the width of the confidence band at x, and $u(\cdot)$ denotes a specified probability measure over the interval (a,b) on which the confidence band is defined. The average width of a confidence band corresponds to the uniform probability measure over (a,b). The weighted average width of a confidence band over an infinite interval is well defined by choosing a suitable $u(\cdot)$.

The width of a one-sided confidence band is often taken as the distance between the band and the estimated regression line. Hence the average width of the one-sided hyperbolic band in (2.32) is given, for example, by

$$\int_a^b c\hat{\sigma}\sqrt{v(1,x)}dx/(b-a). \tag{2.84}$$

Intuitively, a narrower confidence band provides more accurate information about the unknown regression line. The average width of a confidence band indicates its narrowness. Hence the smaller is the average width, the narrower and better is the confidence band. Average width was used explicitly in Gafarian (1964) as an optimality criterion for the first time. Various authors discussed the motivations behind the average width optimality; see for example Naiman (1983). Naiman (1983) compared the hyperbolic and constant width bands for a multiple linear regression model over a particular ellipsoidal region of the covariates under the average width criterion. Naiman (1984a) provided a probability measure $u(\cdot)$ with respect to which the hyperbolic band over the finite interval (a,b) has the smallest weighted average width. Piegorsch (1985b) found the probability measures $u(\cdot)$ for which the hyperbolic band and the two-segment band have the smallest weighted average width over the whole line $(a,b) = (-\infty,\infty)$. Kanoh (1988) considered how to reduce the average width of the band in (2.77) by choosing a suitable value of τ which determines the curvature of the band.

Next we present some numerical comparisons of the hyperbolic band and the constant width band on a finite interval under the average width criterion. Specifically, we consider only the case that $a = \bar{x} - \delta$ and $b = \bar{x} + \delta$, i.e., the interval (a,b) is symmetric about $\bar{x}$. In this case, the constant width band is given by the three-segment band in (2.45) but with $c_1 = c_2$, which is denoted as c_c to indicate it is the critical value of the constant width band. Furthermore, the ratio of the average width of the hyperbolic band and the average width of the constant width band is given by

$$\begin{aligned} e &= \int_a^b 2c\hat{\sigma}\sqrt{v(1,x)}dx \Big/ \int_a^b 2\hat{\sigma}H_3(x)dx \\ &= \frac{c}{c_c}\int_0^{\sqrt{n\delta^2/S_{xx}}} \sqrt{w^2+1}dw \Big/ \left(\sqrt{n\delta^2/S_{xx}}\sqrt{n\delta^2/S_{xx}+1}\right) \end{aligned} \tag{2.85}$$

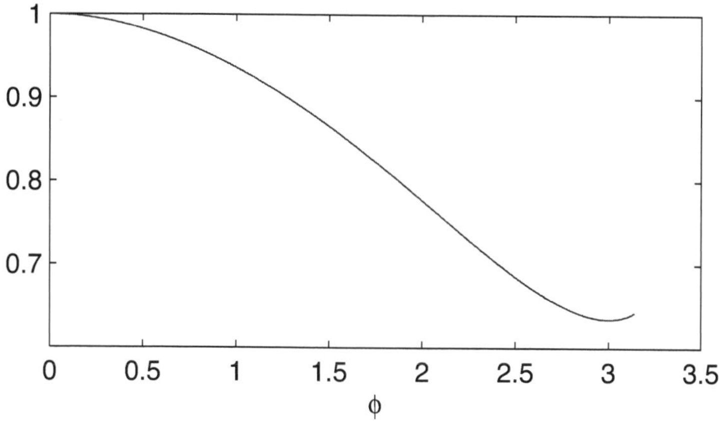

Figure 2.27: *A plot of the function* $e(\phi)$

where c is the critical value of the hyperbolic band, and $S_{xx} = \sum_{i=1}^{n}(x_i - \bar{x})^2$.

Note that the $\cos\phi$ in (2.13) is calculated to be

$$\cos\phi = (1 - n\delta^2/S_{xx})/(1 + n\delta^2/S_{xx})$$

and so $n\delta^2/S_{xx}$ is determined by ϕ. Furthermore, the critical value c of the hyperbolic band is determined by ϕ, degrees of freedom ν and α from (2.20), and the critical value c_c of the constant width band is determined also by ϕ, ν and α from (2.54) by noting that $\xi_1 = \phi/2$ and $\eta_1 = (\pi + \phi)/2$ in this case. Hence the ratio e given by (2.85) depends on ϕ, ν and α only.

Figure 2.27 presents how e changes with ϕ for given values of $\alpha = 0.05$ and $\nu = 10$. This picture is typical for all the combinations of $\alpha = 0.01, 0.05, 0.10$ and $\nu = 6, 10, 20, \infty$ that I have tried. As a function of ϕ, e first decreases from 1 as ϕ increases from 0, and then increases as ϕ increases from about 3.0 to $\pi = 3.1415$. The value of e is always smaller than 1 for $\phi \in (0, \pi)$, which indicates that the constant width band is less efficient than the hyperbolic band under the average width criterion. The smallest value of e over $\phi \in (0, \pi)$ is somewhere about 0.65 and so the constant width band can be considerably less efficient than the hyperbolic band under the average width criterion. But the shape of the constant width band may be more desirable than that of the hyperbolic band in some applications. For example, in Piegorsch *et al.* (2005) where confidence bands are used in risk analysis, it is crucial that a confidence band should be narrower especially near the lower end $x = a$. Hence a constant width or three segment bands should do better than the hyperbolic band that was considered by Piegorsch *et al.* (2005). Figure 2.27 is produced by MATLAB program example020701.m.

Naiman (1983) generalized this to the comparisons of hyperbolic and constant width bands for a multiple linear regression model over a special ellipsoidal co-

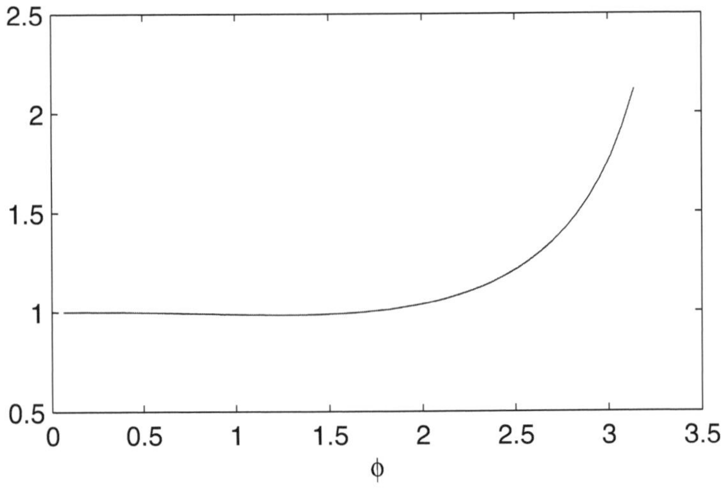

Figure 2.28: *A plot of the function* $E(\phi)$

variate region centered at the mean vector of the observed covariate values given in (3.17) in Chapter 3.

2.7.2 *Minimum area confidence set criterion*

Note that each regression line that is completely inside the confidence band is deemed by the confidence band as a plausible candidate of the true but unknown regression line, and so a better confidence band should contain less regression lines. From Section 2.6.1, each confidence band corresponds to a confidence set for $\boldsymbol{\beta}$, that is, each regression line that is contained inside the confidence band corresponds to a point inside the confidence set for $\boldsymbol{\beta}$. It is therefore desirable that a confidence band has its corresponding confidence set as small as possible in area. This motivates the minimum area confidence set (MACS) criterion proposed in Liu and Hayter (2007). With the confidence level fixed at $1-\alpha$, a confidence band that has a smaller confidence set area is better, and the confidence band with minimum confidence set area will be optimal under this MACS criterion. This criterion is closely related to the classical D-optimality in experiment design in that D-optimal designs minimize the area of the ellipsoidal confidence set for $\boldsymbol{\beta}$ in (2.1); see e.g., Atkinson and Donev (1992).

Under the MACS criterion, it is shown in Liu and Hayter (2007) that among all the $1-\alpha$ level two-sided three-segment confidence bands, the one that has $c_1 = c_2$ is optimal, that among all the $1-\alpha$ level two-sided two-segment confidence bands, the one that has $c_1 = c_2$ is optimal, and that the optimal two-segment band is better than the optimal three-segment band unless $\phi = \pi/2$.

To compare the optimal three-segment band with the hyperbolic band over an interval (a,b), we can compare the ratio of the areas of the corresponding confidence sets

$$E(\phi) = \frac{\text{area of confidence set corresponding to the optimal three} - \text{segment band}}{\text{area of confidence set corresponding to the hyperbolic band}}$$

which turns out to be dependent on ϕ, ν and α only. Figure 2.28 depicts the function $E(\phi)$ for $\alpha = 0.05$ and $\nu = 10$, which is typical for all combinations of $\alpha = 0.01, 0.05, 0.10$ and $\nu = 6, 10, 20, \infty$. The function $E(\phi)$ first decreases from 1 as ϕ increases from 0. The minimum of $E(\phi)$ is attained near $\phi = 1$ and the minimum value of $E(\phi)$ is no less than 0.98. The function $E(\phi)$ then increases as ϕ increases from the minimum point, is equal to 1 at about $\phi = 1.5$, and approaches ∞ as ϕ approaches π. So under the MACS criterion the best three-segment band is just slightly more efficient than the hyperbolic band when ϕ is no larger than about $\pi/2$. But the best three-segment band is less efficient than the hyperbolic band when ϕ is larger than about $\pi/2$, and dramatically less efficient than the hyperbolic band when ϕ is close to π. Figure 2.28 is produced by the MATLAB program example020702.m.

Comparison of the hyperbolic and constant width bands for a multiple linear regression model over a special ellipsoidal covariate region (see Chapter 3) under the minimum volume confidence set criterion is considered in Liu, Hayter, Piegorsch and Ah-Kine (2009), and results similar to the case of $p = 1$ above are observed. Searching for the optimal confidence band under the MACS criterion within certain family of $1-\alpha$ level simultaneous confidence bands is considered in Liu and Ah-Kine (2010). One can further consider how to choose a suitable design (matrix $\mathbf{X}$) to produce a better confidence band under a given criterion. Gafarian (1964) considered the optimal design for the constant width band under the average width criterion and is the only work available on optimal designs for confidence bands.

2.8 Confidence bands for percentiles and tolerance bands

The regression function $\mathbf{x}'\beta$ is a special case of the regression percentile function $\mathbf{x}'\beta + z^\gamma\sigma$, where z^γ denotes the upper γ point of the standard normal distribution. Construction of simultaneous confidence bands for $\mathbf{x}'\beta + z^\gamma\sigma$ has been considered by Steinhorst and Bowden (1971), Kabe (1976), Turner and Bowden (1977, 1979) and Thomas and Thomas (1986) among others.

Lieberman and Miller (1963) considered the construction of simultaneous confidence bands for the central $(1-2\gamma)$-content of the normal distribution $N(\mathbf{x}'\beta, \sigma^2)$:

$$(\mathbf{x}'\beta - z^\gamma\sigma,\ \mathbf{x}'\beta + z^\gamma\sigma).$$

Simultaneous $(1-2\gamma)$-content (not necessarily central) tolerance bands are further studied in Scheffé (1973), Jones *et al.* (1985), Limam and Thomas (1988),

Odeh and Mee (1990), and Mee *et al.* (1991). Many ideas presented in this and the next chapters are useful in the construction of these types of simultaneous bands.

2.9 Bayesian simultaneous credible bands

One may consider the construction of $1-\alpha$ level *Bayesian simultaneous credible bands* for $\mathbf{x}'\boldsymbol{\beta}$, such as

$$\mathrm{P}\left\{\mathbf{x}'\boldsymbol{\beta} \in \mathbf{x}'\mathrm{E}(\boldsymbol{\beta}) \pm c\sqrt{\mathrm{Var}(\mathbf{x}'\boldsymbol{\beta})} \quad \forall\, x \in (a,b)\right\} = 1-\alpha.$$

Here $\boldsymbol{\beta}$ is a random vector, and $\mathrm{P}\{\cdot\}$, $\mathrm{E}(\boldsymbol{\beta})$ and $\mathrm{Var}(\mathbf{x}'\boldsymbol{\beta})$ are with respect to the posterior distribution of $\boldsymbol{\beta}$ given $\mathbf{Y}$. The critical constant c is chosen so that the posterior probability of $\mathbf{x}'\boldsymbol{\beta}$ being contained in the given interval simultaneously for all $x \in (a,b)$ is equal to $1-\alpha$. For this, one needs to postulate prior distribution on $(\boldsymbol{\beta},\sigma)$ (cf. Broemeling, 1985) and evaluate the posterior distribution of $\boldsymbol{\beta}$. Ultimately, one needs to evaluate the posterior distribution of

$$\sup_{x\in(a,b)} \frac{|\mathbf{x}'(\boldsymbol{\beta}-\mathrm{E}(\boldsymbol{\beta}))|}{\sqrt{\mathrm{Var}(\mathbf{x}'\boldsymbol{\beta})}} \tag{2.86}$$

in order to determine the critical constant c. If the prior distribution is such that the posterior distribution of $\boldsymbol{\beta}$ can be determined analytically, then the posterior distribution of the random variable in (2.86) can be evaluated by using the ideas presented in this chapter. Otherwise, Markov Chain Monte Carlo simulation methods (cf. Gamerman and Lopes, 2006) are necessary. To the best of my knowledge, there is no published work on Bayesian simultaneous credible bands.

3

Confidence Bands for One Multiple Regression Model

As pointed out at the beginning of Chapter 2, a simultaneous confidence band for $\mathbf{x}'\boldsymbol{\beta}$ quantifies the plausible range of the true model $\mathbf{x}'\boldsymbol{\beta}$ directly. Any regression model function $\mathbf{x}'\boldsymbol{\beta}_0$ that lies inside the simultaneous confidence band is deemed by this band as a plausible candidate of the true model $\mathbf{x}'\boldsymbol{\beta}$. On the other hand, any regression model $\mathbf{x}'\boldsymbol{\beta}_0$ that does not lie inside the simultaneous confidence band is not a plausible candidate of the true model $\mathbf{x}'\boldsymbol{\beta}$. As a by-product, a confidence band can also be used to test the hypotheses in (2.2).

In this chapter we focus on the construction of simultaneous confidence bands for a linear regression model that has $p \geq 1$ explanatory variables. All the results except those in Sections 3.4 and 3.5 are for a general $p \geq 1$. Let $\mathbf{x}_{(0)} = (x_1, \cdots, x_p)' \in \Re^p$.

3.1 Hyperbolic bands over the whole space

3.1.1 Two-sided band

The most well-known simultaneous confidence band of level $1-\alpha$ for the regression model $\mathbf{x}'\boldsymbol{\beta}$ is for all the $\mathbf{x}_{(0)} \in \Re^p$ and given by

$$\mathbf{x}'\boldsymbol{\beta} \in \mathbf{x}'\hat{\boldsymbol{\beta}} \pm \sqrt{(p+1)f^{\alpha}_{p+1,n-p-1}}\,\hat{\sigma}\sqrt{\mathbf{x}'(\mathbf{X}'\mathbf{X})^{-1}\mathbf{x}} \quad \forall\, \mathbf{x}_{(0)} \in \Re^p. \tag{3.1}$$

This confidence band was given by Hoel (1951) and Scheffé (1953, 1959), and generalizes the band for a simple linear regression model of Working and Hotelling (1929). The lower and upper parts of the band are symmetric about the fitted model $\mathbf{x}'\hat{\boldsymbol{\beta}}$, and the width of the band at each $\mathbf{x}$ is proportional to $\sqrt{\mathrm{Var}(\mathbf{x}'\hat{\boldsymbol{\beta}})} = \sigma\sqrt{\mathbf{x}'(\mathbf{X}'\mathbf{X})^{-1}\mathbf{x}}$. This particular shape of the confidence band is often called hyperbolic or Scheffé type in the statistical literature. The critical value $\sqrt{(p+1)f^{\alpha}_{p+1,n-p-1}}$ in (3.1) is larger than the critical value $t^{\alpha/2}_{n-p-1}$ used in the pointwise confidence band or interval in (1.17). This is to ensure that the confidence band (3.1) has a simultaneous coverage probability of $1-\alpha$ for all $\mathbf{x}_{(0)} = (x_1, \cdots x_p)' \in \Re^p$ as given by the following result.

Theorem 3.1 For any $\beta \in \Re^{p+1}$ and $\sigma > 0$, we have

$$P\left\{ \mathbf{x}'\beta \in \mathbf{x}'\hat{\beta} \pm \sqrt{(p+1)f^{\alpha}_{p+1,n-p-1}}\,\hat{\sigma}\sqrt{\mathbf{x}'(\mathbf{X}'\mathbf{X})^{-1}\mathbf{x}}\ \ \forall\, \mathbf{x}_{(0)} \in \Re^p \right\} = 1-\alpha.$$

Proof 3.1 *Let* $\mathbf{P}$ *be the unique square root matrix of* $(\mathbf{X}'\mathbf{X})^{-1}$ *and so satisfying* $(\mathbf{X}'\mathbf{X})^{-1} = \mathbf{P}^2$, *and let* $\mathbf{N} = \mathbf{P}^{-1}(\hat{\beta} - \beta)/\sigma$. *Then* $\mathbf{N}$ *has the multivariate normal distribution* $\mathcal{N}_{p+1}(\mathbf{0},\mathbf{I})$. *Now we have*

$$\begin{aligned}
& P\left\{ \mathbf{x}'\beta \in \mathbf{x}'\hat{\beta} \pm \sqrt{(p+1)f^{\alpha}_{p+1,n-p-1}}\,\hat{\sigma}\sqrt{\mathbf{x}'(\mathbf{X}'\mathbf{X})^{-1}\mathbf{x}}\ \ \forall\, \mathbf{x}_{(0)} \in \Re^p \right\} \\
= \;& P\left\{ \sup_{\mathbf{x}_{(0)} \in \Re^p} \frac{|\mathbf{x}'(\hat{\beta}-\beta)|}{\hat{\sigma}\sqrt{\mathbf{x}'(\mathbf{X}'\mathbf{X})^{-1}\mathbf{x}}} \le \sqrt{(p+1)f^{\alpha}_{p+1,n-p-1}} \right\} \\
= \;& P\left\{ \sup_{\mathbf{x}_{(0)} \in \Re^p} \frac{|(\mathbf{P}\mathbf{x})'\mathbf{N}|}{(\hat{\sigma}/\sigma)\sqrt{(\mathbf{P}\mathbf{x})'(\mathbf{P}\mathbf{x})}} \le \sqrt{(p+1)f^{\alpha}_{p+1,n-p-1}} \right\} \\
= \;& P\left\{ \frac{\|\mathbf{N}\|}{(\hat{\sigma}/\sigma)} \left(\sup_{\mathbf{x}_{(0)} \in \Re^p} \frac{|(\mathbf{P}\mathbf{x})'\mathbf{N}|}{\|\mathbf{P}\mathbf{x}\|\,\|\mathbf{N}\|} \right) \le \sqrt{(p+1)f^{\alpha}_{p+1,n-p-1}} \right\} \\
= \;& P\left\{ \frac{\|\mathbf{N}\|}{(\hat{\sigma}/\sigma)} \le \sqrt{(p+1)f^{\alpha}_{p+1,n-p-1}} \right\} \qquad (3.2) \\
= \;& P\left\{ \frac{(\hat{\beta}-\beta)'(\mathbf{X}'\mathbf{X})(\hat{\beta}-\beta)}{(p+1)\hat{\sigma}^2} \le f^{\alpha}_{p+1,n-p-1} \right\} = 1-\alpha \qquad (3.3)
\end{aligned}$$

where the equality in (3.2) follows from the Cauchy-Schwarz inequality:

$$|(\mathbf{P}\mathbf{x})'\mathbf{N}| \le \|\mathbf{P}\mathbf{x}\|\,\|\mathbf{N}\| \ \text{ for all } \mathbf{x}_{(0)} \in \Re^p$$

and the equality holds when $\mathbf{P}\mathbf{x} = \lambda\mathbf{N}$ *for some constant* λ, *and the first equality in (3.3) follows from the definition of* $\mathbf{N}$. *Note that the probability statement in (3.3) gives just the confidence region (2.1) for* β *since* $\hat{\sigma}^2 = \|\mathbf{Y} - \mathbf{X}\hat{\beta}\|^2/(n-p-1)$. *The proof is thus complete.* ▮

It is clear from the proof that the simultaneous confidence band (3.1) is just an alternative representation of the confidence region (2.1). This of course agrees with the observations made in Section 2.6.1. The advantage of the confidence band (3.1) is that it provides information on the regression model $\mathbf{x}'\beta$ directly, while the confidence region (2.1) provides information on the unknown coefficients β only. Bohrer (1973a) showed that the confidence band (3.1) is optimal under the average width criterion when certain conditions are satisfied.

Any regression hyper-plane $\mathbf{x}'\beta_0$ that lies between the upper and lower parts of the band is a plausible candidate of the true model $\mathbf{x}'\beta$. In particular, the confidence band (3.1) can be used to test the hypotheses $H_0 : \boldsymbol{\beta} = \boldsymbol{\beta}_0$ against $H_a : \boldsymbol{\beta} \neq \boldsymbol{\beta}_0$ by rejecting H_0

$$\text{if and only if } \mathbf{x}'\boldsymbol{\beta}_0 \text{ is outside the band (3.1) for at least one } \mathbf{x}_{(0)} \in \Re^p. \quad (3.4)$$

This test is of size α since the confidence band has simultaneous confidence level $1-\alpha$. In fact, this test is the same as the test in (2.3) as given by the following.

Theorem 3.2 Tests (2.3) and (3.4) reject and accept H_0 at the same time.

Proof 3.2 *Using the definition of* $\mathbf{N}$ *in the proof of Theorem 3.1 except replacing* $\boldsymbol{\beta}$ *with* $\boldsymbol{\beta}_0$, *it is clear that test (3.4) rejects* H_0 *if and only if*

$$\begin{aligned}
\sqrt{(p+1)f^{\alpha}_{p+1,n-p-1}} < & \sup_{\mathbf{x}_{(0)} \in \Re^p} \frac{|\mathbf{x}'(\hat{\boldsymbol{\beta}} - \boldsymbol{\beta}_0)|}{\hat{\sigma}\sqrt{\mathbf{x}'(\mathbf{X}'\mathbf{X})^{-1}\mathbf{x}}} \\
= & \sup_{\mathbf{x}_{(0)} \in \Re^p} \frac{|(\mathbf{P}\mathbf{x})'\mathbf{N}|}{(\hat{\sigma}/\sigma)\sqrt{(\mathbf{P}\mathbf{x})'(\mathbf{P}\mathbf{x})}} \\
= & \frac{\|\mathbf{N}\|}{(\hat{\sigma}/\sigma)} \left(\sup_{\mathbf{x}_{(0)} \in \Re^p} \frac{|(\mathbf{P}\mathbf{x})'\mathbf{N}|}{\|\mathbf{P}\mathbf{x}\|\,\|\mathbf{N}\|} \right) \\
= & \frac{\|\mathbf{N}\|}{(\hat{\sigma}/\sigma)} \\
= & \sqrt{\frac{(\hat{\boldsymbol{\beta}} - \boldsymbol{\beta}_0)'\mathbf{P}^{-1}\mathbf{P}^{-1}(\hat{\boldsymbol{\beta}} - \boldsymbol{\beta}_0)}{\hat{\sigma}^2}} \\
= & \sqrt{\frac{(\hat{\boldsymbol{\beta}} - \boldsymbol{\beta}_0)'(\mathbf{X}'\mathbf{X})(\hat{\boldsymbol{\beta}} - \boldsymbol{\beta}_0)}{\hat{\sigma}^2}},
\end{aligned}$$

which is clearly the same as the test (2.3). This completes the proof. ∎

To apply test (3.4), one checks whether or not $\mathbf{x}'\boldsymbol{\beta}_0$ is inside the confidence band for all $\mathbf{x}_{(0)} \in \Re^p$ in order to accept or reject H_0 accordingly. To apply the test (2.3), one checks whether or not $\boldsymbol{\beta}_0$ is in the confidence region (2.1) for $\boldsymbol{\beta}$. They are the same test in different appearances according to Theorem 3.2.

However, viewing test (2.3) from the standpoint of confidence band via (3.4) exposes a problem with this test. Let us consider Example 1.1. It is clear that, for this example, the linear regression model cannot be true for a negative x_1 (which is the birth weight of an infant) or for a negative x_2 (which is the age of an infant). When we apply the test (2.3) or its equivalent test (3.4), however, H_0 will be rejected if $\mathbf{x}'\boldsymbol{\beta}_0$ is outside the confidence band (3.1) for some $x_1 < 0$ or $x_2 < 0$ even though $\mathbf{x}'\boldsymbol{\beta}_0$ lies completely inside the confidence band for all $x_1 \geq 0$ and $x_2 \geq 0$. In other words, the test (2.3) requires implicitly that the linear regression model $\mathbf{x}'\boldsymbol{\beta}$ holds over the whole range $(-\infty, \infty)$ of each predictor variable, which is seldom true in any real problems. So the F test (2.3) or (3.4) might reject the null hypothesis H_0 for a wrong reason when the regression model doesn't hold over the whole range $(-\infty, \infty)$ of each predictor variable!

An obvious remedy for this problem is to use only the part of the confidence band (3.1) over the region $x_1 \geq 0$ and $x_2 \geq 0$, while the part of the confidence band over the remaining region is ignored. So H_0 is rejected if and only if $\mathbf{x}'\beta_0$ is outside the confidence band for some $x_1 \geq 0$ or $x_2 \geq 0$. However, since only the part of the confidence band over the region $x_1 \geq 0$ and $x_2 \geq 0$ is used in making this statistical inference, to guarantee $1-\alpha$ simultaneous coverage probability of the confidence band over the entire region $\mathbf{x}_{(0)} \in \Re^2$ is clearly a waste of resource. One should therefore construct a $1-\alpha$ simultaneous confidence band only over the region $x_1 \geq 0$ and $x_2 \geq 0$. This confidence band is narrower and hence allows sharper inferences than the part of confidence band (3.1) over the region $x_1 \geq 0$ and $x_2 \geq 0$. Of course, for this example, it is also sensible to put some upper bounds on x_1 and x_2. This leads to the construction of simultaneous confidence bands over restricted regions of the predictor variables, the focus of this Chapter. Stewart (1991b) discussed potential pitfalls in plotting/visualizing confidence bands over an infinite covariate region. For the special situation that model (1.1) has no intercept term, i.e. $\beta_0 = 0$, Bohrer (1967) considered how to construct two-sided hyperbolic confidence bands over the covariate region $\{x_i \geq 0,\ i = 1, \cdots, p\}$.

3.1.2 *One-sided band*

Sometimes the purpose of inference is to bound the regression model $\mathbf{x}'\beta$ from one side only, either from below or from above. In this case, it is not efficient to use a two-sided simultaneous confidence band, as given in the last subsection, which bounds the regression model $\mathbf{x}'\beta$ from both sides. One-sided simultaneous confidence bands are pertinent for one-sided inferences.

A lower hyperbolic simultaneous confidence band over the whole space of the covariates is given by

$$\mathbf{x}'\beta > \mathbf{x}'\hat{\beta} - c\hat{\sigma}\sqrt{\mathbf{x}'(\mathbf{X}'\mathbf{X})^{-1}\mathbf{x}} \quad \text{for all } \mathbf{x}_{(0)} \in \Re^p \tag{3.5}$$

where the critical constant c is chosen so that the simultaneous confidence level of this one-sided band is equal to $1-\alpha$. The determination of c is given by the following result of Hochberg and Quade (1975).

Theorem 3.3 The confidence band (3.5) has an exact confidence level $1-\alpha$ if the critical constant c is the solution to the equation

$$\frac{1}{2}F_{p+1,\nu}\left(\frac{c^2}{p+1}\right) + \frac{1}{2}F_{p,\nu}\left(\frac{c^2}{p}\right) = 1-\alpha$$

where $F_{m,\nu}(\cdot)$ is the cumulative distribution function of the F distribution with m and $\nu = n-p-1$ degrees of freedom.

This result is a special case of a more general result given in Subsection 3.4.2. From this result, the value of c can be found easily by using any software that is able to calculate the F distribution function, even though a table of c is provided in Hochberg and Quade (1975).

An upper hyperbolic simultaneous confidence band over the whole space of the covariates has the form

$$\mathbf{x}'\beta < \mathbf{x}'\hat{\beta} + c\hat{\sigma}\sqrt{\mathbf{x}'(\mathbf{X}'\mathbf{X})^{-1}\mathbf{x}} \quad \text{for all } \mathbf{x}_{(0)} \in \Re^p. \tag{3.6}$$

It is straightforward to show that the critical constant c required for this exact $1-\alpha$ level upper confidence band is equal to the critical constant of the exact $1-\alpha$ level lower confidence band (3.5). This observation is true for all the corresponding upper and lower confidence bands in this book, and so it is necessary to consider only the lower confidence band for the purpose of computing the critical constant. For the special case that regression model (1.1) has no intercept term, i.e., $\beta_0 = 0$, Bohrer (1969) pointed out that the exact $1-\alpha$ level one-sided hyperbolic band over $\mathbf{x}_{(0)} \in \Re^p$ uses the same critical constant as the exact $1-\alpha$ level two-sided hyperbolic band over $\mathbf{x}_{(0)} \in \Re^p$, while Bohrer and Francis (1972b) considered how to construct exact one-sided hyperbolic confidence bands over the covariate region $\{x_i \geq 0,\ i = 1, \cdots, p\}$.

Example 3.1 For Example 1.1, the 95% simultaneous confidence band is given by

$$\mathbf{x}'\beta \in \mathbf{x}'\hat{\beta} \pm 3.167\hat{\sigma}\sqrt{\mathbf{x}'(\mathbf{X}'\mathbf{X})^{-1}\mathbf{x}} \quad \text{for all } \mathbf{x}_{(0)} = (x_1, x_2)' \in \Re^2,$$

while the 95% pointwise confidence interval is given by

$$\mathbf{x}'\beta \in \mathbf{x}'\hat{\beta} \pm 2.145\hat{\sigma}\sqrt{\mathbf{x}'(\mathbf{X}'\mathbf{X})^{-1}\mathbf{x}} \quad \text{for each } \mathbf{x}_{(0)} = (x_1, x_2)' \in \Re^2.$$

Clearly the simultaneous confidence band is wider than the pointwise confidence interval at each $\mathbf{x}_{(0)} \in \Re^2$.

The critical value of the 95% one-sided lower band (3.5) is given by $c = 2.984$, which is smaller than the $c = 3.167$ of the two-sided band. All these results are computed by using the MATLAB® program example0301.m.

3.2 Hyperbolic bands over a rectangular region

A great deal of research efforts have gone into the construction of simultaneous confidence bands over restricted regions of the predictor variables. It has been argued by several researchers (cf. Casella and Strawderman, 1980, and Naiman, 1987a) that a simultaneous confidence band over a rectangular region of the predictor variables of the form

$$\mathcal{X}_r = \{(x_1, \cdots, x_p)' : a_i \leq x_i \leq b_i, i = 1, \cdots, p\}, \tag{3.7}$$

where $-\infty \le a_i < b_i \le \infty, i = 1, \cdots, p$ are given constants, is often most useful for practical purposes. In many applications involving a linear regression model the experimenter can specify constraints on a predictor variable in terms of a lower and an upper bounds for each of the predictor variables, which results in the region $\mathcal{X}_r$. The central topic of this section is the construction of hyperbolic confidence bands over $\mathcal{X}_r$.

3.2.1 *Two-sided band*

A two-sided hyperbolic confidence band over the rectangular region $\mathcal{X}_r$ in (3.7) has the form

$$\mathbf{x}'\beta \in \mathbf{x}'\hat{\beta} \pm c\hat{\sigma}\sqrt{\mathbf{x}'(\mathbf{X}'\mathbf{X})^{-1}\mathbf{x}} \quad \text{for all } \mathbf{x}_{(0)} \in \mathcal{X}_r \tag{3.8}$$

where c is a critical constant chosen suitably so that the simultaneous confidence level is $1-\alpha$. It is clear that the confidence level of this simultaneous confidence band is given by $P\{S < c\}$, where

$$S = \sup_{\mathbf{x}_{(0)} \in \mathcal{X}_r} \frac{|\mathbf{x}'(\hat{\beta} - \beta)/\sigma|}{(\hat{\sigma}/\sigma)\sqrt{\mathbf{x}'(\mathbf{X}'\mathbf{X})^{-1}\mathbf{x}}}.$$

The distribution of S does not depend on the unknown parameters β and σ, and so S is a pivotal random quantity. However, it depends on the bounds $[a_i, b_i]$ and the design matrix $\mathbf{X}$ in a complicated way in this general setting.

Let $\mathbf{P}$ and $\mathbf{N}$ be as defined in the proof of Theorem 3.1. Denote $\mathbf{T} = \mathbf{N}/(\hat{\sigma}/\sigma)$ which has the $\mathcal{T}_{p+1,\nu}$ distribution. Now S can be expressed as

$$\begin{aligned} S &= \sup_{\mathbf{x}_{(0)} \in \mathcal{X}_r} \frac{\left|(\mathbf{P}\mathbf{x})'\left(\mathbf{P}^{-1}(\hat{\beta} - \beta)/\hat{\sigma}\right)\right|}{\sqrt{(\mathbf{P}\mathbf{x})'(\mathbf{P}\mathbf{x})}} \\ &= \sup_{\mathbf{x}_{(0)} \in \mathcal{X}_r} \frac{|(\mathbf{P}\mathbf{x})'\mathbf{T}|}{\|\mathbf{P}\mathbf{x}\|} \\ &= \sup_{\mathbf{v} \in C(\mathbf{P}, \mathcal{X}_r)} \frac{|\mathbf{v}'\mathbf{T}|}{\|\mathbf{v}\|} \end{aligned} \tag{3.9}$$

where

$$\begin{aligned} C(\mathbf{P}, \mathcal{X}_r) &:= \{\lambda \mathbf{P}\mathbf{x} : \lambda \ge 0, \, \mathbf{x}_{(0)} \in \mathcal{X}_r\} \\ &= \{\lambda(\mathbf{p}_0 + x_1\mathbf{p}_1 + \cdots + x_p\mathbf{p}_p) : \lambda \ge 0, \, x_i \in [a_i, b_i] \, \forall \, i = 1, \cdots, p\} \end{aligned}$$

with $\mathbf{P} = (\mathbf{p}_0, \mathbf{p}_1, \cdots, \mathbf{p}_p)$. For a general $p \ge 1$, it is difficult to derive a useful formula for the distribution of S in order to determine the critical constant c.

Now we give the simulation-based method of Liu, Jamshidian, Zhang and Donnelly (2005) to calculate the critical constant c. This method is 'exact' in the sense that if the simulation can be replicated for a sufficiently large number of times than the critical constant calculated can be as close to the exact value as one requires. In this book we emphasize simulation-based methods for computing the critical constants of simultaneous confidence bands when no exact method is available. These simulation-based methods are preferable to conservative or approximate methods since they are exact in the sense pointed out above and practical with the available software and the computation power of modern computers.

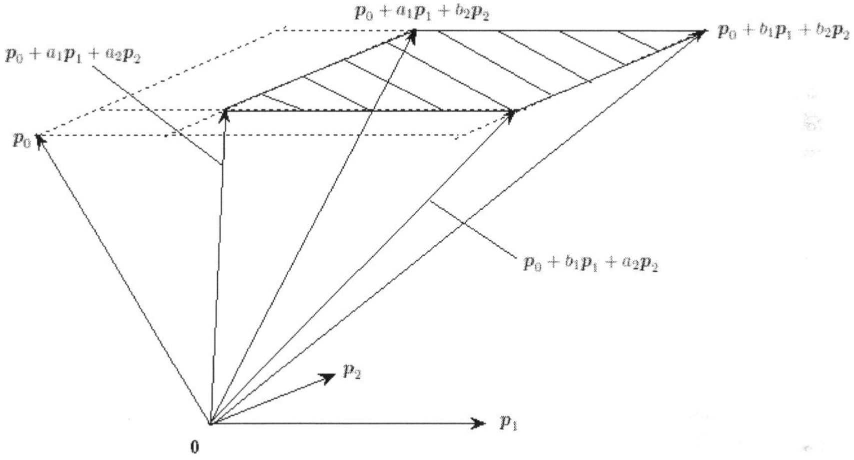

Figure 3.1: *The cone $C(\mathbf{P}, \mathcal{X}_r)$ for $p = 2$*

Note that $C(\mathbf{P}, \mathcal{X}_r)$ is a cone spanned by the vectors

$$\mathbf{p}_0 + c_1\mathbf{p}_1 + \cdots + c_p\mathbf{p}_p \text{ where } c_i \text{ is either } a_i \text{ or } b_i \text{ for } i = 1, \cdots, p.$$

The cone $C(\mathbf{P}, \mathcal{X}_r)$ for the special case of $p = 2$ is depicted in Figure 3.1.

Let $\pi(\mathbf{t}, \mathbf{P}, \mathcal{X}_r)$ denote the projection of $\mathbf{t} \in \Re^{p+1}$ to the cone $C(\mathbf{P}, \mathcal{X}_r)$, i.e., $\pi(\mathbf{t}, \mathbf{P}, \mathcal{X}_r)$ is the $\mathbf{v} \in \Re^{p+1}$ that solves the problem

$$\min_{\mathbf{v} \in C(\mathbf{P}, \mathcal{X}_r)} \|\mathbf{v} - \mathbf{t}\|^2. \tag{3.10}$$

Now it follows from Naiman (1987a, Theorem 2.1) that S in (3.9) is further equal to

$$S = \max\left\{\|\pi(\mathbf{T}, \mathbf{P}, \mathcal{X}_r)\|,\ \|\pi(-\mathbf{T}, \mathbf{P}, \mathcal{X}_r)\|\right\}. \tag{3.11}$$

Also note that the solution of the problem (3.10), $\pi(\mathbf{t},\mathbf{P},\mathcal{X}_r)$, can be found accurately and efficiently in a finite number of steps for any given $\mathbf{t}$, $\mathbf{P}$ and $\mathcal{X}_r$ by using quadratic programming under linear constraints; more details on this are provided in Appendix B.

The random variable S can therefore be simulated exactly and efficiently in the following way. First, independent $\mathbf{N}$ and $\hat{\sigma}/\sigma$ are generated from $\mathcal{N}_{p+1}(\mathbf{0},\mathbf{I})$ and $\sqrt{\chi^2_\nu/\nu}$ respectively. Then $\pi(\mathbf{T},\mathbf{P},\mathcal{X}_r)$ and $\pi(-\mathbf{T},\mathbf{P},\mathcal{X}_r)$ are computed separately. Finally, S is calculated by using (3.11).

Repeat the process for R times to simulate R independent replicates $S_1,\cdots,S_R$ of S. The exact critical constant c we are looking for is the $100(1-\alpha)$ percentile of the population distribution of S. It can be approximated by the $100(1-\alpha)$ sample percentile of the sample $S_1,\cdots,S_R$, $\hat{c}$, which is the $\langle R(1-\alpha)\rangle$-th largest S_i's. Here $\langle a\rangle$ denotes the integer part of a.

The accuracy of $\hat{c}$ can be assessed in several ways. One is to estimate its standard error since $\hat{c}$ is a random quantity. This can be estimated by using the large sample result (see e.g., Serfling, 1980) or bootstrap methods (e.g., Efron and Tibshirani, 1993). Another method, due to Edwards and Berry (1987), looks at the difference between the conditional probability $\bar{G}(\hat{c}) = P\{S > \hat{c}|\hat{c}\}$, which is a random quantity, and the target tail probability $\bar{G}(c) = P\{S > c\} = \alpha$. More details on these are provided in Appendix A.

As R becomes larger, $\hat{c}$ becomes more accurate as an estimator of c. The accuracy measurements above provide guidance on how large R should be set, depending on the accuracy measurement used and accuracy required. From our experience, if $R = 50,000$ then $\hat{c}$ varies only at the second decimal place, and if $R = 100,000$ then $\hat{c}$ is often accurate to the second decimal place.

Liu, Jamshidian, Zhang and Donnelly (LJZD, 2005) provided a specific quadratic-programming-based method to compute the S in (3.9); this method is slightly more efficient than the built-in quadratic programming function `quadprog` in MATLAB that is used in this book. Jamshidian *et al.* (2005) provided a MATLAB program which implements LJZD's (2005) simulation procedure to compute $\hat{c}$ and its standard error.

Example 3.2 Consider the infant data set given in Table 1.1. The response variable (Y) is systolic blood pressure. The two predictor variables are birth weight in oz (x_1) and age in days (x_2). There are seventeen data points. So $p = 2$, $n = 17$ and $\nu = n - p - 1 = 14$. The fitted linear regression model is given by $Y = 47.583 + 0.181x_1 + 5.313x_2$, $\hat{\sigma} = 1.314$, and $R^2 = 0.95$. The observed values of x_1 range from 92 to 149, and the observed values of x_2 range from 2 to 5. If the 95% confidence band over the region

$$\mathcal{X}_r = \{(x_1,x_2)' : \; 92 \le x_1 \le 149, \; 2 \le x_2 \le 5\}$$

is required then the critical constant $\hat{c}$ is computed to be 3.116 (s.e. 0.0081) by using our MATLAB program example0302.m based on 100,000 simulations. This critical value is marginally smaller than the critical value $c = 3.167$ of the band over the whole range of the two explanatory variables. LJZD (2005) also provided one example with $p = 6$ explanatory variables.

Naiman (1987a) proposed a simulation-based method to compute a conservative critical constant c, which can be substantially larger than the exact critical constant as shown by Donnelly (2003). Naiman (1990) considered an approach based on the volume of tubular neighborhoods to provide a conservative critical constant. The method of Sun and Loader (1994) can be used to compute an approximate critical constant for $p \leq 2$. In particular, their Proposition 2 (Sun and Loader, 1994, pp.1330) gives the following approximation to the simultaneous confidence level when $p = 2$:

$$\begin{aligned} P\{S \leq c\} &\approx 1 - \frac{\kappa_0}{\pi^{3/2}} \frac{\Gamma((\nu+1)/2)}{\Gamma(\nu/2)} \frac{c}{\sqrt{\nu}} \left(1 + \frac{c^2}{\nu}\right)^{-(\nu+1)/2} \\ &\quad - \frac{\xi_0}{2\pi} \left(1 + \frac{c^2}{\nu}\right)^{-\nu/2} - P\{|t_\nu| > c\} \end{aligned}$$

where κ_0 is the area of $\mathcal{M}$ and ξ_0 is the length of the boundary of $\mathcal{M}$ where

$$\mathcal{M} = \left\{ \frac{\mathbf{X}(\mathbf{X}'\mathbf{X})^{-1}\mathbf{x}'}{\|\mathbf{X}(\mathbf{X}'\mathbf{X})^{-1}\mathbf{x}'\|} : \mathbf{x}_{(0)} \in \mathcal{X}_r \right\}$$

is a manifold. Calculation of κ_0 requires two-dimensional numerical integration; formulae for computing κ_0 and ξ_0 are given in Sun and Loader (1994, pp.1334-1335); also see Loader (2004).

3.2.2 *One-sided band*

An upper hyperbolic simultaneous confidence band over the rectangular region $\mathcal{X}_r$ of the covariates is given by

$$\mathbf{x}'\beta < \mathbf{x}'\hat{\beta} + c\hat{\sigma}\sqrt{\mathbf{x}'(\mathbf{X}'\mathbf{X})^{-1}\mathbf{x}} \quad \text{for all } \mathbf{x}_{(0)} \in \mathcal{X}_r \tag{3.12}$$

where the critical constant c is chosen so that the simultaneous confidence level of this one-sided band is equal to $1 - \alpha$. A lower hyperbolic simultaneous confidence band over the rectangular region $\mathcal{X}_r$ can be defined in an obvious way and requires the same critical constant as the upper band.

Note that the simultaneous confidence level of the band (3.12) is given by $P\{S < c\}$ where

$$S = \sup_{\mathbf{x}_{(0)} \in \mathcal{X}_r} \frac{(\mathbf{P}\mathbf{x})'(-\mathbf{T})}{\|\mathbf{P}\mathbf{x}\|} = \sup_{\mathbf{v} \in C(\mathbf{P}, \mathcal{X}_r)} \frac{\mathbf{v}'(-\mathbf{T})}{\|\mathbf{v}\|}$$

where $\mathbf{T}$ and $C(\mathbf{P}, \mathcal{X}_r)$ are defined in Section 3.2.1. Even though the last expression of S may have a negative value and so is not equal to $\|\pi(-\mathbf{T},\mathbf{P},\mathcal{X}_r)\|$, it follows from Naiman (1987a, Theorem 2.1) that, for $c > 0$,

$$P\{S < c\} = P\{\|\pi(-\mathbf{T},\mathbf{P},\mathcal{X}_r)\| < c\}.$$

An estimator $\hat{c}$ of c can therefore be computed by simulating a sample of R independent replicates of $\|\pi(-\mathbf{T},\mathbf{P},\mathcal{X}_r)\|$ as in the two-sided case.

Example 3.3 For the infant data set considered in Example 1.1, the 95% upper confidence band over the rectangular region $\mathcal{X}_r$ in Example 3.2 has its critical constant $\hat{c}$ computed to be 2.767 (s.e. 0.0079) by our MATLAB program example0303.m based on 100,000 simulations. This critical value is smaller than the critical value $c = 3.116$ of the two-sided band over the region $\mathcal{X}_r$ given in Example 3.2 and the critical value $c = 2.984$ of the one-sided band over the whole space $\mathbf{x}_{(0)} \in \Re^2$ in Example 3.1 as expected.

3.3 Constant width bands over a rectangular region

The hyperbolic band discussed in the last section has a width at $\mathbf{x}_{(0)}$ proportional to the standard error $\sigma\sqrt{\mathbf{x}'(\mathbf{X}'\mathbf{X})^{-1}\mathbf{x}}$ of $\mathbf{x}'\hat{\beta}$. This means that the width of the band is smaller when $\mathbf{x}_{(0)}$ is nearer the center of the covariates $\bar{\mathbf{x}}_{(0)} = (\bar{x}_{\cdot 1},\cdots,\bar{x}_{\cdot p})'$ and becomes larger when $\mathbf{x}_{(0)}$ is further away from the center, where $\bar{x}_{\cdot j} = \frac{1}{n}\sum_{i=1}^{n} x_{ij}$ is the mean of the observations on the jth predictor variable $(1 \le j \le p)$. Sometimes one may want to have a band that has the same width for all $\mathbf{x}_{(0)} \in \mathcal{X}_r$. A confidence band that has the same width throughout the covariate region $\mathcal{X}_r$ is called a constant width band. This is the generalization of the constant width bands considered first by Gafarian (1964) for the case of $p = 1$, which is a special case of the three-segment band given in Chapter 2.

For a general p, the calculation of the simultaneous confidence level of a constant width confidence band involves a multivariate t probability. While it is difficult to evaluate a multivariate t probability in general, especially when p is large, methods are available for computing multivariate t probabilities; see e.g., Genz and Bretz (1999, 2002), Bretz *et al.* (2001), Somerville (1997, 1998, 1999) and Somerville and Bretz (2001). All these methods involve Monte Carlo simulations. Genz and Bretz (2009) provided an excellent review on the topic of computation of multivariate normal and t probabilities. Liu, Jamshidian, Zhang and Bretz (2005) used a simulation method as in the last section to compute the critical constant c of a constant width band for a general p, which is discussed below.

3.3.1 *Two-sided band*

A two-sided constant width confidence band has the form

$$\mathbf{x}'\beta \in \mathbf{x}'\hat{\beta} \pm c\hat{\sigma} \quad \text{for all } \mathbf{x}_{(0)} \in \mathcal{X}_r \tag{3.13}$$

where c is a critical constant chosen so that the simultaneous confidence level of the band is equal to $1-\alpha$. It is clear that the width of the band is given by $2c\hat{\sigma}$ throughout the covariate region $\mathcal{X}_r$.

The confidence level of the two-sided constant width band in (3.13) is given by $P\{S<c\}$ where

$$S = \sup_{\mathbf{x}_{(0)} \in \mathcal{X}_r} |\mathbf{x}'(\hat{\beta}-\beta)/\sigma|/(\hat{\sigma}/\sigma). \tag{3.14}$$

The distribution of S depends on the design matrix $\mathbf{X}$ and the covariate region $\mathcal{X}_r$, but not on the unknown parameters β and σ. Let $\mathbf{N}=(\hat{\beta}-\beta)/\sigma \sim \mathcal{N}_{p+1}(\mathbf{0},(\mathbf{X}'\mathbf{X})^{-1})$ and $\mathbf{T}=\mathbf{N}/(\hat{\sigma}/\sigma)$ which has the multivariate t distribution $\mathcal{T}_{p+1,\nu}(\mathbf{0},(\mathbf{X}'\mathbf{X})^{-1})$ (cf. Dunnett and Sobel, 1954b, and Genz and Bretz, 2009). Then S in (3.14) can be expressed as

$$S = \sup_{\mathbf{x}_{(0)} \in \mathcal{X}_r} |\mathbf{x}'\mathbf{T}|.$$

Since $\mathbf{x}'\mathbf{T}$ is a linear function of $\mathbf{x}_{(0)}=(x_1,\cdots,x_p)'$, it must attain both its maximum and minimum over $\mathbf{x}_{(0)} \in \mathcal{X}_r$ at some vertices of $\mathcal{X}_r$. Obviously $\mathcal{X}_r$ has 2^p vertices given by

$$V = \{(v_1,\cdots,v_p)' : \text{ each } v_i \text{ is either } a_i \text{ or } b_i \text{ for } 1 \le i \le p\} \subset \Re^p. \tag{3.15}$$

So

$$S = \sup_{\mathbf{x}_{(0)} \in V} |\mathbf{x}'\mathbf{T}|. \tag{3.16}$$

To simulate one realization of S, one $\mathbf{N} \sim \mathcal{N}_{p+1}(\mathbf{0},(\mathbf{X}'\mathbf{X})^{-1})$ and one independent $\hat{\sigma}/\sigma \sim \sqrt{\chi^2_\nu/\nu}$ are simulated. Then S is easily calculated from (3.16) since the maximization is only over 2^p points. Now that we know how to simulate one S, the critical constant c can be determined in a similar way as in the last section: simulate R replicates of the random variable S, and set the $\langle(1-\alpha)R\rangle$th largest simulated value, $\hat{c}$, as the critical constant c.

Example 3.4 For the infant data set, the critical constant c of the two-sided constant width band is found from example0304.m to be 1.850 (with s.e. 0.0052) based on 100,000 simulations when $[a_1,b_1]\times[a_2,b_2]=[92,149]\times[2,5]$ and $\alpha=.05$. Figure 3.2 plots the 95% two-sided hyperbolic and constant width bands over the given rectangular covariate region. Similar to the case of $p=1$, the hyperbolic band tends to be narrower than the constant width band when $(x_1,x_2)'$ is near $(\bar{x}_{\cdot 1},\bar{x}_{\cdot 2})'=(113.7,3.9)'$, and wider than the constant width band when $(x_1,x_2)'$ is far away from $(\bar{x}_{\cdot 1},\bar{x}_{\cdot 2})'=(113.7,3.9)'$.

It is noteworthy from (3.16) that if any one of the bounds a_i's or b_i's is infinite then $S=\infty$ with probability one. So constant width confidence bands can be constructed only over a finite rectangular region $\mathcal{X}_r$. But hyperbolic bands can be constructed over the whole space $\mathbf{x}_{(0)} \in \Re^p$.

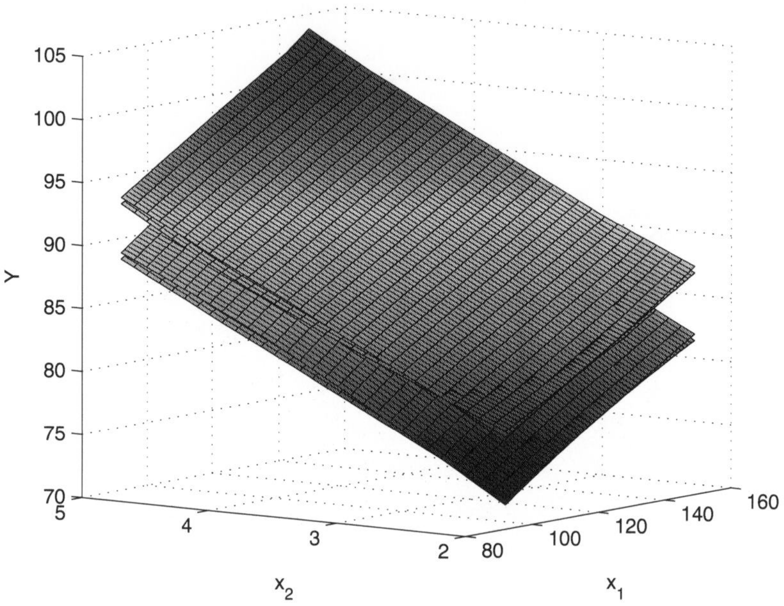

Figure 3.2: *The two-sided hyperbolic and constant width bands*

3.3.2 One-sided band

An upper constant width confidence band is given by

$$\mathbf{x}'\beta < \mathbf{x}'\hat{\beta} + c\hat{\sigma} \quad \text{for all } \mathbf{x}_{(0)} \in \mathcal{X}_r$$

where c is a critical constant chosen so that the simultaneous confidence level of the band is equal to $1-\alpha$. A lower constant width confidence band can be defined in a similar way and uses the same critical constant as an upper constant width band.

The confidence level of this one-sided constant width band is given by $P\{S < c\}$ where

$$S = \sup_{\mathbf{x}_{(0)} \in \mathcal{X}_r} \left(-\mathbf{x}'(\hat{\beta} - \beta)/\hat{\sigma} \right) = \sup_{\mathbf{x}_{(0)} \in V} \mathbf{x}'(-\mathbf{T})$$

where $\mathbf{T} = (\hat{\beta} - \beta)/\hat{\sigma}$ and V are the same as in the two-sided case. For a general p one can use a simulation method to find the critical constant as in the two-sided case.

Example 3.5 For the infant data set, the critical constant c of the one-sided constant width band is found from program example0305.m to be 1.639 (with s.e.

0.0051) based on 100,000 simulations when $[a_1,b_1] \times [a_2,b_2] = [92,149] \times [2,5]$ and $\alpha = .05$. Figure 3.3 plots the 95% one-sided hyperbolic and constant width bands over the given rectangular covariate region. The hyperbolic band tends to be lower than the constant width band when $(x_1,x_2)'$ is near $(\bar{x}_{\cdot 1},\bar{x}_{\cdot 2})' = (113.7,3.9)'$, and higher than the constant width band when $(x_1,x_2)'$ is far away from $(\bar{x}_{\cdot 1},\bar{x}_{\cdot 2})' = (113.7,3.9)'$.

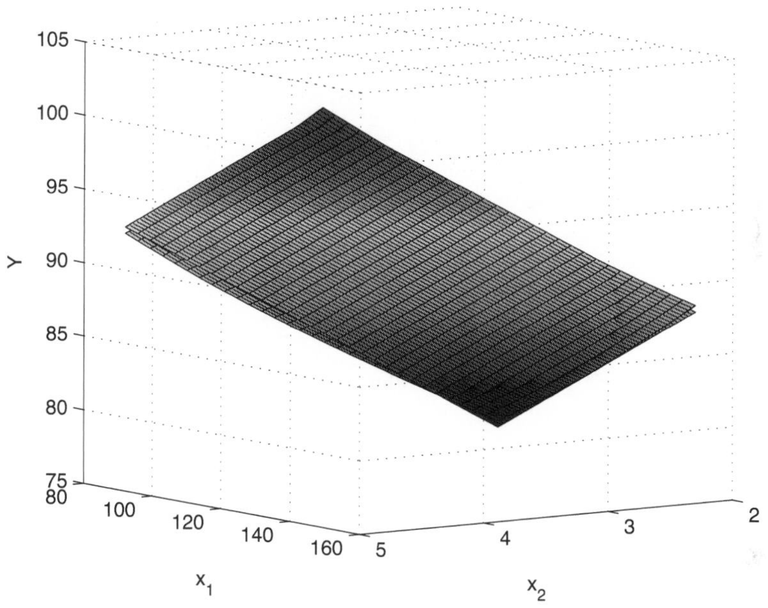

Figure 3.3: *The one-sided hyperbolic and constant width bands*

3.4 Hyperbolic bands over an ellipsoidal region

Another finite predictor variable region, over which construction of confidence bands is of interest when $p \geq 2$, is an ellipsoidal region defined in the following way. Let $\mathbf{X}_{(0)}$ be the $n \times p$ matrix produced from the design matrix $\mathbf{X}$ by deleting the first column of 1's from $\mathbf{X}$. Define a $p \times p$ matrix

$$\mathbf{V} = \frac{1}{n}\left(\mathbf{X}_{(0)} - \mathbf{1}\bar{\mathbf{x}}_{(0)}'\right)'\left(\mathbf{X}_{(0)} - \mathbf{1}\bar{\mathbf{x}}_{(0)}'\right) = \frac{1}{n}\left(\mathbf{X}_{(0)}'\mathbf{X}_{(0)} - n\bar{\mathbf{x}}_{(0)}\bar{\mathbf{x}}_{(0)}'\right)$$

where $\mathbf{1}$ is an n-vector of 1's, and $\bar{\mathbf{x}}_{(0)} = (\bar{x}_{\cdot 1},\cdots,\bar{x}_{\cdot p})'$. Note that $\mathbf{V}$ is just the sample variance-covariance matrix of the p predictor variables. Matrix $\mathbf{V}$ is non-singular since $\mathbf{X}$ is assumed to be of full column-rank. Now the ellipsoidal region is defined to be

$$\mathcal{X}_e = \left\{ \mathbf{x}_{(0)} : \left(\mathbf{x}_{(0)} - \bar{\mathbf{x}}_{(0)}\right)' \mathbf{V}^{-1} \left(\mathbf{x}_{(0)} - \bar{\mathbf{x}}_{(0)}\right) \le a^2 \right\} \tag{3.17}$$

where $a > 0$ is a given constant. It is clear that this region is centered at $\bar{\mathbf{x}}_{(0)}$ and has an ellipsoidal shape. See Figure 3.9 below for some examples of $\mathcal{X}_e$ when $p = 2$.

Note that

$$\begin{aligned} \mathbf{X}'\mathbf{X} &= \begin{pmatrix} n, & n\bar{\mathbf{x}}'_{(0)} \\ n\bar{\mathbf{x}}_{(0)}, & \mathbf{X}'_{(0)}\mathbf{X}_{(0)} \end{pmatrix} \quad \text{and} \\ (\mathbf{X}'\mathbf{X})^{-1} &= \begin{pmatrix} \frac{1}{n}(1 + \bar{\mathbf{x}}'_{(0)}\mathbf{V}^{-1}\bar{\mathbf{x}}_{(0)}), & -\frac{1}{n}\bar{\mathbf{x}}'_{(0)}\mathbf{V}^{-1} \\ -\frac{1}{n}\mathbf{V}^{-1}\bar{\mathbf{x}}_{(0)}, & \frac{1}{n}\mathbf{V}^{-1} \end{pmatrix}. \end{aligned}$$

Direct calculation shows that the variance of the fitted regression model at $\mathbf{x}_{(0)}$ is given by

$$\mathrm{Var}(\mathbf{x}'\hat{\beta}) = \frac{\sigma^2}{n}\left[1 + \left(\mathbf{x}_{(0)} - \bar{\mathbf{x}}_{(0)}\right)' \mathbf{V}^{-1} \left(\mathbf{x}_{(0)} - \bar{\mathbf{x}}_{(0)}\right)\right]. \tag{3.18}$$

So, for all the $\mathbf{x}_{(0)}$ on the surface of the ellipsoidal region $\mathcal{X}_e$, $\mathrm{Var}(\mathbf{x}'\hat{\beta})$ are equal and given by $\frac{\sigma^2}{n}\left[1 + a^2\right]$; the minimum value of $\mathrm{Var}(\mathbf{x}'\hat{\beta})$ is attained at $\mathbf{x}_{(0)} = \bar{\mathbf{x}}_{(0)}$. All the $\mathbf{x}_{(0)}$ on the surface of $\mathcal{X}_e$ can therefore be regarded as of equal 'distance', in terms of $\mathrm{Var}(\mathbf{x}'\hat{\beta})$, from $\bar{\mathbf{x}}_{(0)}$. Hence it is of interest to learn via a simultaneous confidence band about the regression model over $\mathcal{X}_e$ for a pre-specified a^2 value. It is clear that the size of $\mathcal{X}_e$ increases with the value of a. If the axes of $\mathcal{X}_e$ coincide with the axes of the coordinates then the design is called orthogonal and, in particular, if $\mathcal{X}_e$ is a sphere then the design is called rotatable; see e.g., Atkinson and Donev (1992, page 48).

Halperin and Gurian (1968) constructed a conservative two-sided hyperbolic confidence band over $\mathcal{X}_e$ by using a result of Halperin *et al.* (1967). Casella and Strawderman (1980) were able to construct an exact two-sided hyperbolic confidence band over a region that is more general than $\mathcal{X}_e$. Bohrer (1973b) constructed an exact one-sided hyperbolic confidence band over $\mathcal{X}_e$. The exposition of this section follows largely Liu and Lin (2009).

The problems are to construct a two-sided confidence band of the form

$$\mathbf{x}'\beta \in \mathbf{x}'\hat{\beta} \pm c\hat{\sigma}\sqrt{\mathbf{x}'(\mathbf{X}'\mathbf{X})^{-1}\mathbf{x}} \quad \text{for all } \mathbf{x}_{(0)} = (x_1, \cdots, x_p)' \in \mathcal{X}_e, \tag{3.19}$$

and to construct a one-sided lower confidence band of the form

$$\mathbf{x}'\beta \ge \mathbf{x}'\hat{\beta} - c\hat{\sigma}\sqrt{\mathbf{x}'(\mathbf{X}'\mathbf{X})^{-1}\mathbf{x}} \quad \text{for all } \mathbf{x}_{(0)} = (x_1, \cdots, x_p)' \in \mathcal{X}_e. \tag{3.20}$$

In order to determine the critical constants c in (3.19) and (3.20) so that each confidence band has a confidence level equal to pre-specified $1 - \alpha$, the key is to calculate the confidence level of the band for a given constant $c > 0$: $\mathrm{P}\{S_1 \le c\}$ and $\mathrm{P}\{S_2 \le c\}$, where

$$S_1 = \sup_{\mathbf{x}_{(0)} \in \mathcal{X}_e} \frac{\mathbf{x}'(\hat{\beta} - \beta)}{\hat{\sigma}\sqrt{\mathbf{x}'(\mathbf{X}'\mathbf{X})^{-1}\mathbf{x}}}, \quad S_2 = \sup_{\mathbf{x}_{(0)} \in \mathcal{X}_e} \frac{|\mathbf{x}'(\hat{\beta} - \beta)|}{\hat{\sigma}\sqrt{\mathbf{x}'(\mathbf{X}'\mathbf{X})^{-1}\mathbf{x}}}. \tag{3.21}$$

Let $\mathbf{z} = \sqrt{n}(1, \bar{\mathbf{x}}'_{(0)})'$. Note that $\mathbf{z}'(\mathbf{X}'\mathbf{X})^{-1}\mathbf{z} = 1$ and hence $\mathbf{z}$ is of length one in the $(p+1)$-dimensional vector space with inner product defined by $\langle \mathbf{z}_1, \mathbf{z}_2 \rangle = \mathbf{z}_1'(\mathbf{X}'\mathbf{X})^{-1}\mathbf{z}_2$. One can find another p vectors of this space which, together with $\mathbf{z}$, form an orthonormal basis of this space. Denoting the $(p+1) \times p$ matrix formed by these p vector by $\mathbf{Z}$, then $(\mathbf{z}, \mathbf{Z})'(\mathbf{X}'\mathbf{X})^{-1}(\mathbf{z}, \mathbf{Z}) = \mathbf{I}_{p+1}$. It follows therefore that $\mathbf{N} = (\mathbf{z}, \mathbf{Z})^{-1}(\mathbf{X}'\mathbf{X})(\hat{\beta} - \beta)/\sigma \sim \mathcal{N}_{p+1}(\mathbf{0}, \mathbf{I})$.

Also note that $\mathbf{z}'(\mathbf{X}'\mathbf{X})^{-1}\mathbf{x} = 1/\sqrt{n}$. Denote $\mathbf{w} = (\mathbf{z}, \mathbf{Z})'(\mathbf{X}'\mathbf{X})^{-1}\mathbf{x} = (1/\sqrt{n}, \mathbf{w}'_{(0)})'$ where $\mathbf{w}_{(0)} = (w_1, \cdots, w_p)' = \mathbf{Z}'(\mathbf{X}'\mathbf{X})^{-1}\mathbf{x}$. Then $\mathbf{x}'(\mathbf{X}'\mathbf{X})^{-1}\mathbf{x} = \mathbf{w}'\mathbf{w} = \|\mathbf{w}\|^2$. From this and the fact that the region $\mathcal{X}_e$ in (3.17) can also be expressed as

$$\mathcal{X}_e = \left\{ \mathbf{x}_{(0)} : \mathbf{x}'(\mathbf{X}'\mathbf{X})^{-1}\mathbf{x} \le \frac{1+a^2}{n} \right\} \tag{3.22}$$

by using (3.18), all the possible values of $\mathbf{w}_{(0)}$, determined from the relationship $\mathbf{w} = (\mathbf{z}, \mathbf{Z})'(\mathbf{X}'\mathbf{X})^{-1}\mathbf{x}$, when $\mathbf{x}_{(0)}$ varies over the region $\mathcal{X}_e$, form the set

$$\mathcal{W}_e = \left\{ \mathbf{w}_{(0)} : \|\mathbf{w}\|^2 \le \frac{1+a^2}{n} \right\}. \tag{3.23}$$

The random variable S_1 in (3.21) can now be expressed as

$$\begin{aligned} S_1 &= \sup_{\mathbf{x}_{(0)} \in \mathcal{X}_e} \frac{\left((\mathbf{z}, \mathbf{Z})'(\mathbf{X}'\mathbf{X})^{-1}\mathbf{x}\right)' \left((\mathbf{z}, \mathbf{Z})^{-1}(\mathbf{X}'\mathbf{X})(\hat{\beta} - \beta)/\sigma\right)}{(\hat{\sigma}/\sigma)\sqrt{\left((\mathbf{z}, \mathbf{Z})'(\mathbf{X}'\mathbf{X})^{-1}\mathbf{x}\right)' \left((\mathbf{z}, \mathbf{Z})'(\mathbf{X}'\mathbf{X})^{-1}\mathbf{x}\right)}} \\ &= \sup_{\mathbf{w}_{(0)} \in \mathcal{W}_e} \frac{\mathbf{w}'\mathbf{N}}{(\hat{\sigma}/\sigma)\|\mathbf{w}\|}. \end{aligned} \tag{3.24}$$

Furthermore, note that $\mathbf{w}'\mathbf{N}/\|\mathbf{w}\|$ is invariant if $\mathbf{w}$ is replaced with $\mathbf{v} = u\mathbf{w}$ for any $u > 0$, and that

$$\begin{aligned} \mathcal{V}_e &:= \left\{ \mathbf{v} = u\mathbf{w} = (v_0, \cdots, v_p)' : u > 0, \mathbf{w}_{(0)} \in \mathcal{W}_e \right\} \\ &= \left\{ \mathbf{v} = (v_0, \cdots, v_p)' : \|\mathbf{v}\| \le v_0\sqrt{1+a^2} \right\} \\ &= \left\{ \mathbf{v} : v_0 \ge r\|\mathbf{v}\| \right\} \subset \Re^{p+1} \end{aligned} \tag{3.25}$$

where $r = 1/\sqrt{1+a^2}$. Note that the set $\mathcal{V}_e$ is a circular cone in $\Re^{p+1}$ with its vertex at the origin and its central direction given by the v_0-axis. The half angle

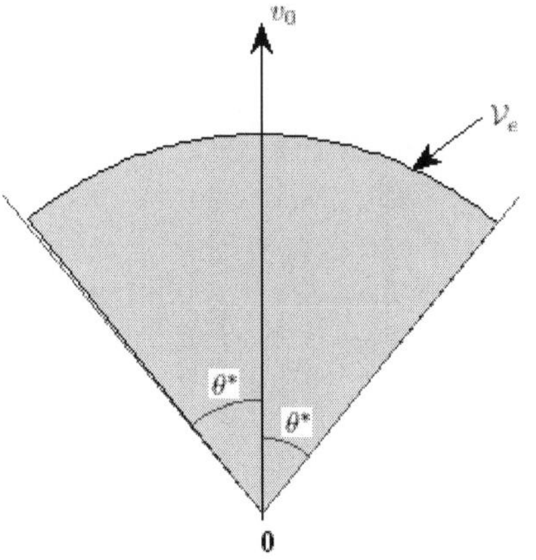

Figure 3.4: *The circular cone* $\mathcal{V}_e$

of this circular cone, i.e., the angle between any ray on the boundary of the cone and the v_0-axis, is $\theta^* = \arccos(r)$. This cone is depicted in Figure 3.4.

We therefore have from (3.24) and (3.25) that

$$\begin{aligned} P\{S_1 \le c\} &= P\left\{ \sup_{\mathbf{v}\in\mathcal{V}_e} \frac{\mathbf{v}'\mathbf{N}}{(\hat{\sigma}/\sigma)\|\mathbf{v}\|} \le c \right\} & (3.26) \\ &= P\left\{ \mathbf{v}'\left(\frac{\mathbf{N}}{(\hat{\sigma}/\sigma)}\right) \le c\|\mathbf{v}\| \;\forall\, \mathbf{v}\in\mathcal{V}_e \right\} & (3.27) \end{aligned}$$

where $\mathcal{V}_e$ is given in (3.25). Three methods are given in Section 3.4.2 for computing this probability.

Similarly, we have

$$\begin{aligned} P\{S_2 \le c\} &= P\left\{ \sup_{\mathbf{v}\in\mathcal{V}_e} \frac{|\mathbf{v}'\mathbf{N}|}{(\hat{\sigma}/\sigma)\|\mathbf{v}\|} \le c \right\} & (3.28) \\ &= P\left\{ \left|\mathbf{v}'\left(\frac{\mathbf{N}}{(\hat{\sigma}/\sigma)}\right)\right| \le c\|\mathbf{v}\| \;\forall\, \mathbf{v}\in\mathcal{V}_e \right\}. & (3.29) \end{aligned}$$

Three methods are given in Section 3.4.1 for computing this probability.

Polar coordinates of $(p+1)$-dimension are used in several places below and so reviewed briefly here. For a $(p+1)$-dimensional vector $\mathbf{v} = (v_0, \cdots, v_p)'$, define its polar coordinates $(R_{\mathbf{v}}, \theta_{\mathbf{v}1}, \ldots, \theta_{\mathbf{v},p})'$ by

$$\begin{aligned}
v_0 &= R_{\mathbf{v}} \cos\theta_{\mathbf{v}1} \\
v_1 &= R_{\mathbf{v}} \sin\theta_{\mathbf{v}1} \cos\theta_{\mathbf{v}2} \\
v_2 &= R_{\mathbf{v}} \sin\theta_{\mathbf{v}1} \sin\theta_{\mathbf{v}2} \cos\theta_{\mathbf{v}3} \\
&\cdots \quad \cdots \\
v_{p-1} &= R_{\mathbf{v}} \sin\theta_{\mathbf{v}1} \sin\theta_{\mathbf{v}2} \cdots \sin\theta_{\mathbf{v},p-1} \cos\theta_{\mathbf{v},p} \\
v_p &= R_{\mathbf{v}} \sin\theta_{\mathbf{v}1} \sin\theta_{\mathbf{v}2} \cdots \sin\theta_{\mathbf{v},p-1} \sin\theta_{\mathbf{v},p}
\end{aligned}$$

where

$$\begin{aligned}
&0 \le \theta_{\mathbf{v}1} \le \pi \\
&0 \le \theta_{\mathbf{v}2} \le \pi \\
&\cdots \quad \cdots \\
&0 \le \theta_{\mathbf{v},p-1} \le \pi \\
&0 \le \theta_{\mathbf{v},p} \le 2\pi \\
&R_{\mathbf{v}} \ge 0.
\end{aligned}$$

The Jacobian of the transformation is given by

$$|J| = R_{\mathbf{v}}^p \sin^{p-1}\theta_{\mathbf{v}1} \sin^{p-2}\theta_{\mathbf{v}2} \cdots \sin\theta_{\mathbf{v},p-1}.$$

Polar coordinates were used by Tamhankar (1967) to characterize a multivariate normal distribution.

For $\mathbf{T} = \mathbf{N}/(\hat{\sigma}/\sigma)$, which is a standard $(p+1)$-dimensional t random vector, one can directly find the joint density function of $(R_{\mathbf{T}}, \theta_{\mathbf{T}1}, \ldots, \theta_{\mathbf{T},p})'$ by using the Jacobian above. In particular, all the polar coordinates are independent random variables, the marginal density of $\theta_{\mathbf{T}1}$ is given by

$$f(\theta) = k \sin^{p-1}\theta, \;\; 0 \le \theta \le \pi \tag{3.30}$$

where $k = 1/(\int_0^\pi \sin^{p-1}\theta d\theta)$ is the normalizing constant, and the marginal distribution of $R_{\mathbf{T}}$ is given by

$$R_{\mathbf{T}} = \|\mathbf{N}/(\hat{\sigma}/\sigma)\| \;\sim\; \sqrt{(p+1)F_{p+1,\nu}} \tag{3.31}$$

where $F_{p+1,\nu}$ denotes an F random variable that has $p+1$ and ν degrees of freedom.

3.4.1 Two-sided band

3.4.1.1 A method based on Bohrer's (1973b) approach

This method can be regarded as the generalization of the method of Wynn and Bloomfield (1971) given in Chapter 2 and is similar to Bohrer's (1973b) method for the one-sided band given in Section 3.4.2 below. From (3.29), the confidence level is given by

$$P\{\mathbf{N}/(\hat{\sigma}/\sigma) \in A_{r,2}\} = P\{\mathbf{T} \in A_{r,2}\} \qquad (3.32)$$

where

$$A_{r,2} = A_{r,2}(c) = \{\mathbf{t} = (t_0, \cdots, t_p)' : \ |\mathbf{v}'\mathbf{t}| \le c\|\mathbf{v}\| \ \ \forall \ \mathbf{v} \text{ in } \mathcal{V}_e\}, \qquad (3.33)$$

where $\mathcal{V}_e$ is the circular cone given in (3.25).

In the definition of $A_{r,2}$ in (3.33), each $|\mathbf{v}'\mathbf{t}| \le c\|\mathbf{v}\|$ restricts $\mathbf{t}$ to the origin-containing stripe that is bounded by the two hyper-planes which are perpendicular to the vector $\mathbf{v}$ and c-distance away from the origin. So when $\mathbf{v}$ varies over $\mathcal{V}_e$ the resultant set $A_{r,2}$ has the shape given in Figure 3.5. The set $A_{r,2}$ can be partitioned into four sets, also depicted in Figure 3.5, which can be expressed easily using the polar coordinates, as given by the following lemma.

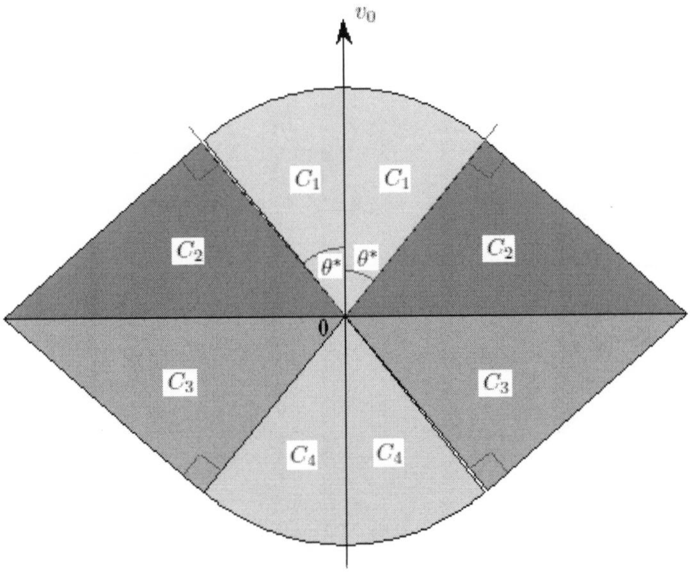

Figure 3.5: *The set $A_{r,2}$ and its partition $A_{r,2} = C_1 + C_2 + C_3 + C_4$*

Lemma 3.4 *We have $A_{r,2} = C_1 + C_2 + C_3 + C_4$ where*

$$C_1 \;=\; \{\mathbf{t} : 0 \le \theta_{\mathbf{t}1} \le \theta^*, \ R_{\mathbf{t}} \le c\},$$

$$
\begin{aligned}
C_2 &= \{\mathbf{t} : \theta^* < \theta_{\mathbf{t}1} \le \frac{\pi}{2},\ R_{\mathbf{t}} \cos(\theta_{\mathbf{t}1} - \theta^*) \le c\}, \\
C_3 &= \{\mathbf{t} : \frac{\pi}{2} < \theta_{\mathbf{t}1} \le \pi - \theta^*,\ R_{\mathbf{t}} \cos(\pi - \theta^* - \theta_{\mathbf{t}1}) \le c\}, \\
C_4 &= \{\mathbf{t} : \pi - \theta^* < \theta_{\mathbf{t}1} \le \pi,\ R_{\mathbf{t}} \le c\}.
\end{aligned}
$$

The lemma is clear from Figure 3.5 and can be proved using basic calculus, the details of which is omitted here. Now the four probabilities can be calculated in the following way. Firstly,

$$
\begin{aligned}
\mathrm{P}\{\mathbf{T} \in C_1\} = \mathrm{P}\{\mathbf{T} \in C_4\} &= \int_0^{\theta^*} k \sin^{p-1}\theta d\theta \cdot \mathrm{P}\{R_{\mathbf{T}} \le c\} \\
&= \int_0^{\theta^*} k \sin^{p-1}\theta d\theta \cdot \mathrm{P}\{(p+1)F_{p+1,\nu} \le c^2\} \\
&= \int_0^{\theta^*} k \sin^{p-1}\theta d\theta \cdot F_{p+1,\nu}\left(\frac{c^2}{p+1}\right).
\end{aligned}
$$

Secondly,

$$
\begin{aligned}
\mathrm{P}\{\mathbf{T} \in C_2\} = \mathrm{P}\{\mathbf{T} \in C_3\} &= \int_{\theta^*}^{\frac{\pi}{2}} k \sin^{p-1}\theta \cdot \mathrm{P}\{R_{\mathbf{T}} \cos(\theta - \theta^*) \le c\} d\theta \\
&= \int_0^{\frac{\pi}{2}-\theta^*} k \sin^{p-1}(\theta + \theta^*) \cdot \mathrm{P}\left\{R_{\mathbf{T}} \le \frac{c}{\cos\theta}\right\} d\theta \\
&= \int_0^{\frac{\pi}{2}-\theta^*} k \sin^{p-1}(\theta + \theta^*) \cdot F_{p+1,\nu}\left(\frac{c^2}{(p+1)\cos^2\theta}\right) d\theta.
\end{aligned}
$$

The confidence level is therefore given by

$$
\begin{aligned}
& \int_0^{\theta^*} 2k \sin^{p-1}\theta d\theta \cdot F_{p+1,\nu}\left(\frac{c^2}{p+1}\right) \\
+ \quad & \int_0^{\pi/2-\theta^*} 2k \sin^{p-1}(\theta + \theta^*) \cdot F_{p+1,\nu}\left(\frac{c^2}{(p+1)\cos^2\theta}\right) d\theta. \qquad (3.34)
\end{aligned}
$$

3.4.1.2 *The method of Casella and Strawderman (1980)*

The key idea of this method is to find the supreme in (3.28) explicitly, as given in Lemma 3.5 below, from which the confidence level can be evaluated.

Lemma 3.5 *We have*

$$
\sup_{\mathbf{v} \in \mathcal{V}_e} \frac{|\mathbf{v}'\mathbf{N}|}{\|\mathbf{v}\|} = \begin{cases} \|\mathbf{N}\| & \text{if } \pm\mathbf{N} \in \mathcal{V}_e \\ \frac{q|N_0| + \|\mathbf{N}_{(0)}\|}{\sqrt{q^2+1}} & \text{if } \pm\mathbf{N} \notin \mathcal{V}_e \end{cases}
$$

where $\mathbf{N} = (N_0, \cdots, N_p)' = (N_0, \mathbf{N}'_{(0)})'$, *and* $q = \sqrt{r^2/(1-r^2)} = 1/a$.

This can be proved by using basic calculus and the detail is available in Casella and Strawderman (1980). From this lemma and by denoting $s = \hat{\sigma}/\sigma$, the confidence level of the band is given by

$$
\begin{aligned}
& P\left\{\sup_{\mathbf{v}\in\mathcal{V}_e} \frac{|\mathbf{v}'\mathbf{N}|}{\|\mathbf{v}\|} \le cs\right\} \\
= \; & P\{\pm\mathbf{N} \in \mathcal{V}_e,\ \|\mathbf{N}\| \le cs\} + P\left\{\pm\mathbf{N} \notin \mathcal{V}_e,\ \frac{q|N_0| + \|\mathbf{N}_{(0)}\|}{\sqrt{q^2+1}} \le cs\right\}. \\
= \; & P\left\{|N_0| \ge q\|\mathbf{N}_{(0)}\|, |N_0|^2 + \|\mathbf{N}_{(0)}\|^2 \le c^2s^2\right\} \\
& + P\left\{|N_0| < q\|\mathbf{N}_{(0)}\|, \frac{q|N_0| + \|\mathbf{N}_{(0)}\|}{\sqrt{q^2+1}} \le cs\right\}.
\end{aligned}
$$

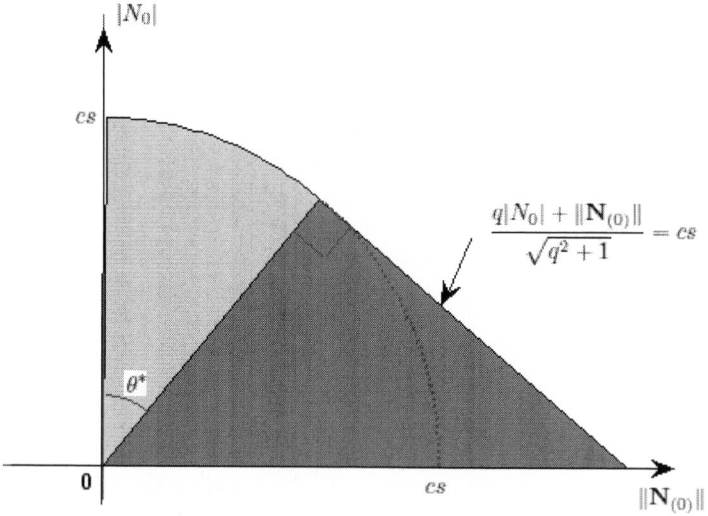

Figure 3.6: *The two regions and the new partition*

The two regions

$$
\begin{aligned}
& \left\{|N_0| \ge q\|\mathbf{N}_{(0)}\|,\ |N_0|^2 + \|\mathbf{N}_{(0)}\|^2 \le c^2s^2\right\} \\
\text{and} \quad & \left\{|N_0| < q\|\mathbf{N}_{(0)}\|,\ \frac{q|N_0|+\|\mathbf{N}_{(0)}\|}{\sqrt{q^2+1}} \le cs\right\}
\end{aligned}
$$

are depicted in Figure 3.6 in the $(\|\mathbf{N}_{(0)}\|, |N_0|)$-coordinate system. The union of these two regions can be re-partitioned into two regions. The first region is the quarter disc

$$\{R^2 \le c^2 s^2\} \text{ where } R^2 = |N_0|^2 + \|\mathbf{N}_{(0)}\|^2.$$

The second region is the remaining part which can be expressed as

$$\left\{ c^2 s^2 < R^2 < (q^2+1)c^2 s^2,\ 0 \le \frac{|N_0|}{\|\mathbf{N}_{(0)}\|} \le \frac{qcs - \sqrt{R^2 - c^2 s^2}}{cs + q\sqrt{R^2 - c^2 s^2}} \right\}.$$

This expression comes from the fact that, when the point $(\|\mathbf{N}_{(0)}\|, |N_0|)$ varies on the segment of the circle $|N_0|^2 + \|\mathbf{N}_{(0)}\|^2 = R^2 \in (c^2 s^2, (q^2+1)c^2 s^2)$ that is within the second region, $|N_0|/\|\mathbf{N}_{(0)}\|$ attains its minimum value 0 at the right end of the circle-segment $(\|\mathbf{N}_{(0)}\|, |N_0|) = (R, 0)$ and attains its maximum value

$$(qcs - \sqrt{R^2 - c^2 s^2})/(cs + q\sqrt{R^2 - c^2 s^2})$$

at the left end of the circle-segment whose coordinates can be solved from the simultaneous equations $|N_0|^2 + \|\mathbf{N}_{(0)}\|^2 = R^2$ and $(q|N_0| + \|\mathbf{N}_{(0)}\|)/\sqrt{q^2+1} = cs$. From this new partition, the confidence level is further equal to

$$\begin{aligned} &\mathrm{P}\{R^2 \le c^2 s^2\} \\ + \ &\mathrm{P}\left\{ c^2 s^2 < R^2 < (q^2+1)c^2 s^2,\ 0 < \frac{|N_0|}{\|\mathbf{N}_{(0)}\|} \le \frac{qcs - \sqrt{R^2 - c^2 s^2}}{cs + q\sqrt{R^2 - c^2 s^2}} \right\}. \end{aligned} \quad (3.35)$$

Now note that $R^2 = |N_0|^2 + \|\mathbf{N}_{(0)}\|^2$ and $N_0/\|\mathbf{N}_{(0)}\|$ are independent random variables by considering the polar coordinates of $\mathbf{N}$, and both are independent of s. So the first probability in (3.35) is equal to $F_{p+1,v}\{c^2/(p+1)\}$, and the second probability in (3.35) is equal to

$$\begin{aligned} &\mathrm{P}\left\{ c^2 < \tfrac{R^2}{s^2} < (q^2+1)c^2,\ 0 < \tfrac{|N_0|}{\|\mathbf{N}_{(0)}\|} \le \tfrac{qc - \sqrt{(R^2/s^2) - c^2}}{c + q\sqrt{(R^2/s^2) - c^2}} \right\} \\ = \ &\int_{c^2/(p+1)}^{(q^2+1)c^2/(p+1)} F_{1,p}\left(p \left(\tfrac{qc - \sqrt{(p+1)w - c^2}}{c + q\sqrt{(p+1)w - c^2}} \right)^2 \right) dF_{p+1,v}(w). \end{aligned}$$

The simultaneous confidence level is therefore given by

$$\begin{aligned} &F_{p+1,v}\left(\frac{c^2}{p+1}\right) + \\ &\int_{c^2/(p+1)}^{(q^2+1)c^2/(p+1)} F_{1,p}\left(p \left(\frac{qc - \sqrt{(p+1)w - c^2}}{c + q\sqrt{(p+1)w - c^2}} \right)^2 \right) dF_{p+1,v}(w). \end{aligned} \quad (3.36)$$

In fact, Casella and Strawderman (1980) considered the construction of an exact two-sided hyperbolic confidence band over a region of the predictor variables whose simultaneous confidence level can be reduced to the form

$$P\left\{ \sup_{\mathbf{v}\in\mathcal{V}^*(m)} \frac{|\mathbf{v}'\mathbf{N}|}{(\hat{\sigma}/\sigma)\parallel \mathbf{v} \parallel} \le c \right\} \text{ where } \mathcal{V}^*(m) = \left\{ \mathbf{v} : \sum_{i=0}^{m} v_i^2 \ge \frac{r^2}{1-r^2} \sum_{i=m+1}^{p} v_i^2 \right\}$$

where $0 \le m \le p$ is a given integer. Note that $\mathcal{V}_e = \mathcal{V}^*(0)$. Seppanen and Uusipaikka (1992) provided an explicit form of the predictor variable region over which the two-sided hyperbolic confidence band has its simultaneous confidence level given by this form. For $1 \le m \le p$, the predictor variable region corresponding to $\mathcal{V}^*(m)$ is not bounded however, and so a confidence band over such a predictor variable region is of less interest considering that a regression model holds most likely only over a finite region of the predictor variables in real problems. The main purpose of studying $\mathcal{V}^*(m)$ for $1 \le m \le p$ in Casella and Strawderman (1980) is to find a conservative two-sided hyperbolic band over the rectangular predictor variable region $\mathcal{X}_r$: use the predictor variable regions corresponding to $\mathcal{V}^*(m)$ for $0 \le m \le p$ to bound the given $\mathcal{X}_r$, calculate the critical values for these $\mathcal{V}^*(m)$'s, and use the smallest calculated critical values as a conservative critical value for the confidence band over $\mathcal{X}_r$.

3.4.1.3 A method based on the volume of tubular neighborhoods

This method is based on the volume of tubular neighborhoods of the circular cone $\mathcal{V}_e$ and similar to those used by Naiman (1986, 1990), Sun (1993) and Sun and Loader (1994), among others. Due to the special form of the cone $\mathcal{V}_e$, the exact volume of its tubular neighborhoods can be calculated easily. From (3.28), the confidence level is given by

$$P\left\{ \sup_{\mathbf{v}\in\mathcal{V}_e} \frac{|\mathbf{v}'\mathbf{N}|}{\parallel \mathbf{v} \parallel\parallel \mathbf{N} \parallel} \le c\frac{(\hat{\sigma}/\sigma)}{\parallel \mathbf{N} \parallel} \right\}. \tag{3.37}$$

Note that $\mathbf{N}/\|\mathbf{N}\|$ depends only on the $\theta_{\mathbf{N},i}$'s and that $\|\mathbf{N}\| = R_{\mathbf{N}}$. Hence $\mathbf{N}/\|\mathbf{N}\|$ is independent of $\|\mathbf{N}\|$ and so $(\hat{\sigma}/\sigma)/\parallel \mathbf{N} \parallel$. Furthermore, the supreme in (3.37) is no larger than one, and $c/\sqrt{(p+1)w} < 1$ if and only if $w > c^2/(p+1)$. So the confidence level is equal to

$$\begin{aligned} & 1 - \int_0^\infty P\left\{ \sup_{\mathbf{v}\in\mathcal{V}_e} \frac{|\mathbf{v}'\mathbf{N}|}{\parallel \mathbf{v} \parallel\parallel \mathbf{N} \parallel} > \frac{c}{\sqrt{(p+1)w}} \right\} dF_{p+1,\nu}(w) \\ = \; & 1 - \int_{c^2/(p+1)}^\infty P\left\{ \sup_{\mathbf{v}\in\mathcal{V}_e} \frac{|\mathbf{v}'\mathbf{N}|}{\parallel \mathbf{v} \parallel\parallel \mathbf{N} \parallel} > \frac{c}{\sqrt{(p+1)w}} \right\} dF_{p+1,\nu}(w). \end{aligned} \tag{3.38}$$

The key to this method is to find the probability in (3.38). This hinges on the following result, which is a generalization of the result in Section 2.2.1.3.

Lemma 3.6 *Let* $0 < h = c/\sqrt{(p+1)w} < 1$, $\alpha = \arccos(h) \in (0, \pi/2)$, $(R_{\mathbf{N}}, \theta_{\mathbf{N}1}, \ldots, \theta_{\mathbf{N},p})'$ *be the polar coordinates of* $\mathbf{N}$, *and* $\theta^* = \arccos(r)$. *We have*

$$\left\{\mathbf{N}:\ \sup_{\mathbf{v}\in\mathcal{V}_e} \frac{|\mathbf{v}'\mathbf{N}|}{\|\mathbf{v}\|\|\mathbf{N}\|} > h\right\}$$
$$= \begin{cases} \{\mathbf{N}:\ \theta_{\mathbf{N}1} \in [0,\ \theta^*+\alpha] \cup [\pi-\theta^*-\alpha,\ \pi]\} & \text{if } \theta^*+\alpha < \frac{\pi}{2} \\ \{\mathbf{N}:\ \theta_{\mathbf{N}1} \in [0,\ \pi]\} & \text{if } \theta^*+\alpha \geq \frac{\pi}{2}. \end{cases}$$

The proof involves only basic calculus and the detail is omitted here. From this lemma, the fact that $\theta^* + \arccos(c/\sqrt{(p+1)w}) < \pi/2$ if and only if $w < (q^2+1)c^2/(p+1)$, and that the pdf of $\theta_{\mathbf{N}1}$ is given in (3.30), we have, for $c^2/(p+1) \leq w < (q^2+1)c^2/(p+1)$,

$$\mathrm{P}\left\{\sup_{\mathbf{v}\in\mathcal{V}_e} \frac{|\mathbf{v}'\mathbf{N}|}{\|\mathbf{v}\|\|\mathbf{N}\|} > \frac{c}{\sqrt{(p+1)w}}\right\}$$
$$= \mathrm{P}\left\{\ \theta_{\mathbf{N}1} \in [0,\ \theta^*+\arccos(c/\sqrt{(p+1)w})] \cup [\pi-\theta^*-\arccos(c/\sqrt{(p+1)w}),\ \pi]\right\}$$
$$= 2\int_0^{\theta^*+\arccos(c/\sqrt{(p+1)w})} k\sin^{p-1}\theta d\theta$$

and, for $w \geq (q^2+1)c^2/(p+1)$,

$$\mathrm{P}\left\{\sup_{\mathbf{v}\in\mathcal{V}_e} \frac{|\mathbf{v}'\mathbf{N}|}{\|\mathbf{v}\|\|\mathbf{N}\|} > \frac{c}{\sqrt{(p+1)w}}\right\} = \mathrm{P}\{\ \theta_{\mathbf{N}1} \in [0,\ \pi]\} = 1.$$

Substituting these two expressions into (3.38), the confidence level is equal to

$$1 - \int_{c^2/(p+1)}^{(q^2+1)c^2/(p+1)} \int_0^{\theta^*+\arccos(c/\sqrt{(p+1)w})} 2k\sin^{p-1}\theta d\theta dF_{p+1,\nu}(w)$$
$$- \int_{(q^2+1)c^2/(p+1)}^{\infty} 1dF_{(p+1),\nu}(w). \tag{3.39}$$

It is straightforward to show by changing the order of integrations that the double integral above is equal to

$$\int_0^{\theta^*} 2k\sin^{p-1}\theta d\theta \cdot \left(F_{p+1,\nu}\left(\frac{(q^2+1)c^2}{p+1}\right) - F_{p+1,\nu}\left(\frac{c^2}{(p+1)}\right)\right)$$
$$+ \int_{\theta^*}^{\frac{\pi}{2}} 2k\sin^{p-1}\theta \cdot \left(F_{p+1,\nu}\left(\frac{(q^2+1)c^2}{p+1}\right) - F_{p+1,\nu}\left(\frac{c^2}{(p+1)\cos^2(\theta-\theta^*)}\right)\right) d\theta.$$

By substituting this into (3.39), it is clear that the confidence level expression (3.39) is equal to the expression given in (3.34).

Numerical results have been computed from expressions (3.34) and (3.36) and these agree with the entries of Seppanen and Uusipaikka (1992, Table 1 for $r = 1$). Expression (3.34) is slightly easier to use than the expression (3.36) for numerical computation.

3.4.2 One-sided band

3.4.2.1 The method of Bohrer (1973b)

This method was given by Bohrer (1973b). From (3.27), we have

$$P\{S_1 \le c\} = P\{\mathbf{N}/(\hat{\sigma}/\sigma) \in A_{r,1}\} = P\{\mathbf{T} \in A_{r,1}\} \tag{3.40}$$

where

$$A_{r,1} = A_{r,1}(c) = \{\mathbf{t} = (t_0, \cdots, t_p)' : \ \mathbf{v}'\mathbf{t} \le c\|\mathbf{v}\| \ \ \forall \ \mathbf{v} \text{ in } \mathcal{V}_e\}, \tag{3.41}$$

where $\mathcal{V}_e$ is the circular cone given in (3.25).

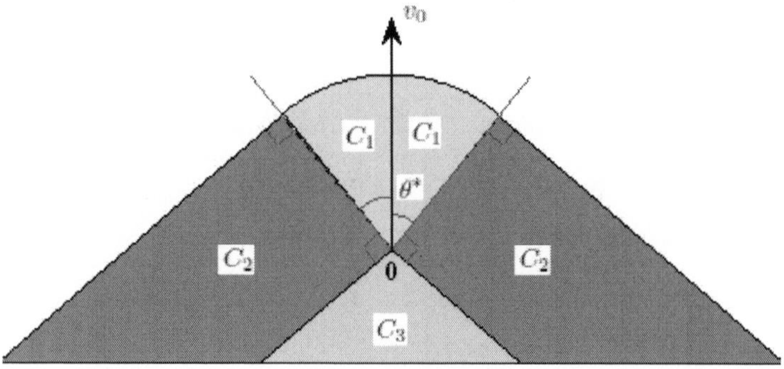

Figure 3.7: *The set $A_{r,1}$ and its partition $A_{r,1} = C_1 + C_2 + C_3$*

Assume $c > 0$. Note that each $\mathbf{v}'\mathbf{t} \le c\|\mathbf{v}\|$ in the definition of $A_{r,1}$ restricts $\mathbf{t}$ to the origin-containing side of the plane that is perpendicular to the vector $\mathbf{v}$ and c-distance away from the origin in the direction of $\mathbf{v}$. So when $\mathbf{v}$ varies over $\mathcal{V}_e$ the resultant set $A_{r,1}$ has the shape given in Figure 3.7. The set $A_{r,1}$ can be partitioned into three sets which can be expressed easily using the polar coordinates.

Lemma 3.7 *We have $A_{r,1} = C_1 + C_2 + C_3$ where*

$$\begin{aligned} C_1 &= \{\mathbf{t} : 0 \le \theta_{\mathbf{t}1} \le \theta^*, \ R_{\mathbf{t}} \le c\}, \\ C_2 &= \{\mathbf{t} : \theta_{\mathbf{t}1} - \theta^* \in (0, \pi/2], \ R_{\mathbf{t}} \cos(\theta_{\mathbf{t}1} - \theta^*) \le c\}, \\ C_3 &= \{\mathbf{t} : \theta^* + \pi/2 < \theta_{\mathbf{t}1} \le \pi\}. \end{aligned}$$

The lemma is clear from Figure 3.7 and can be proved by introducing the notation f_j for $j = 1, \ldots, p$ as in Bohrer (1973b); the detail is omitted here. Note that C_1 is the intersection of the circular cone $\mathcal{V}_e$ and the $(p+1)$-dimension ball centered at the origin with radius c, C_3 is the dual cone of $\mathcal{V}_e$, and $C_2 = C_1 \oplus C_3$. These three sets are also depicted in Figure 3.7. Wynn (1975) generalized this partition to the situation where $\mathcal{V}_e$ may not be a circular cone; but the computation of the probabilities involved is not trivial at all.

Now the three probabilities can be calculated as follows. Firstly,

$$\begin{aligned} P\{\mathbf{T} \in C_1\} &= \int_0^{\theta^*} k \sin^{p-1}\theta d\theta \cdot P\{R_{\mathbf{T}} \le c\} \\ &= \int_0^{\theta^*} k \sin^{p-1}\theta d\theta \cdot P\{(p+1)F_{p+1,v} \le c^2\} \\ &= \int_0^{\theta^*} k \sin^{p-1}\theta d\theta \cdot F_{p+1,v}\left(c^2/(p+1)\right). \end{aligned}$$

Secondly,

$$P\{\mathbf{T} \in C_3\} = \int_{\theta^*+\frac{\pi}{2}}^{\pi} k \sin^{p-1}\theta d\theta = \int_0^{\frac{\pi}{2}-\theta^*} k \sin^{p-1}\theta d\theta.$$

Thirdly,

$$\begin{aligned} P\{\mathbf{T} \in C_2\} &= \int_{\theta^*}^{\theta^*+\frac{\pi}{2}} k \sin^{p-1}\theta \cdot P\{R_{\mathbf{T}}\cos(\theta-\theta^*) \le c\} d\theta \\ &= \int_0^{\frac{\pi}{2}} k \sin^{p-1}(\theta+\theta^*) \cdot P\left\{R_{\mathbf{T}} \le \frac{c}{\cos\theta}\right\} d\theta \\ &= \int_0^{\frac{\pi}{2}} k \sin^{p-1}(\theta+\theta^*) \cdot F_{p+1,v}\left(\frac{c^2}{(p+1)\cos^2\theta}\right) d\theta. \end{aligned}$$

The confidence level can therefore be calculated from

$$P\{S_1 \le r\} = P\{\mathbf{T} \in C_1\} + P\{\mathbf{T} \in C_2\} + P\{\mathbf{T} \in C_3\}. \tag{3.42}$$

One can express $P\{\mathbf{T} \in C_2\}$ as a linear combination of several F probabilities by following the idea of Bohrer (1973b). While this is interesting mathematically, expression (3.42) is easier for numerical calculation.

When $a = \infty$ it is clear that $r = 0$ and $\theta^* = \pi/2$. In this special case, C_1 is a half ball in $\Re^{p+1}$ and so

$$P\{\mathbf{T} \in C_1\} = \frac{1}{2} F_{p+1,v}\left\{c^2/(p+1)\right\},$$

C_2 is a half cylinder that has the expression

$$C_2 = \{\mathbf{t} :\ t_0 < 0,\ \|\mathbf{t}_{(0)}\| \le c\}, \ \text{where } \mathbf{t}_{(0)} = (t_1, \cdots, t_p)'$$

and so

$$\mathrm{P}\{\mathbf{T} \in C_2\} = \frac{1}{2} F_{p,\nu}(c^2/p),$$

and C_3 is empty. Hence the confidence level is given by

$$\frac{1}{2} F_{p+1,\nu}\left\{c^2/(p+1)\right\} + \frac{1}{2} F_{p,\nu}(c^2/p),$$

which agrees with the result of Hochberg and Quade (1975, expression (2.4)) as given in Theorem 3.3.

3.4.2.2 An algebraic method

This approach is similar to that of Casella and Strawderman (1980) for the two-sided case given in Section 3.4.1 above. The key idea is to find the supreme in (3.26) explicitly, as given in Lemma 3.8 below, from which the confidence level can be evaluated.

Lemma 3.8 *We have*

$$\sup_{\mathbf{v} \in \mathcal{V}_e} \frac{\mathbf{v}'\mathbf{N}}{\|\mathbf{v}\|} = \begin{cases} \|\mathbf{N}\| & \text{if } \mathbf{N} \in \mathcal{V}_e \\ \frac{qN_0 + \|\mathbf{N}_{(0)}\|}{\sqrt{q^2+1}} & \text{if } \mathbf{N} \notin \mathcal{V}_e \end{cases}$$

where N_0 is the first element of $\mathbf{N}$, $\mathbf{N}_{(0)} = (N_1, \cdots, N_p)'$, *and* $q = \sqrt{r^2/(1-r^2)} = 1/a$.

This can be proved by using basic calculus and the detail is omitted here. From this lemma and by denoting $s = \hat{\sigma}/\sigma$, the confidence level of the band is given by

$$\begin{aligned} & \mathrm{P}\left\{\sup_{\mathbf{v} \in \mathcal{V}_e} \frac{\mathbf{v}'\mathbf{N}}{\|\mathbf{v}\|} \leq cs\right\} \\ = & \mathrm{P}\{\mathbf{N} \in \mathcal{V}_e,\ \|\mathbf{N}\| \leq cs\} + \mathrm{P}\left\{\mathbf{N} \notin \mathcal{V}_e,\ \frac{qN_0 + \|\mathbf{N}_{(0)}\|}{\sqrt{q^2+1}} \leq cs\right\}. \\ = & \mathrm{P}\left\{N_0 \geq q\|\mathbf{N}_{(0)}\|,\ |N_0|^2 + \|\mathbf{N}_{(0)}\|^2 \leq c^2 s^2\right\} \\ & + \mathrm{P}\left\{N_0 < q\|\mathbf{N}_{(0)}\|,\ \frac{qN_0 + \|\mathbf{N}_{(0)}\|}{\sqrt{q^2+1}} \leq cs\right\}. \end{aligned}$$

The two regions

$$\left\{N_0 \geq q\|\mathbf{N}_{(0)}\|,\ \|N_0\|^2 + \|\mathbf{N}_{(0)}\|^2 \leq c^2 s^2\right\}$$

$$\text{and} \qquad \left\{N_0 < q\|\mathbf{N}_{(0)}\|,\ \frac{qN_0 + \|\mathbf{N}_{(0)}\|}{\sqrt{q^2+1}} \leq cs\right\}$$

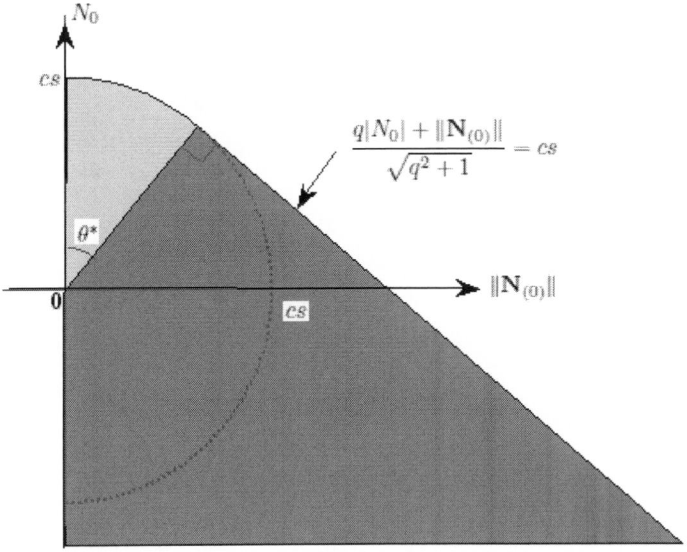

Figure 3.8: *The two regions and the new partition*

are depicted in Figure 3.8 in the $(\|\mathbf{N}_{(0)}\|, N_0)$-coordinate system. The union of these two regions can be re-partitioned into two regions. The first region is the half disc

$$\{R^2 \le c^2 s^2\} \text{ where } R^2 = N_0^2 + \|\mathbf{N}_{(0)}\|^2.$$

The second region is the remaining part which has the expression

$$\left\{ c^2 s^2 < R^2 < \infty, \ -\infty < \frac{N_0}{\|\mathbf{N}_{(0)}\|} \le \frac{qcs - \sqrt{R^2 - c^2 s^2}}{cs + q\sqrt{R^2 - c^2 s^2}} \right\}.$$

This expression comes from the fact that, when the point $(\|\mathbf{N}_{(0)}\|, N_0)$ varies on the segment of the circle $N_0^2 + \|\mathbf{N}_{(0)}\|^2 = R^2$ $(> c^2 s^2)$ that is within the second region, $N_0/\|\mathbf{N}_{(0)}\|$ attains its minimum value $-\infty$ at the lower-left end of the circle-segment $(\|\mathbf{N}_{(0)}\|, N_0) = (0, -R)$ and attains its maximum value

$$(qcs - \sqrt{R^2 - c^2 s^2})/(cs + q\sqrt{R^2 - c^2 s^2})$$

at the upper-right end of the circle-segment whose coordinates can be solved from the simultaneous equations $N_0^2 + \|\mathbf{N}_{(0)}\|^2 = R^2$ and $(qN_0 + \|\mathbf{N}_{(0)}\|)/\sqrt{q^2+1} = cs$

as in the two-sided case. From this new partition, the confidence level is further equal to

$$\begin{aligned} & \mathrm{P}\{R^2 \le c^2 s^2\} \\ + \quad & \mathrm{P}\left\{ c^2 s^2 < R^2 < \infty,\ -\infty < \frac{N_0}{\|\mathbf{N}_{(0)}\|} \le \frac{qcs - \sqrt{R^2 - c^2 s^2}}{cs + q\sqrt{R^2 - c^2 s^2}} \right\} \end{aligned} \tag{3.43}$$

similar to the two-sided case.

Now the first probability in (3.43) is equal to $F_{p,v}(c^2/(p+1))$ as in the two-sided case, and the second probability in (3.43) is equal to

$$\begin{aligned} & \mathrm{P}\left\{ c^2 < \frac{R^2}{s^2} < \infty,\ -\infty < \frac{N_0}{\|\mathbf{N}_{(0)}\|} \le \frac{qc - \sqrt{(R^2/s^2) - c^2}}{c + q\sqrt{(R^2/s^2) - c^2}} \right\} \\ = \quad & \int_{c^2/(p+1)}^{\infty} g(w) dF_{p+1,v}(w) \end{aligned}$$

where

$$g(w) = \mathrm{P}\left\{ -\infty < \frac{N_0}{\|\mathbf{N}_{(0)}\|} \le \frac{qc - \sqrt{(p+1)w - c^2}}{c + q\sqrt{(p+1)w - c^2}} \right\}.$$

Next we express $g(w)$ in term of the cdf of an F distribution. Note that

$$qc - \sqrt{(p+1)w - c^2} < 0 \Longleftrightarrow w > (q^2+1)c^2/(p+1).$$

Hence for $w > (q^2+1)c^2/(p+1)$ we have

$$(qc - \sqrt{(p+1)w - c^2})/(c + q\sqrt{(p+1)w - c^2}) < 0$$

and so

$$\begin{aligned} g(w) &= \frac{1}{2}\mathrm{P}\left\{ \frac{N_0^2}{\|\mathbf{N}_{(0)}\|^2} \ge \left(\frac{qc - \sqrt{(p+1)w - c^2}}{c + q\sqrt{(p+1)w - c^2}} \right)^2 \right\} \\ &= \frac{1}{2} - \frac{1}{2} F_{1,p}\left(p \left(\frac{qc - \sqrt{(p+1)w - c^2}}{c + q\sqrt{(p+1)w - c^2}} \right)^2 \right), \end{aligned} \tag{3.44}$$

and for $c^2/(p+1) < w \le (q^2+1)c^2/(p+1)$ we have

$$(qc - \sqrt{(p+1)w - c^2})/(c + q\sqrt{(p+1)w - c^2}) \ge 0$$

and so

$$\begin{aligned} g(w) &= \mathrm{P}\{N_0 \le 0\} + \mathrm{P}\left\{ 0 \le \frac{N_0}{\|\mathbf{N}_{(0)}\|} \le \frac{qc - \sqrt{(p+1)w - c^2}}{c + q\sqrt{(p+1)w - c^2}} \right\} \\ &= \frac{1}{2} + \frac{1}{2} F_{1,p}\left(p \left(\frac{qc - \sqrt{(p+1)w - c^2}}{c + q\sqrt{(p+1)w - c^2}} \right)^2 \right). \end{aligned} \tag{3.45}$$

Finally, the confidence level is given by

$$F_{p+1,\nu}\left(\frac{r^2}{p+1}\right) + \int_{c^2/(p+1)}^{\infty} g(w)dF_{p+1,\nu}(w) \quad (3.46)$$

with the function $g(w)$ being given by (3.44) and (3.45).

3.4.2.3 A method based on the volume of tubular neighborhoods

This method is similar to the two-sided case given in Section 3.4.1. From (3.26), the confidence level is given by

$$P\left\{\sup_{\mathbf{v}\in\mathcal{V}_e} \frac{\mathbf{v}'\mathbf{N}}{\|\mathbf{v}\|\|\mathbf{N}\|} \le c\frac{(\hat{\sigma}/\sigma)}{\|\mathbf{N}\|}\right\} \quad (3.47)$$

$$= 1 - \int_{c^2/(p+1)}^{\infty} P\left\{\sup_{\mathbf{v}\in\mathcal{V}_e} \frac{\mathbf{v}'\mathbf{N}}{\|\mathbf{v}\|\|\mathbf{N}\|} > \frac{c}{\sqrt{(p+1)w}}\right\} dF_{p+1,\nu}(w). \quad (3.48)$$

Now the probability in (3.48) can be easily evaluated by using the following result, which is the generalization of the result in Section 2.2.2.3.

Lemma 3.9 *Let* $0 < h < 1$, $\alpha = \arccos(h) \in (0, \pi/2)$, $\theta^* = \arccos(r)$ *as before, and* $(R_{\mathbf{N}}, \theta_{\mathbf{N}1}, \ldots, \theta_{\mathbf{N},p})$ *be the polar coordinates of* $\mathbf{N}$. *We have*

$$\left\{\mathbf{N}:\ \sup_{\mathbf{v}\in\mathcal{V}_e} \frac{\mathbf{v}'\mathbf{N}}{\|\mathbf{v}\|\|\mathbf{N}\|} > h\right\} = \{\mathbf{N}:\ \theta_{\mathbf{N}1} < \theta^* + \alpha\}.$$

Again the proof involves only calculus and the detail is omitted here. From this lemma and the fact that the pdf of $\theta_{\mathbf{N}1}$ is given in (3.30), we have for $0 < h < 1$

$$P\left\{\sup_{\mathbf{v}\in\mathcal{V}_e} \frac{\mathbf{v}'\mathbf{N}}{\|\mathbf{v}\|\|\mathbf{N}\|} > h\right\} = P\{\theta_{\mathbf{N}1} < \theta^* + \arccos(h)\} = \int_0^{\theta^*+\arccos(h)} k\sin^{p-1}\theta d\theta.$$

Substituting this expression for the probability in (3.48) with $h = c/\sqrt{(p+1)w}$ and changing the order of the double integration give the confidence level equal to

$$\begin{aligned}
& 1 - \int_{c^2/(p+1)}^{\infty} \int_0^{\theta^*+\arccos(h)} k\sin^{p-1}\theta d\theta dF_{p+1,\nu}(w) \\
= \ & 1 - \int_0^{\theta^*} \int_{c^2/(p+1)}^{\infty} k\sin^{p-1}\theta dF_{p+1,\nu}(w) d\theta \\
& - \int_{\theta^*}^{\theta^*+\frac{\pi}{2}} \int_{c^2/\{(p+1)\cos^2(\theta-\theta^*)\}}^{\infty} k\sin^{p-1}\theta dF_{p+1,\nu}(w) d\theta \\
= \ & 1 - \int_0^{\theta^*} k\sin^{p-1}\theta d\theta \cdot P\left\{F_{p+1,\nu} > \frac{c^2}{p+1}\right\} \\
& - \int_{\theta^*}^{\theta^*+\frac{\pi}{2}} k\sin^{p-1}\theta \cdot P\left\{F_{p+1,\nu} > \frac{c^2}{(p+1)\cos^2(\theta-\theta^*)}\right\} d\theta.
\end{aligned}$$

Now by replacing the 1 in the last expression above by

$$\int_0^{\pi} k\sin^{p-1}\theta d\theta = \int_0^{\theta^*} k\sin^{p-1}\theta d\theta + \int_{\theta^*}^{\theta^*+\frac{\pi}{2}} k\sin^{p-1}\theta d\theta + \int_0^{\frac{\pi}{2}-\theta^*} k\sin^{p-1}\theta d\theta$$

and straightforward manipulation, the confidence level is finally given by

$$\int_0^{\theta^*} k\sin^{p-1}\theta d\theta \cdot F_{p+1,\nu}\left(c^2/(p+1)\right)$$
$$+ \int_0^{\frac{\pi}{2}} k\sin^{p-1}(\theta+\theta^*) \cdot F_{p+1,\nu}\left(\frac{c^2}{(p+1)\cos^2\theta}\right) d\theta + \int_0^{\frac{\pi}{2}-\theta^*} k\sin^{p-1}\theta d\theta,$$

which is the same as the expression in (3.42).

Numerical computations for various parameter values of a, r, p and ν do confirm that the two expressions (3.42) and (3.46) are equal. For example, both expressions are equal to 0.77887 for $a = 2.0$, $r = 2.5$, $p = 6$ and $\nu = \infty$, and equal to 0.95620 for $a = 1.5$, $r = 3.0$, $p = 4$ and $\nu = 20$. For numerical computation, expression (3.42) is easier to use than expression (3.46).

Example 3.6 For the infant data set , the observed values of x_1 range from 92 to 149 with average $\bar{x}_{\cdot 1} = 113.71$, and the observed values of x_2 range from 2 to 5 with average $\bar{x}_{\cdot 2} = 3.88$. So the ellipsoidal region $\mathcal{X}_e$ is centered at $(\bar{x}_{\cdot 1}, \bar{x}_{\cdot 2})' = (113.71, 3.88)'$. The size of $\mathcal{X}_e$ increases with the value of a. Figure 3.9 gives four $\mathcal{X}_e$'s corresponding to various a-values indicated in the picture. The rectangular region indicates the observed range $[92, 149] \times [2, 5]$ of the predictor variables $(x_1, x_2)'$.

For a chosen $\mathcal{X}_e$, one can use a confidence band to quantify the plausible range of the unknown regression model over $\mathcal{X}_e$. Suppose $a = 1.3$ and so $\mathcal{X}_e$ is given by the second largest ellipse in Figure 3.9, and simultaneous confidence level is $1 - \alpha = 95\%$. Then the critical constants of the two-sided and one-sided hyperbolic bands are given by $c = 3.044$ and $c = 2.670$, respectively. These confidence bands are plotted in Figure 3.10: the bands are given only by the parts that are inside the cylinder which is in the y-direction and has the $\mathcal{X}_e$ as the cross-section in the $(x_1, x_2)'$-plane. The upper one-sided confidence band is slightly below the upper part of the two-sided band. Note from the discussion immediately below expression (3.18) that the width of the two-sided confidence band on the boundary of the $\mathcal{X}_e$ is a constant and given by

$$2c\hat{\sigma}\sqrt{\mathbf{x}^T(\mathbf{X}^T\mathbf{X})^{-1}\mathbf{x}} = 2c\hat{\sigma}\sqrt{(1+a^2)/n}$$
$$= 2 * 3.044 * 1.314 * \sqrt{(1+1.3^2)/17} = 3.187.$$

Of course the upper confidence band quantifies how high the unknown regression model can go over $\mathcal{X}_e$. All the computations in this example are done using MATLAB program example0306.m.

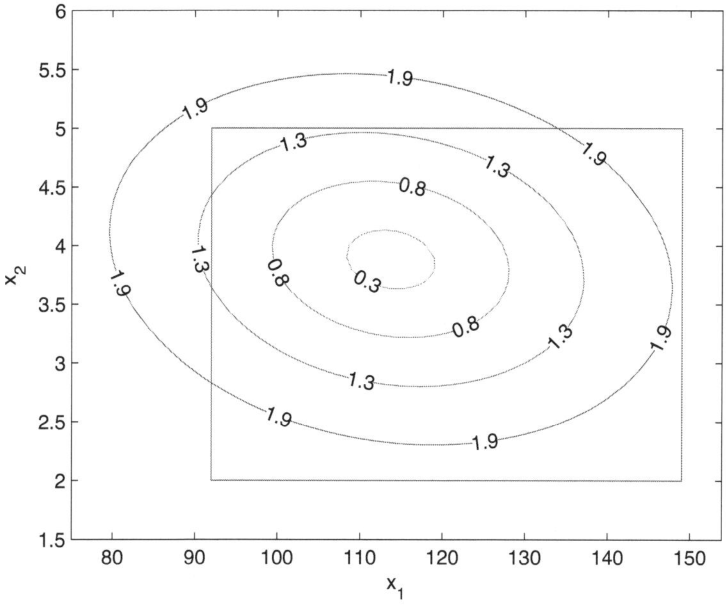

Figure 3.9: *Several ellipsoidal region $\mathcal{X}_e$'s*

3.5 Constant-width bands over an ellipsoidal region

Now we consider the construction of constant width bands over the ellipsoidal region $\mathcal{X}_e$.

3.5.1 *Two-sided band*

A two-sided constant width band has the form

$$\mathbf{x}'\beta \in \mathbf{x}'\hat{\beta} \pm c\sqrt{(1+a^2)/n}\,\hat{\sigma} \quad \text{for all } \mathbf{x}_{(0)} = (x_1, \cdots, x_p)' \in \mathcal{X}_e, \tag{3.49}$$

where c is chosen so that the simultaneous confidence level of the band is equal to $1-\alpha$. The confidence level is given by

$$\begin{aligned} & P\left\{ \sup_{\mathbf{x}_{(0)} \in \mathcal{X}_e} |\mathbf{x}'(\hat{\beta} - \beta)|/\hat{\sigma} \le c\sqrt{(1+a^2)/n} \right\} \\ = \; & P\left\{ \sup_{\mathbf{x}_{(0)} \in \mathcal{X}_e} \left| \{(\mathbf{z}, \mathbf{Z})'(\mathbf{X}'\mathbf{X})^{-1}\mathbf{x}\}' \mathbf{T} \right| \le c\sqrt{(1+a^2)/n} \right\} \end{aligned}$$

where $\mathbf{T} = (\mathbf{z}, \mathbf{Z})^{-1}(\mathbf{X}'\mathbf{X})(\hat{\beta} - \beta)/\hat{\sigma}$ is a standard $(p+1)$-dimensional t random vector as in the last section. Now note that, similar to (3.24) in the hyperbolic

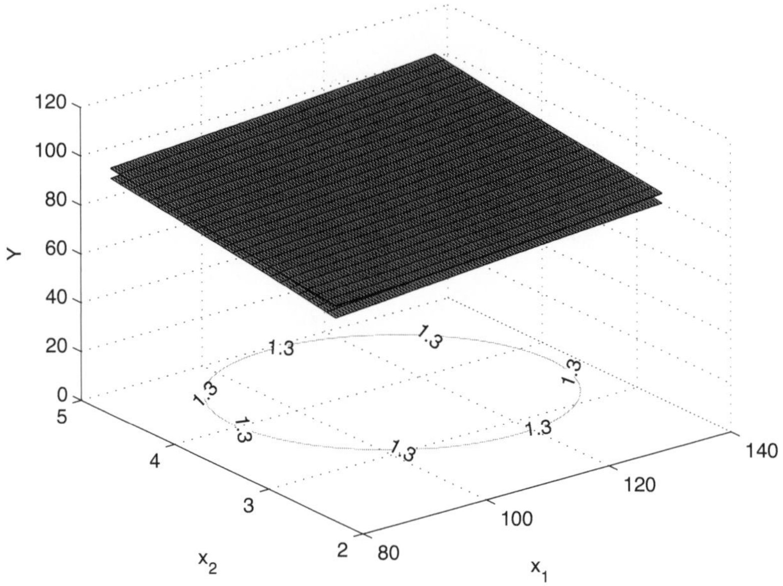

Figure 3.10: *The two-sided and upper hyperbolic bands over the $\mathcal{X}_e$ with $a = 1.3$*

band case,

$$\sup_{\mathbf{x}_{(0)} \in \mathcal{X}_e} \left| \left((\mathbf{z}, \mathbf{Z})'(\mathbf{X}'\mathbf{X})^{-1}\mathbf{x} \right)' \mathbf{T} \right| = \sup_{\mathbf{w}_{(0)} \in \mathcal{W}_e} |\mathbf{w}'\mathbf{T}|$$

where $\mathcal{W}_e$ is given in (3.23). Hence the confidence level is further equal to

$$P\left\{ \sup_{\mathbf{w}_{(0)} \in \mathcal{W}_e} |\mathbf{w}'\mathbf{T}| \leq c\sqrt{(1+a^2)/n} \right\}. \tag{3.50}$$

Let $\mathbf{T} = (T_0, \mathbf{T}'_{(0)})'$. Note that

$$\sup_{\mathbf{w}_{(0)} \in \mathcal{W}_e} |\mathbf{w}'\mathbf{T}| = \sup_{\mathbf{w}_{(0)} \in \mathcal{W}_e} \left| T_0/\sqrt{n} + \mathbf{w}'_{(0)}\mathbf{T}_{(0)} \right| \leq |T_0|/\sqrt{n} + \sqrt{a^2/n}\, \|\mathbf{T}_{(0)}\|$$

and the upper bound above is attained at $\mathbf{w}_{(0)} = \text{sign}(T_0)\sqrt{a^2/n}\, \mathbf{T}_{(0)}/\|\mathbf{T}_{(0)}\| \in \mathcal{W}_e$. Hence the confidence level in (3.50) can further be expressed as

$$P\left\{ |T_0|/\sqrt{n} + \sqrt{a^2/n}\, \|\mathbf{T}_{(0)}\| \leq c\sqrt{(1+a^2)/n} \right\}.$$

Now, using the polar coordinates of $\mathbf{T}$, this probability is equal to

$$\begin{aligned}
& \mathrm{P}\left\{ |R_{\mathbf{T}}\cos\theta_{\mathbf{T}1}|/\sqrt{n} + \sqrt{a^2/n}\,|R_{\mathbf{T}}\sin\theta_{\mathbf{T}1}| \le c\sqrt{(1+a^2)/n} \right\} \\
= \ & \mathrm{P}\{\, 0 \le \theta_{\mathbf{T}1} \le \pi/2,\ R_{\mathbf{T}}\cos(\theta_{\mathbf{T}1} - \theta^*) \le c\} \\
& + \mathrm{P}\{\, \pi/2 \le \theta_{\mathbf{T}1} \le \pi,\ R_{\mathbf{T}}\cos(\pi - \theta_{\mathbf{T}1} - \theta^*) \le c\} \\
= \ & 2\mathrm{P}\{\, 0 \le \theta_{\mathbf{T}1} \le \pi/2,\ R_{\mathbf{T}}\cos(\theta_{\mathbf{T}1} - \theta^*) \le c\} \\
= \ & \int_0^{\pi/2} 2k\sin^{p-1}\theta F_{p+1,\nu}\left(\frac{c^2}{(p+1)\cos^2(\theta - \theta^*)}\right) d\theta
\end{aligned}$$

where $\theta^* = \arccos(1/\sqrt{1+a^2}) = \arccos(r)$ as before.

3.5.2 *One-sided band*

A one-sided lower constant width band has the form

$$\mathbf{x}'\beta \ge \mathbf{x}'\hat{\beta} - c\sqrt{(1+a^2)/n}\,\hat{\sigma} \quad \text{for all } \mathbf{x}_{(0)} = (x_1, \cdots, x_p)' \in \mathcal{X}_e, \tag{3.51}$$

where c is the critical constant. Similar to the two-sided case, the confidence level is given by

$$\begin{aligned}
& \mathrm{P}\left\{ \sup_{\mathbf{x}_{(0)} \in \mathcal{X}_e} \mathbf{x}'(\hat{\beta} - \beta)/\hat{\sigma} \le c\sqrt{(1+a^2)/n} \right\} \\
= \ & \mathrm{P}\left\{ \sup_{\mathbf{x}_{(0)} \in \mathcal{X}_e} ((\mathbf{z}, \mathbf{Z})'(\mathbf{X}'\mathbf{X})^{-1}\mathbf{x})'\,\mathbf{T} \le c\sqrt{(1+a^2)/n} \right\} \\
= \ & \mathrm{P}\left\{ \sup_{\mathbf{w}_{(0)} \in \mathcal{W}_e} \mathbf{w}'\mathbf{T} \le c\sqrt{(1+a^2)/n} \right\}.
\end{aligned} \tag{3.52}$$

Now note that

$$\sup_{\mathbf{w}_{(0)} \in \mathcal{W}_e} \mathbf{w}'\mathbf{T} = \sup_{\mathbf{w}_{(0)} \in \mathcal{W}_e} \left\{ T_0/\sqrt{n} + \mathbf{w}_{(0)}'\mathbf{T}_{(0)} \right\} = T_0/\sqrt{n} + \sqrt{a^2/n}\,\|\mathbf{T}_{(0)}\|,$$

where the supreme above is attained at $\mathbf{w}_{(0)} = \sqrt{a^2/n}\,\mathbf{T}_{(0)}/\|\mathbf{T}_{(0)}\| \in \mathcal{W}_e$. Hence the confidence level in (3.52) is further equal to

$$\begin{aligned}
& \mathrm{P}\left\{ T_0/\sqrt{n} + \sqrt{a^2/n}\,\|\mathbf{T}_{(0)}\| \le c\sqrt{(1+a^2)/n} \right\} \\
= \ & \mathrm{P}\left\{ R_{\mathbf{T}}\cos\theta_{\mathbf{T}1}/\sqrt{n} + \sqrt{a^2/n}\,R_{\mathbf{T}}\sin\theta_{\mathbf{T}1} \le c\sqrt{(1+a^2)/n} \right\} \\
= \ & \mathrm{P}\{\, R_{\mathbf{T}}\cos(\theta_{\mathbf{T}1} - \theta^*) \le c\}
\end{aligned}$$

$$= \mathrm{P}\{\, 0 \le \theta_{\mathrm{T}1} \le \theta^* + \pi/2,\ R_{\mathrm{T}} \cos(\theta_{\mathrm{T}1} - \theta^*) \le c\}$$
$$+ \mathrm{P}\{\, \theta^* + \pi/2 \le \theta_{\mathrm{T}1} \le \pi\}$$
$$= \int_0^{\theta^*+\pi/2} k \sin^{p-1}\theta F_{p+1,\nu}\left(\frac{c^2}{(p+1)\cos^2(\theta-\theta^*)}\right) d\theta$$
$$+ \int_{\theta^*+\pi/2}^{\pi} k \sin^{p-1}\theta d\theta.$$

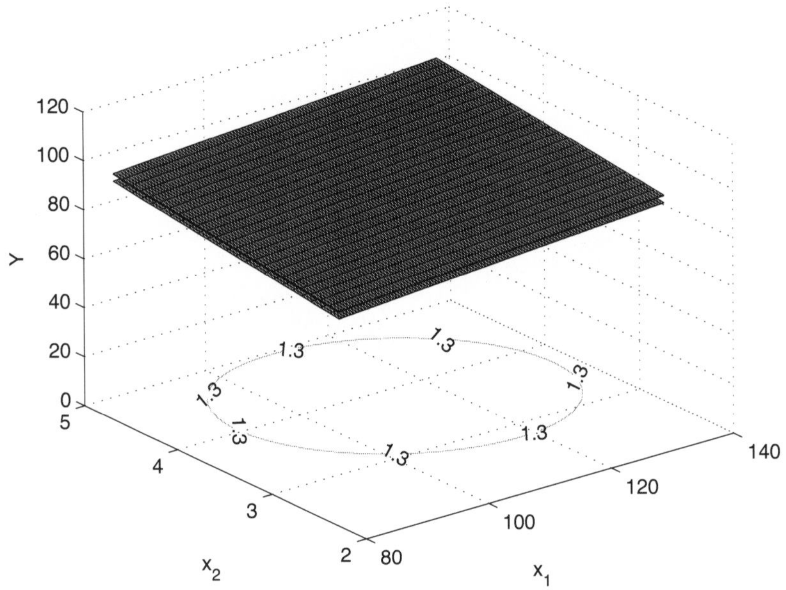

Figure 3.11: *The two-sided and upper constant width bands over the $\mathcal{X}_e$ with $a = 1.3$*

Example 3.7 For the infant data set, we can construct constant width two-sided and one-sided confidence bands over a given covariate region $\mathcal{X}_e$. Suppose that $a = 1.3$ and simultaneous confidence level is $1-\alpha = 95\%$. Then the critical constants of the two-sided and one-sided constant width bands are given by $c = 2.963$ and $c = 2.603$, respectively. These confidence bands are plotted in Figure 3.11: again the bands are given only by the parts that are inside the cylinder which is in the y-direction and has the $\mathcal{X}_e$ as the cross-section in the $(x_1,x_2)'$-plane. The upper one-sided confidence band is slightly lower than the upper part of the two-sided band. For this example, there is little difference between the hyperbolic confidence bands in Figure 3.10 and constant width bands in Figure 3.11. All the computations in this example are done using MATLAB program example0307.m.

Liu, Hayter, Piegorsch and Ah-kine (2009) compared the hyperbolic and constant width bands over $\mathcal{X}_e$ under the minimum volume confidence set criterion, a generalization of the minimum area confidence set criterion proposed in Liu and Hayter (2007), and results similar to the case of $p = 1$ given in Section 2.7.2 were observed.

3.6 Other confidence bands

In this chapter, we have focused on the hyperbolic and constant width bands. They are direct generalizations of the hyperbolic and constant width bands for a simple regression line considered in Chapter 2. The three-segment and two-segment bands of Chapter 2 can also be generalized to a multiple linear regression model. The key idea in the construction of a three-segment confidence band is to bound the simple linear regression model at two covariate points. For a multiple linear regression model, one needs to bound the regression model at several points of the covariates $\mathbf{x}_{(0)}$. Hayter, Wynn and Liu (2006b) considered how to choose several points of $\mathbf{x}_{(0)}$ suitably so that the estimators $\mathbf{x}'\hat{\beta}$ at these $\mathbf{x}_{(0)}$-points are independent random variables. Hence the determination of critical constants for bounding the multiple linear regression model at these points involves just one-dimensional numerical integration. The key idea in the construction of a two-segment band for a simple regression line is to bound the regression line at the mean of the observed values of the only covariate and to bound the gradient of the regression line. For a multiple linear regression model, one can bound the regression model at $\mathbf{x}_{(0)} = \bar{\mathbf{x}}_{(0)}$ and bound the gradients of the regression model in p directions. Hayter *et al.* (2009) considered how to choose the p directions suitably so that the estimators of the gradients in these directions and the estimator of the regression model at $\mathbf{x}_{(0)} = \bar{\mathbf{x}}_{(0)}$ are independent random variables. Hence the computation of the critical constants requires only one-dimensional numerical integration. Dalal (1983) constructed a simultaneous confidence band for the population regression function when the intercept of the linear regression model is assumed to be random by converting a confidence set for the regression coefficients. For a special linear regression model, see Spurrier (1983) who showed that the hyperbolic and constant width bands are actually identical. Naiman (1984b) considered confidence bands that guarantee the expected coverage measure instead of the simultaneous confidence level.

4

Assessing Part of a Regression Model

One important inferential problem for the model

$$\mathbf{Y} = \mathbf{X}\beta + \mathbf{e} \tag{4.1}$$

is to assess whether some of the coefficients β_i's are equal to zero and so the corresponding covariates x_i's have no effect on the response variable Y. The model can therefore be simplified. To be specific, let $\beta = (\beta_1', \beta_2')'$ where $\beta_1' = (\beta_0, \cdots, \beta_{p-k})$ and $\beta_2' = (\beta_{p-k+1}, \cdots, \beta_p)$ with $1 \le k \le p$ being a given integer, and the corresponding partitions of $\mathbf{x}$ are $\mathbf{x}_1 = (1, x_1, \cdots, x_{p-k})'$ and $\mathbf{x}_2 = (x_{p-k+1}, \cdots, x_p)'$. If β_2 is a zero vector then the covariates $x_{p-k+1}, \cdots, x_p$ have no effect on the response variable Y and so model (4.1) reduces to

$$\mathbf{Y} = \mathbf{X}_1\beta_1 + \mathbf{e} \tag{4.2}$$

where $\mathbf{X}_1$ is formed by the first $p-k+1$ columns of the matrix $\mathbf{X}$.

4.1 Partial F test approach

One commonly used statistical approach to assessing whether β_2 is a zero vector, found in most text books on multiple linear regression (see e.g., Kleinbaum *et al.*, 1998, Draper and Smith, 1998, and Dielman, 2001), is to test the hypotheses

$$H_0 : \beta_2 = \mathbf{0} \quad \text{against} \quad H_a : \beta_2 \neq \mathbf{0}. \tag{4.3}$$

Applying the partial F test (1.7), H_0 is rejected if and only if

$$\frac{(SS_R \text{ of model (4.1)} - SS_R \text{ of model (4.2)})/k}{\text{MS residual of model (4.1)}} > f_{k,\nu}^{\alpha} \tag{4.4}$$

where $f_{k,\nu}^{\alpha}$ is the upper α point of the F distribution with k and $\nu = n-(p+1)$ degrees of freedom. Equivalently, from (1.11), H_0 is rejected if and only if

$$\frac{(\mathbf{A}\hat{\beta})'(\mathbf{A}(\mathbf{X}'\mathbf{X})^{-1}\mathbf{A}')^{-1}(\mathbf{A}\hat{\beta})/k}{\text{MS residual of model (4.1)}} > f_{k,\nu}^{\alpha}$$

where $\mathbf{A}$ is a $k \times (p+1)$ matrix given by $\mathbf{A} = (\mathbf{0}, \mathbf{I}_k)$. In fact, we have

$$\begin{aligned} & SS_R \text{ of model (4.1)} - SS_R \text{ of model (4.2)} \\ = \quad & (\mathbf{A}\hat{\beta})'(\mathbf{A}(\mathbf{X}'\mathbf{X})^{-1}\mathbf{A}')^{-1}(\mathbf{A}\hat{\beta}) \end{aligned}$$

from Theorems 1.3 and 1.5.

The inferences that can be drawn from this partial F test are that if H_0 is rejected then β_2 is deemed to be a non-zero vector and so at least some of the covariates $x_{p-k+1}, \cdots, x_p$ affect the response variable Y, and that if H_0 is not rejected then there is not enough statistical evidence to conclude that β_2 is not equal to zero. Unfortunately, this latter case is often falsely interpreted as β_2 is equal to zero and so model (4.2) is accepted as more appropriate than model (4.1). Whether $\beta_2'\mathbf{x}_2$ can be deleted from model (4.1) is better judged based on the magnitude of $\beta_2'\mathbf{x}_2$. The partial F test above provides no information on the magnitude of $\beta_2'\mathbf{x}_2$, irrespective whether H_0 is rejected or not. On the other hand, a confidence band on $\beta_2'\mathbf{x}_2$ provides useful information on the magnitude of $\beta_2'\mathbf{x}_2$.

4.2 Hyperbolic confidence bands

The next theorem provides a hyperbolic confidence band for $\beta_2'\mathbf{x}_2$ over the whole space of $\mathbf{x}_2 \in \Re^k$. Let $\hat{\beta}_2 = \mathbf{A}\hat{\beta}$ denote the estimator of β_2, which is made of the last k components of $\hat{\beta}$. Clearly $\hat{\beta}_2$ has the distribution $\mathcal{N}_k(\beta_2, \sigma^2\mathbf{V})$, where $\mathbf{V}$ is the $k \times k$ partition matrix formed from the last k rows and the last k columns of $(\mathbf{X}'\mathbf{X})^{-1}$, i.e. $\mathbf{V} = \mathbf{A}(\mathbf{X}'\mathbf{X})^{-1}\mathbf{A}'$.

Theorem 4.1 We have

$$P\left\{ \mathbf{x}_2'\beta_2 \in \mathbf{x}_2'\hat{\beta}_2 \pm \sqrt{kf^{\alpha}_{k,n-p-1}}\,\hat{\sigma}\sqrt{\mathbf{x}_2'\mathbf{V}\mathbf{x}_2} \;\; \forall\, \mathbf{x}_2 \in \Re^k \right\} = 1 - \alpha. \tag{4.5}$$

This theorem can be proved similarly to Theorem 3.1. The $1-\alpha$ simultaneous confidence band given in this theorem provides the plausible range of the true model $\mathbf{x}_2'\beta_2$. In particular, if β_2 is a zero vector then $\mathbf{x}_2'\beta_2$ is the zero hyper-plane $\mathbf{x}_2'\mathbf{0}$ in the $\Re^{k+1}$ space and it is contained in the confidence band (4.5) for all $\mathbf{x}_2 \in \Re^k$ with probability $1-\alpha$. So an exact size α test of the hypotheses in (4.3) implied by the confidence band (4.5) is to

$$\text{reject } H_0 \;\Leftrightarrow\; \text{the band (4.5) does not contain } \mathbf{x}_2'\mathbf{0} \text{ for some } \mathbf{x}_2 \in \Re^k. \tag{4.6}$$

In fact, this test is just the partial F test (4.4) as asserted by the next theorem, which can be proved in a similar way as Theorem 3.2.

Theorem 4.2 Tests (4.4) and (4.6) reject and accept H_0 at the same time and so are the same.

So the partial F test (4.4) will not reject H_0 so long as the confidence band (4.5) contains $\mathbf{x}_2'\mathbf{0}$ even though the confidence band may be very wide and so the true model $\mathbf{x}_2'\beta_2$ can potentially be very different from $\mathbf{x}_2'\mathbf{0}$. Hence the non-rejection of the null hypothesis H_0 can mean anything but that $\mathbf{x}_2'\beta_2$ is equal to zero. But the confidence band provides information on the magnitude of $\mathbf{x}_2'\beta_2$ for every $\mathbf{x}_2 \in R^k$, which is suitable for judging whether $\mathbf{x}_2'\beta_2$ is small enough to be deleted from the original model (4.1) and hence to conclude that model (4.2) is more appropriate.

As argued in Chapter 3, in most real problems, the model (4.1) holds only over a certain region of the covariates. Hence it does not make sense that the partial F may reject H_0 due to the behavior of the fitted model $\mathbf{x}_2'\hat{\beta}_2$ outside this covariate region. Also the model (4.1) is often established to make inferences only in a certain range of the covariates. Hence it is only necessary to assess whether the true model $\mathbf{x}_2'\beta_2$ is small enough over a given region of interest of the covariates $\mathbf{x}_2$ to be deleted from the overall model (4.1) when used over this particular region of $\mathbf{x}_2$. For this purpose, a confidence band for $\mathbf{x}_2'\beta_2$ over a restricted region of $\mathbf{x}_2$ is useful and discussed next following largely Jamshidian, Jenrich and Liu (JJL, 2007).

We focus on the rectangular region

$$\mathcal{X}_2 = \{\mathbf{x}_2 : x_i \in [a_i, b_i],\ i = p-k+1, \cdots, p\}$$

where $-\infty \le a_i \le b_i \le \infty$, $i = p-k+1, \cdots, p$ are given constants. A hyperbolic confidence band over $\mathcal{X}_2$ is given by

$$\mathbf{x}_2'\beta_2 \ \in\ \mathbf{x}_2'\hat{\beta}_2 \pm c\hat{\sigma}\sqrt{\mathbf{x}_2'\mathbf{V}\mathbf{x}_2} \quad \text{for all } \mathbf{x}_2 \in \mathcal{X}_2 \tag{4.7}$$

where the critical constant c is determined so that the simultaneous confidence level of this band is equal to $1-\alpha$.

Let $\mathbf{W}$ be the square-root matrix of $\mathbf{V}$ and so $\mathbf{V} = \mathbf{W}^2$. Denote $\mathbf{N}_2 = \mathbf{W}^{-1}(\hat{\beta}_2 - \beta_2)/\sigma \sim \mathcal{N}_k(\mathbf{0}, \mathbf{I})$ and $\mathbf{T}_2 = \mathbf{N}_2/(\hat{\sigma}/\sigma) \sim \mathcal{T}_{k,\nu}$. Then the confidence level of the band (4.7) is given by $\mathrm{P}\{S < c\}$ where

$$\begin{aligned} S &= \sup_{\mathbf{x}_2 \in \mathcal{X}_2} \frac{|\mathbf{x}_2'(\hat{\beta}_2 - \beta_2)|}{\hat{\sigma}\sqrt{\mathbf{x}_2'\mathbf{V}\mathbf{x}_2}} \\ &= \sup_{\mathbf{x}_2 \in \mathcal{X}_2} \frac{|(\mathbf{W}\mathbf{x}_2)'\mathbf{W}^{-1}(\hat{\beta}_2 - \beta_2)/\hat{\sigma}|}{\sqrt{(\mathbf{W}\mathbf{x}_2)'(\mathbf{W}\mathbf{x}_2)}} \\ &= \sup_{\mathbf{v} \in C(\mathbf{W}, \mathcal{X}_2)} \frac{|\mathbf{v}'\mathbf{T}_2|}{\|\mathbf{v}\|} \end{aligned} \tag{4.8}$$

where

$$C(\mathbf{W}, \mathcal{X}_2) = \left\{\lambda \mathbf{W}\mathbf{x}_2 = \lambda(x_{p-k+1}\mathbf{w}_1 + \cdots + x_p\mathbf{w}_k) : \lambda > 0 \text{ and } \mathbf{x}_2 \in \mathcal{X}_2\right\} \tag{4.9}$$

with $\mathbf{W} = (\mathbf{w}_1, \cdots, \mathbf{w}_k)$. The distribution of S does not depend on the unknown parameters $\boldsymbol{\beta}$ and σ, but it does depend on the region $\mathcal{X}_2$ and $\mathbf{W}$ through $C(\mathbf{W}, \mathcal{X}_2)$.

Note that for the special case of $k = 1$, which corresponds to assessing whether the coefficient of one particular covariate of the model (4.1) is equal to zero, S in (4.8) is clearly equal to $\|\mathbf{T}_2\|$ and so the critical constant c is given by the usual t-value $t_\nu^{\alpha/2}$ and the confidence band is just the usual two-sided t confidence interval for one coefficient β_i. Also, for the special case that the set $C(\mathbf{W}, \mathcal{X}_2) \subset \Re^k$, which is a polyhedral cone, contains the origin $\mathbf{0}$ as an inner point, S in (4.8) is clearly equal to $\|\mathbf{T}_2\|$ (by choosing a $\mathbf{v} \in C(\mathbf{W}, \mathcal{X}_2)$ to be in the same or opposite directions of $\mathbf{T}_2$) and so the critical constant c is given by $\sqrt{k f_{k,\nu}^{\alpha}}$ as in confidence band (4.5). That is, no improvement over the confidence band (4.5) can be achieved by restricting the covariates $\mathbf{x}_2$ to such a region $\mathcal{X}_2$. It is clear that the set $C(\mathbf{W}, \mathcal{X}_2)$ contains the origin $\mathbf{0}$ as an inner point if and only if $\mathcal{X}_2$ contains the origin $\mathbf{0}$ as an inner point, that is, zero is an inner point of the interval $[a_i, b_i]$ for each $i = p - k + 1, \cdots, p$.

In the general setting considered, the critical constant c can be approximated by simulation as before. To simulate one S, a pair of independent $\mathbf{N}_2 \sim \mathcal{N}_k(\mathbf{0}, \mathbf{I})$ and $\hat{\sigma}/\sigma \sim \sqrt{\chi_\nu^2/\nu}$ are simulated first, and the value of S is then calculated from (4.8). Repeat this to generate R replicates of S: $S_1, \cdots, S_R$, and then use the $\langle(1 - \alpha)R\rangle$th largest S_i's as an approximation of the critical constant c.

The key in the simulation of each S is to evaluate the supreme in (4.8). Realizing that $|\mathbf{v}'\mathbf{t}_2| / \| \mathbf{v} \|$ is the length of the projection of $\mathbf{t}_2$ on $\mathbf{v}$, it can be shown that

$$S = \max\{\|\pi^*(\mathbf{T}_2, \mathbf{W}, \mathcal{X}_2)\|, \|\pi^*(-\mathbf{T}_2, \mathbf{W}, \mathcal{X}_2)\|\}$$

similar to the result of Naiman (1987, Theorem 2.1) as given in Section 3.2.1. Here $\pi^*(\mathbf{t}_2, \mathbf{W}, \mathcal{X}_2)$ is the projection of $\mathbf{t}_2 \in \Re^k$ to the polyhedral cone $C(\mathbf{W}, \mathcal{X}_2)$, i.e., $\pi^*(\mathbf{t}_2, \mathbf{W}, \mathcal{X}_2)$ is the $\mathbf{v} \in \Re^k$ that solves the problem

$$\min_{\mathbf{v} \in C(\mathbf{W}, \mathcal{X}_2)} \|\mathbf{v} - \mathbf{t}_2\|^2.$$

Hence $\pi^*(\mathbf{t}_2, \mathbf{W}, \mathcal{X}_2)$ can be solved from a quadratic programming problem in a finite number of steps; see Appendix C for more details on this.

Similar to the test (4.4) implied by the confidence band (4.5) over the whole space $\mathbf{x}_2 \in \Re^k$, a size α test of the hypotheses (4.3) is also induced by the confidence band (4.7): it rejects H_0 if and only if the zero hyper-plane $\mathbf{x}_2\mathbf{0}$ is excluded from the band at one $\mathbf{x}_2 \in \mathcal{X}_2$ at least, that is,

$$H_0 \text{ is rejected} \Leftrightarrow \sup_{\mathbf{x}_2 \in \mathcal{X}_2} \frac{|\mathbf{x}_2'\hat{\boldsymbol{\beta}}_2|}{\hat{\sigma}\sqrt{\mathbf{x}_2'\mathbf{V}\mathbf{x}_2}} > c. \tag{4.10}$$

This test is an improvement to the partial F test (4.4) in the sense that the partial F test may reject H_0 because the zero hyper-plane $\mathbf{x}_2'\mathbf{0}$ is excluded from the band

(4.5) at some $\mathbf{x}_2$ outside $\mathcal{X}_2$ where the assumed model (4.1) is false and hence the consequential inferences may be invalid.

The p-value of this new test (4.10) can be defined in the usual manner. Let $\hat{\beta}_2$ and $\hat{\sigma}$ be the respective estimates of β_2 and σ based on the observed data, and so the observed statistic is given by

$$\begin{aligned} S^* &= \sup_{\mathbf{x}_2 \in \mathcal{X}_2} \frac{|\mathbf{x}_2' \hat{\beta}_2|}{\hat{\sigma}\sqrt{\mathbf{x}_2' \mathbf{V} \mathbf{x}_2}} \\ &= \max\left\{ \|\pi^*(\mathbf{W}^{-1}\hat{\beta}_2/\hat{\sigma}, \mathbf{W}, \mathcal{X}_2)\|, \|\pi^*(-\mathbf{W}^{-1}\hat{\beta}_2/\hat{\sigma}, \mathbf{W}, \mathcal{X}_2)\| \right\}. \end{aligned}$$

The observed p-value is then given by

$$P\{\max\{\|\pi^*(\mathbf{T}_2, \mathbf{W}, \mathcal{X}_2)\|, \|\pi^*(-\mathbf{T}_2, \mathbf{W}, \mathcal{X}_2)\|\} > S^*\}$$

where $\mathbf{T}_2 \sim \mathcal{T}_{k,\nu}$, the standard multivariate t distribution with parameters k and ν. This p-value can be approximated by simulation: simulate R replicates $S_1, \cdots, S_R$ of the random variable

$$S = \max\{\|\pi^*(\mathbf{T}_2, \mathbf{W}, \mathcal{X}_2)\|, \|\pi^*(-\mathbf{T}_2, \mathbf{W}, \mathcal{X}_2)\|\};$$

the proportion of the S_i's that are larger than S^* is used as an approximation to the p-value. An extensive numerical study of the power of test (4.10) is given in JJL (2007).

Example 4.1 In this example we use the aerobic fitness data from the SAS/STAT User's Guide (1990, page 1443) to illustrate our methodology. In particular we will show that at significance level $\alpha = 5\%$ the test (4.10) with bounds on the covariates rejects the H_0 in (4.3), whereas the partial F test (4.4) does not reject the H_0. The data set is given in Table 4.1.

The aerobic fitness data set consists of measurements on men involved in a physical fitness course at North Carolina State University. The variables measured are age (years), weight (kg), oxygen intake rate (ml per kg body weight per minute), time to run 1.5 miles (minutes), heart rate while resting, heart rate while running (same time oxygen rate measured), and maximum heart rate recorded while running. In a regression analysis of the data, the effect of age, weight, time to run 1.5 miles, heart rate while running and maximum heart rate on the oxygen intake rate was of interest. The SAS output for fitting the corresponding model

$$\texttt{OXY} = \beta_0 + \beta_1\,\texttt{age} + \beta_2\,\texttt{weight} + \beta_3\,\texttt{runtime} + \beta_4\,\texttt{runpulse} + \beta_5\,\texttt{maxpulse} + \varepsilon \quad (4.11)$$

is given in Table 4.2.

The coefficients of `weight` (β_2) and `maxpulse` (β_5) are the two least significant parameters. If no restriction is imposed on the covariates, then the critical value for the confidence band of β_2 `weight` $+ \beta_5$ `maxpulse` with 95% confidence

Table 4.1: *Aerobic data from* SAS/STAT User's Guide *(1990, page 1443)*

i	Age	Weight	Oxygen	RunTime	RestPulse	RunPulse	MaxPulse
1	44	89.47	44.609	11.37	62	178	182
2	44	85.84	54.297	8.65	45	156	168
3	38	89.02	49.874	9.22	55	178	180
4	40	75.98	45.681	11.95	70	176	180
5	44	81.42	39.442	13.08	63	174	176
6	44	73.03	50.541	10.13	45	168	168
7	45	66.45	44.754	11.12	51	176	176
8	54	83.12	51.855	10.33	50	166	170
9	51	69.63	40.836	10.95	57	168	172
10	48	91.63	46.774	10.25	48	162	164
11	57	73.37	39.407	12.63	58	174	176
12	52	76.32	45.441	9.63	48	164	166
13	51	67.25	45.118	11.08	48	172	172
14	51	73.71	45.790	10.47	59	186	188
15	49	76.32	48.673	9.40	56	186	188
16	52	82.78	47.467	10.50	53	170	172
17	40	75.07	45.313	10.07	62	185	185
18	42	68.15	59.571	8.17	40	166	172
19	47	77.45	44.811	11.63	58	176	176
20	43	81.19	49.091	10.85	64	162	170
21	38	81.87	60.055	8.63	48	170	186
22	45	87.66	37.388	14.03	56	186	192
23	47	79.15	47.273	10.60	47	162	164
24	49	81.42	49.156	8.95	44	180	185
25	51	77.91	46.672	10.00	48	162	168
26	49	73.37	50.388	10.08	67	168	168
27	54	79.38	46.080	11.17	62	156	165
28	50	70.87	54.625	8.92	48	146	155
29	54	91.63	39.203	12.88	44	168	172
30	57	59.08	50.545	9.93	49	148	155
31	48	61.24	47.920	11.50	52	170	176

level is $c = \sqrt{2f_{2,25}^{0.05}} = 2.602$, with the upper and lower bands given by

$$-.0723 \times \texttt{weight} + .3049 \times \texttt{maxpulse} \pm 2.602\hat{\sigma} \times \tag{4.12}$$
$$\sqrt{\mathbf{V}(1,1) \times \texttt{weight}^2 + \mathbf{V}(2,2) \times \texttt{maxpulse}^2 + 2 * \mathbf{V}(1,2) \times \texttt{weight} \times \texttt{maxpulse}}.$$

This band along with the zero plane are depicted in Figure 4.1. From Figure 4.1, the band contains the zero plane, and so the corresponding partial F test (4.4) does

Table 4.2: *SAS output for Model (4.11)*

Dependent Variable: Oxygen consumption

Analysis of Variance

Source	DF	Sum of Squares	Mean Square	F Value	Pr > F
Model	5	721.97309	144.39462	27.90	<.0001
Error	25	129.40845	5.17634		
Corrected Total	30	851.38154			

Root MSE	2.27516	R-Square	0.8480
Dependent Mean	47.37581	Adj R-Sq	0.8176
Coeff Var	4.80236		

Parameter Estimates

Variable	Label	DF	Parameter Estimate	Standard Error	t Value	Pr > |t|
Intercept	Intercept	1	102.20428	11.97929	8.53	<.0001
age	Age in years	1	-0.21962	0.09550	-2.30	0.0301
weight	Weight in kg	1	-0.07230	0.05331	-1.36	0.1871
runtime	Min. to run 1.5 miles	1	-2.68252	0.34099	-7.87	<.0001
runpulse	Heart rate while running	1	-0.37340	0.11714	-3.19	0.0038
maxpulse	Maximum heart rate	1	0.30491	0.13394	2.28	0.0316

not reject the null hypothesis $H_0: \beta_2 = \beta_5 = 0$ against $H_a:$ not H_0. Obviously, negative values for `weight` and `maxpulse` have no physical meaning, and there is no reason to consider them in the analysis. In fact, the observed values for `weight` are in the interval [59, 91.6] and those for `maxpulse` are in the interval [155,192]. Figure 4.2 shows the portion of Figure 4.1 restricted to this range of the observed values. Clearly the zero plane sits between the lower and the upper parts of the band without intersecting them.

Now we restrict the covariates to their reasonable range, and calculate the confidence band for β_2 `weight` + β_5 `maxpulse`, using the simulation-based procedure outlined above. More specifically, we restricted the covariates `weight` and `maxpulse` to their respective observed intervals mentioned above. The required critical value with 95% confidence, using 100,000 simulations, is $c = 2.105$. Thus the lower and upper parts of the confidence band are the same as those in (4.12) except that the critical value 2.602 is replaced by 2.105 and that the covariates are restricted to their observed range. Figure 4.3 depicts this band along with the zero plane. Interestingly, the lower part of the confidence band intersects the zero plane at values of `weight` ranging from about 59 to 72, and values of `maxpulse` ranging from about 157 to 192. So the restricted confidence band rejects the null hypothesis $H_0: \beta_2 = \beta_5 = 0$, whereas the partial F test (4.4) and equivalently the unrestricted bands do not reject H_0 at 5% level.

It is noteworthy that one may obtain the p-values corresponding to the partial F test and the test induced by the restricted band. In fact, when $k \geq 3$, plots of confidence bands are not available even though we may plot 2-dimensional slices of a band by fixing the values of certain covariates. The p-value provides a way

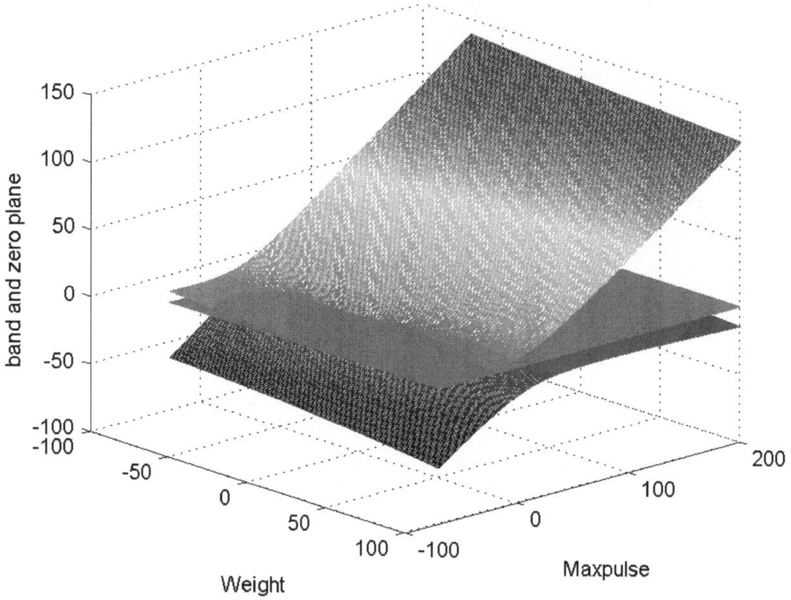

Figure 4.1: *Band (4.12) and the zero plane*

to determine the result of the hypotheses tests. For this example, the p-value corresponding to the partial F test is 0.066 and that for the restricted band is 0.047. These of course agree with our observations from the confidence bands plotted.

The restricted confidence band in Figure 4.3 provides a plausible range of β_2 `weight` $+ \beta_5$ `maxpulse` over $\mathcal{X}_2$: any β_2 `weight` $+ \beta_5$ `maxpulse` that falls completely inside the band is a plausible candidate. The largest (vertical) distance from this confidence band to the zero plane is calculated to be 107.6 (attained at `weight` $= 59$ and `maxpulse` $= 192$. Hence β_2 `weight` $+ \beta_5$ `maxpulse` over $\mathcal{X}_2$ can be as large as 107.6 and is unlikely to be negligible from the overall model (4.11).

The largest (vertical) distance from the band (4.5) to the zero plane over $\mathcal{X}_2$ is calculated to be 120.1 (attained also at `weight` $= 59$ and `maxpulse` $= 192$), which is even larger than 107.6 calculated from the restricted confidence band (4.7), even though the F test does not reject the null hypothesis $H_0 : \beta_2 = \beta_5 = 0$. All the computations in this example are done using the MATLAB® program example0401.m.

The hyperbolic shape of the confidence band (4.7) implies that the band may be unduely wide for those $\mathbf{x}_2$ on the boundary of $\mathcal{X}_2$. This is a disadvantage if one

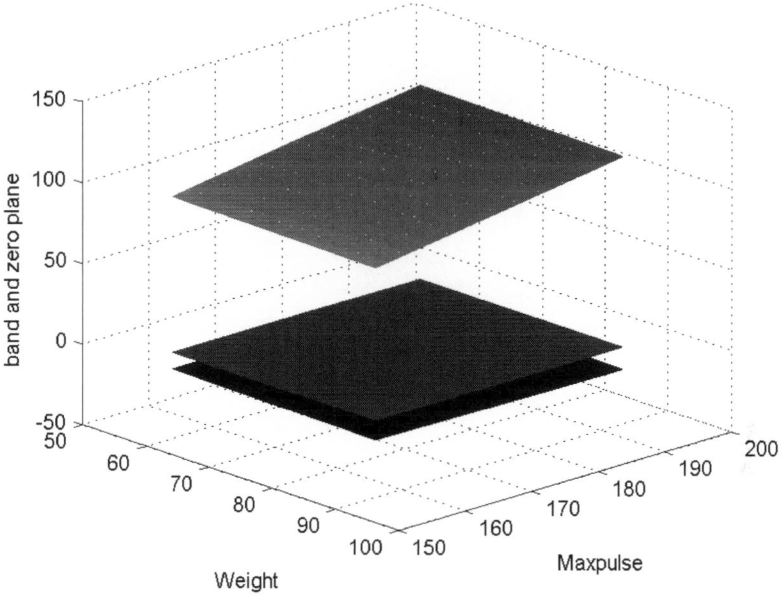

Figure 4.2: *Part of the band and zero plane in Figure 4.1*

hopes to show, by using the confidence band (4.7), that the model $\mathbf{x}_2'\beta_2$ is close to the zero hyper-plane $\mathbf{x}_2'\mathbf{0}$ over the region $\mathbf{x}_2 \in \mathcal{X}_2$. On the other hand, a constant width confidence band has the same width throughout $\mathbf{x}_2 \in \mathcal{X}_2$ and therefore may be advantageous for showing that the model $\mathbf{x}_2'\beta_2$ is close to the zero hyper-plane $\mathbf{x}_2'\mathbf{0}$ over the region $\mathbf{x}_2 \in \mathcal{X}_2$.

A two-sided constant width confidence band over $\mathcal{X}_2$ is given by

$$\mathbf{x}_2'\beta_2 \in \mathbf{x}_2'\hat{\beta}_2 + c\hat{\sigma} \quad \text{for all } \mathbf{x}_2 \in \mathcal{X}_2 \tag{4.13}$$

where the critical constant c can be determined by using simulation as in Section 3.3 so that the simultaneous confidence level of the band is equal to $1-\alpha$. Instead of pursuing this further, in the next section, a hypotheses testing approach is provided if one's only interest is to show that $\mathbf{x}_2'\beta_2$ is sufficiently close to the zero function over $\mathbf{x}_2 \in \mathcal{X}_2$. Note, however, that the simultaneous confidence band approach discussed above allows quite versatile inferences in addition to assessing whether $\mathbf{x}_2'\beta_2$ is sufficiently close to zero over $\mathcal{X}_2$. For example, the plot in Figure 4.3 of the confidence band tells that β_2 `weight` $+\beta_5$ `maxpulse` is positive when $($`weight`,`maxpulse`$) \in \mathcal{X}_2$ is near the point $(59, 192)$.

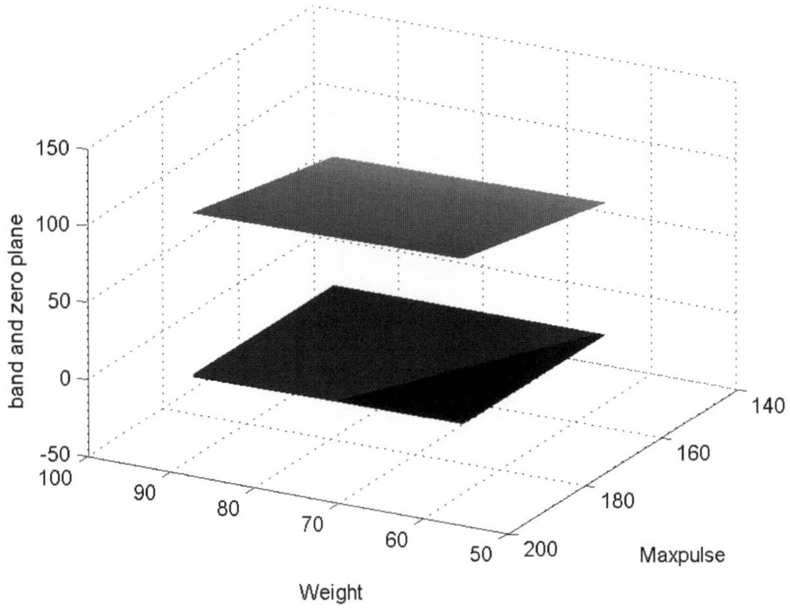

Figure 4.3: *The restricted band (4.7)*

4.3 Assessing equivalence to the zero function

As discussed above, whether the part $\mathbf{x}_2'\beta_2$ can be deleted from the overall model (4.1) over a given region of interest $\mathbf{x}_2 \in \mathcal{X}_2$ is better based on the magnitude of $\mathbf{x}_2'\beta_2$ over $\mathbf{x}_2 \in \mathcal{X}_2$, rather than the non-rejection of the null hypothesis in (4.3). Let us assume that $\delta > 0$ is a pre-specified threshold so that if $|\mathbf{x}_2'\beta_2|$ is strictly less than δ for all $\mathbf{x}_2 \in \mathcal{X}_2$ then $\mathbf{x}_2'\beta_2$ is deemed to be practically equivalent to zero and hence can be deleted from the overall model (4.1) over $\mathbf{x}_2 \in \mathcal{X}_2$.

Since one hopes to demonstrate that $\mathbf{x}_2'\beta_2$ over $\mathbf{x}_2 \in \mathcal{X}_2$ is equivalent to zero, one may use a hypotheses testing approach by setting up the hypotheses as

$$H_0^E : \max_{\mathbf{x}_2 \in \mathcal{X}_2} |\mathbf{x}_2'\beta_2| \geq \delta \quad \text{against} \quad H_a^E : \max_{\mathbf{x}_2 \in \mathcal{X}_2} |\mathbf{x}_2'\beta_2| < \delta. \tag{4.14}$$

The equivalence of $\mathbf{x}_2'\beta_2$ to zero over $\mathbf{x}_2 \in \mathcal{X}_2$ is claimed if and only if the null hypothesis H_0^E is rejected. As a hypotheses test guarantees the size at α, it is necessary to set up the null and alternative hypotheses as in (4.14) in order to guarantee that the chance of falsely claiming equivalence is no more than α.

Equivalence problems started largely in the pharmaceutical industries for the problem of assessing bio-equivalence of a new generic version and the primary formulation of a drug, as discussed in Chow and Liu (1992) for example. In a recent book, Wellek (2003) demonstrated that equivalence problems can potentially come from many scientific areas. The key in determining which setup should be used in a given problem depends upon the purpose of the study. If one wishes to establish a difference between $\mathbf{x}_2'\beta_2$ and the zero function, then the traditional setup such as (4.3) can be used. On the other hand, if one hopes to demonstrate the equivalence of $\mathbf{x}_2'\beta_2$ to the zero function by using a hypotheses testing approach, then the hypotheses should be set up as in (4.14).

To judge whether $\max_{\mathbf{x}_2\in\mathcal{X}_2}|\mathbf{x}_2'\beta_2|$ is less than δ or not, one possibility is to use a confidence band over $\mathcal{X}_2$ to bound the highest and the lowest values of $\mathbf{x}_2'\beta_2$ over $\mathcal{X}_2$ and hence $\max_{\mathbf{x}_2\in\mathcal{X}_2}|\mathbf{x}_2'\beta_2|$. For example, if the hyperbolic confidence band (4.7) is used then one may deduce that

$$\max_{\mathbf{x}_2\in\mathcal{X}_2}|\mathbf{x}_2'\beta_2| \le \max\left\{\max_{\mathbf{x}_2\in\mathcal{X}_2}|\mathbf{x}_2'\hat{\beta}_2 + c\hat{\sigma}\sqrt{\mathbf{x}_2'\mathbf{V}\mathbf{x}_2}|,\ \max_{\mathbf{x}_2\in\mathcal{X}_2}|\mathbf{x}_2'\hat{\beta}_2 - c\hat{\sigma}\sqrt{\mathbf{x}_2'\mathbf{V}\mathbf{x}_2}|\right\}. \tag{4.15}$$

If and only if this upper bound is less than the threshold δ then H_0^E is rejected and equivalence is claimed. The size of this test is α since it is derived from the $1-\alpha$ level simultaneous confidence band (4.7). One may also construct an upper confidence bound on $\max_{\mathbf{x}_2\in\mathcal{X}_2}|\mathbf{x}_2'\beta_2|/\sigma$ by using the method of Liu, Hayter and Wynn (2007). Tseng (2002) provided an upper confidence bound on $|\mu|/\sigma$ for a normal distribution $N(\mu,\sigma^2)$.

Next we give a size α test of the equivalence hypotheses in (4.14) which has a greater chance than the test based on the upper bound in (4.15) in establishing equivalence. This test is an intersection-union (IU) test which is reviewed very briefly in Appendix D. An IU test is particularly suitable for testing equivalence hypotheses as those in (4.14).

Now we construct an IU test for the hypotheses in (4.14). Note that the hypotheses can be expressed as

$$H_0^E = \cup_{\mathbf{x}_2\in\mathcal{X}_2}\left\{H_{0\mathbf{x}_2}^{E+} \cup H_{0\mathbf{x}_2}^{E-}\right\}$$

and

$$H_a^E = \cap_{\mathbf{x}_2\in\mathcal{X}_2}\left\{H_{a\mathbf{x}_2}^{E+} \cap H_{a\mathbf{x}_2}^{E-}\right\}$$

where the individual hypotheses are given by

$$H_{0\mathbf{x}_2}^{E+} : \mathbf{x}_2'\beta_2 \ge \delta, \quad H_{0\mathbf{x}_2}^{E-} : \mathbf{x}_2'\beta_2 \le -\delta$$
$$H_{a\mathbf{x}_2}^{E+} : \mathbf{x}_2'\beta_2 < \delta, \quad H_{a\mathbf{x}_2}^{E-} : \mathbf{x}_2'\beta_2 > -\delta$$

for each $\mathbf{x}_2 \in \mathcal{X}_2$. The individual hypotheses

$$H_{0\mathbf{x}_2}^{E+} : \mathbf{x}_2'\beta_2 \ge \delta \iff H_{a\mathbf{x}_2}^{E+} : \mathbf{x}_2'\beta_2 < \delta$$

can be tested with rejection region

$$\mathbf{x}_2'\hat{\beta}_2 + t_\nu^\alpha\hat{\sigma}\sqrt{\mathbf{x}_2'\mathbf{V}\mathbf{x}_2} < \delta.$$

The size of this individual test is α since, when $H_{0\mathbf{x}_2}^{E+}$ is true,

$$\begin{aligned}
& \mathrm{P}\{ \text{ rejection of } H_{0\mathbf{x}_2}^{E+} \} \\
= \; & \mathrm{P}\{ \mathbf{x}_2'\hat{\beta}_2 + t_\nu^\alpha \hat{\sigma}\sqrt{\mathbf{x}_2'\mathbf{V}\mathbf{x}_2} < \delta \} \\
= \; & \mathrm{P}\{ \mathbf{x}_2'\hat{\beta}_2 - \mathbf{x}_2'\beta_2 + t_\nu^\alpha \hat{\sigma}\sqrt{\mathbf{x}_2'\mathbf{V}\mathbf{x}_2} < \delta - \mathbf{x}_2'\beta_2 \} \\
\leq \; & \mathrm{P}\{ \mathbf{x}_2'\hat{\beta}_2 - \mathbf{x}_2'\beta_2 + t_\nu^\alpha \hat{\sigma}\sqrt{\mathbf{x}_2'\mathbf{V}\mathbf{x}_2} < 0 \} \;\; (\text{since } \delta - \mathbf{x}_2'\beta_2 \leq 0 \text{ under } H_{0\mathbf{x}_2}^{E+}) \\
= \; & \mathrm{P}\left\{ (\mathbf{x}_2'\hat{\beta}_2 - \mathbf{x}_2'\beta_2)/(\hat{\sigma}\sqrt{\mathbf{x}_2'\mathbf{V}\mathbf{x}_2}) < -t_\nu^\alpha \right\} = \alpha.
\end{aligned}$$

Similarly, the individual hypotheses

$$H_{0\mathbf{x}_2}^{E-} : \mathbf{x}_2'\beta_2 \leq -\delta \iff H_{a\mathbf{x}_2}^{E-} : \mathbf{x}_2'\beta_2 > -\delta$$

can be tested with rejection region

$$\mathbf{x}_2'\hat{\beta}_2 - t_\nu^\alpha \hat{\sigma}\sqrt{\mathbf{x}_2'\mathbf{V}\mathbf{x}_2} > -\delta,$$

and the size of this individual test is α. Hence an IU test of H_0^E against H_a^E rejects H_0^E if and only if

$$-\delta < \min_{\mathbf{x}_2 \in \mathcal{X}_2} \{\mathbf{x}_2'\hat{\beta}_2 - t_\nu^\alpha \hat{\sigma}\sqrt{\mathbf{x}_2'\mathbf{V}\mathbf{x}_2}\} \text{ and } \max_{\mathbf{x}_2 \in \mathcal{X}_2} \{\mathbf{x}_2'\hat{\beta}_2 + t_\nu^\alpha \hat{\sigma}\sqrt{\mathbf{x}_2'\mathbf{V}\mathbf{x}_2}\} < \delta. \tag{4.16}$$

The only difference between this rejection region and the rejection region based on the upper bound in (4.15) is that the critical constant c in (4.15) is based on the two-sided simultaneous confidence band (4.7), while the critical constant t_ν^α in (4.16) is based on the one-sided pointwise confidence band. Since $c > t_\nu^\alpha$, whenever H_0^E is rejected by using the upper bound in (4.15) H_0^E is always rejected by the IU test in (4.16), but not vise versa. So the IU test is more efficient than the simultaneous confidence (4.7) for establishing equivalence in terms of testing the hypotheses in (4.14).

Example 4.2 For the aerobic data set given in Example 4.1, one can easily compute the 95% pointwise confidence band with $t_{25}^{0.05} = 1.708$ to find that

$$\min_{\mathbf{x}_2 \in \mathcal{X}_2} \{\mathbf{x}_2'\hat{\beta}_2 - t_\nu^\alpha \hat{\sigma}\sqrt{\mathbf{x}_2'\mathbf{V}\mathbf{x}_2}\} = 5.82 \text{ and } \max_{\mathbf{x}_2 \in \mathcal{X}_2} \{\mathbf{x}_2'\hat{\beta}_2 + t_\nu^\alpha \hat{\sigma}\sqrt{\mathbf{x}_2'\mathbf{V}\mathbf{x}_2}\} = 97.51.$$

Hence H_0^E is rejected and equivalence established by the size 5% IU test, based on the observed data, if and only if the pre-specified threshold δ is larger than 97.51.

5

Comparison of Two Regression Models

One frequently encountered problem in scientific research is to assess whether two regression models, which describe the relationship of a same response variable Y on a same set of predict variables $x_1, \cdots, x_p$, for two groups or two treatments, etc., are the same or not. For example, the relationship between human systolic blood pressure (Y) and age (x_1) can be well described by a linear regression model for a certain range of age. Suppose that a linear regression model of Y on x_1 is set up for each of the two gender groups, male and female. It is interesting to assess whether the two models are different and so two separate models are necessary for the two gender groups, or the two models are similar and therefore only one model is required to describe how blood pressure changes with age for both gender groups.

In general, suppose the two linear regression models are given by

$$\mathbf{Y}_i = \mathbf{X}_i \boldsymbol{\beta}_i + \mathbf{e}_i, \quad i = 1, 2 \tag{5.1}$$

where $\mathbf{Y}_i = (Y_{i,1}, \cdots, Y_{i,n_i})'$ is a vector of random observations, $\mathbf{X}_i$ is an $n_i \times (p+1)$ full column-rank design matrix with the first column given by $(1, \cdots, 1)'$ and the lth $(2 \le l \le p+1)$ column given by $(x_{i,1,l-1}, \cdots, x_{i,n_i,l-1})'$, $\boldsymbol{\beta}_i = (\beta_{i,0}, \cdots, \beta_{i,p})'$ is a vector of unknown regression coefficients, and $\mathbf{e}_i = (e_{i,1}, \cdots, e_{i,n_i})'$ is a vector of random errors with all the $\{e_{i,j} : j = 1, \cdots, n_i, i = 1, 2\}$ being iid $N(0, \sigma^2)$ random variables, where σ^2 is an unknown parameter. Here $\mathbf{Y}_1$ and $\mathbf{Y}_2$ are two groups of observations on a response variable Y that depends on the same p covariates $x_1, \cdots, x_p$ via the classical linear regression models. The two groups may be two gender groups or two treatments, etc.

An important question is whether the two models in (5.1) are similar or different. If the two models are similar then only one model is necessary to describe the relationship between the response variable Y and the p covariates $x_1, \cdots, x_p$ for both groups. On the other hand, if the two models are different then one may want to assess the differences between the two models.

5.1 Partial F test approach

One frequently used approach to this problem of comparing two linear regression models is to test the hypotheses

$$H_0: \boldsymbol{\beta}_1 = \boldsymbol{\beta}_2 \quad \text{against} \quad H_a: \boldsymbol{\beta}_1 \neq \boldsymbol{\beta}_2. \tag{5.2}$$

The partial F test (1.7) can be implemented in the following way. Define a dummy predictor variable

$$z = \begin{cases} 1 & \text{if } Y \text{ is from the model 1} \\ 0 & \text{if } Y \text{ is from the model 2} \end{cases}$$

and set up an overall model for the aggregated data from the two individual models as

$$Y = \mathbf{x}'\mathbf{c}_1 + z\mathbf{x}'\mathbf{c}_2 + e = (\mathbf{x}',\ z\mathbf{x}')\begin{pmatrix}\mathbf{c}_1\\ \mathbf{c}_2\end{pmatrix} + e \tag{5.3}$$

where $\mathbf{x} = (1, x_1, \cdots, x_p)'$ as before, $\mathbf{c}_1 = \boldsymbol{\beta}_2$ and $\mathbf{c}_2 = \boldsymbol{\beta}_1 - \boldsymbol{\beta}_2$. This overall model implies the two original individual models in (5.1):

$$Y = \mathbf{x}'(\mathbf{c}_1 + \mathbf{c}_2) + e = \mathbf{x}'\boldsymbol{\beta}_1 + e$$

for a Y from the individual model 1, and

$$Y = \mathbf{x}'\mathbf{c}_1 + e = \mathbf{x}'\boldsymbol{\beta}_2 + e \tag{5.4}$$

for a Y from the individual model 2. In terms of the overall model (5.3), the hypotheses in (5.2) of the coincidence of the two individual models become

$$H_0: \mathbf{c}_2 = \boldsymbol{\beta}_1 - \boldsymbol{\beta}_2 = \mathbf{0} \quad \text{against} \quad H_a: \mathbf{c}_2 \neq \mathbf{0}.$$

Under H_0, the overall model (5.3) is reduced to the model

$$Y = \mathbf{x}'\mathbf{c}_1 + e \tag{5.5}$$

for the aggregated data. The hypotheses can therefore be tested by using the partial F test (1.7): H_0 is rejected if and only if

$$\frac{(SS_R \text{ of model (5.3)} - SS_R \text{ of model (5.5)})/(p+1)}{\text{MS residual of model (5.3)}} > f^{\alpha}_{p+1,\nu} \tag{5.6}$$

where $f^{\alpha}_{p+1,\nu}$ is the upper α point of an F distribution with $p+1$ and $\nu = n_1 + n_2 - 2(p+1)$ degrees of freedom.

The alternative form (1.11) of the test (1.7) in this case is given by

$$\text{reject } H_0 \text{ if and only if } \frac{(\hat{\boldsymbol{\beta}}_1 - \hat{\boldsymbol{\beta}}_2)'\Delta^{-1}(\hat{\boldsymbol{\beta}}_1 - \hat{\boldsymbol{\beta}}_2)/(p+1)}{\text{MS residual of model (5.3)}} > f^{\alpha}_{p+1,\nu} \tag{5.7}$$

where $\hat{\beta}_i = (\mathbf{X}_i'\mathbf{X}_i)^{-1}\mathbf{X}_i\mathbf{Y}_i$ is the least squares estimator of β_i from the ith individual model ($i = 1,2$), and $\Delta = (\mathbf{X}_1'\mathbf{X}_1)^{-1} + (\mathbf{X}_2'\mathbf{X}_2)^{-1}$. To see this, note that the design matrix of the overall model (5.3) is given by

$$\mathbf{X} = \begin{pmatrix} \mathbf{X}_1 & \mathbf{X}_1 \\ \mathbf{X}_2 & 0 \end{pmatrix}, \text{ and } \mathbf{Y} = \begin{pmatrix} \mathbf{Y}_1 \\ \mathbf{Y}_2 \end{pmatrix}.$$

So direct manipulation shows that

$$\begin{aligned}
\mathbf{X}'\mathbf{X} &= \begin{pmatrix} \mathbf{X}_1'\mathbf{X}_1 + \mathbf{X}_2'\mathbf{X}_2, & \mathbf{X}_1\mathbf{X}_1 \\ \mathbf{X}_1'\mathbf{X}_1, & \mathbf{X}_1'\mathbf{X}_1 \end{pmatrix}, \\
(\mathbf{X}'\mathbf{X})^{-1} &= \begin{pmatrix} (\mathbf{X}_2'\mathbf{X}_2)^{-1}, & -(\mathbf{X}_2'\mathbf{X}_2)^{-1} \\ -(\mathbf{X}_2'\mathbf{X}_2)^{-1}, & (\mathbf{X}_1'\mathbf{X}_1)^{-1} + (\mathbf{X}_2'\mathbf{X}_2)^{-1} \end{pmatrix}, \\
\hat{\mathbf{c}} &= (\mathbf{X}'\mathbf{X})^{-1}\mathbf{X}'\mathbf{Y} = \begin{pmatrix} (\mathbf{X}_2'\mathbf{X}_2)^{-1}\mathbf{X}_2'\mathbf{Y}_2 \\ (\mathbf{X}_1'\mathbf{X}_1)^{-1}\mathbf{X}_1'\mathbf{Y}_1 - (\mathbf{X}_2'X_2)^{-1}\mathbf{X}_2'\mathbf{Y}_2 \end{pmatrix} = \begin{pmatrix} \hat{\beta}_2 \\ \hat{\beta}_1 - \hat{\beta}_2 \end{pmatrix}, \\
\| \mathbf{Y} - \mathbf{X}\hat{\mathbf{c}} \|^2 &= \| \mathbf{Y} - \begin{pmatrix} \mathbf{X}_1\hat{\beta}_1 \\ \mathbf{X}_2\hat{\beta}_2 \end{pmatrix} \|^2 = \| \mathbf{Y}_1 - \mathbf{X}_1\hat{\beta}_1 \|^2 + \| \mathbf{Y}_2 - \mathbf{X}_2\hat{\beta}_2 \|^2.
\end{aligned}$$

Furthermore, note that in this case the matrix $\mathbf{A}$ in test (1.11) is given by the $(p+1) \times 2(p+1)$ matrix $(\mathbf{0}, \mathbf{I}_{p+1})$ and the vector $\mathbf{b}$ in test (1.11) is given by the zero-vector. So the numerator of the test statistic in (1.11) is given by

$$(\mathbf{A}\hat{\mathbf{c}} - \mathbf{b})'(\mathbf{A}(\mathbf{X}'\mathbf{X})^{-1}\mathbf{A}')^{-1}(\mathbf{A}\hat{\mathbf{c}} - \mathbf{b})/r = (\hat{\beta}_1 - \hat{\beta}_2)'\Delta^{-1}(\hat{\beta}_1 - \hat{\beta}_2)/(p+1)$$

as given in (5.7), and the denominator of the test statistic in (1.11), i.e., the mean square residual of the model (5.3), is given by

$$\begin{aligned}
\hat{\sigma}^2 &= \frac{\| \mathbf{Y} - \mathbf{X}\hat{\mathbf{c}} \|^2}{n_1 + n_2 - 2(p+1)} = \frac{\| \mathbf{Y}_1 - \mathbf{X}_1\hat{\beta}_1 \|^2 + \| \mathbf{Y}_2 - \mathbf{X}_2\hat{\beta}_2 \|^2}{n_1 + n_2 - 2(p+1)} \\
&= \frac{n_1 - p - 1}{n_1 + n_2 - 2(p+1)} \frac{\| \mathbf{Y}_1 - \mathbf{X}_1\hat{\beta}_1 \|^2}{(n_1 - p - 1)} + \frac{n_2 - p - 1}{n_1 + n_2 - 2(p+1)} \frac{\| \mathbf{Y}_2 - \mathbf{X}_2\hat{\beta}_2 \|^2}{(n_2 - p - 1)} \\
&= \frac{n_1 - p - 1}{n_1 + n_2 - 2(p+1)} \hat{\sigma}_1^2 + \frac{n_2 - p - 1}{n_1 + n_2 - 2(p+1)} \hat{\sigma}_2^2
\end{aligned}$$

where $\hat{\sigma}_1^2$ and $\hat{\sigma}_2^2$ are the respective mean square residuals of the two individual models in (5.1).

The inferences that can be drawn from the partial F test (5.6) or the equivalent form (5.7) are that if H_0 is rejected then the two regression models are deemed to be different, and that if H_0 is not rejected then there is not sufficient statistical evidence to conclude that the two regression models are different. This latter case is often falsely taken to mean the two regression models are the same. Regardless whether H_0 is rejected or not, however, no information on the magnitude of difference between the two models is provided directly by this approach of hypotheses testing.

Table 5.1: *Data adapted from Kleinbaum et al. (1998, pages 49 and 192)*

Obs	Y	x_1	z	Obs	Y	x_1	z	Obs	Y	x_1	z
1	144	39	1	24	160	44	1	47	128	19	0
2	138	45	1	25	158	53	1	48	130	22	0
3	145	47	1	26	144	63	1	49	138	21	0
4	162	65	1	27	130	29	1	50	150	38	0
5	142	46	1	28	125	25	1	51	156	52	0
6	170	67	1	29	175	69	1	52	134	41	0
7	124	42	1	30	158	41	0	53	134	18	0
8	158	67	1	31	185	60	0	54	174	51	0
9	154	56	1	32	152	41	0	55	174	55	0
10	162	64	1	33	159	47	0	56	158	65	0
11	150	56	1	34	176	66	0	57	144	33	0
12	140	59	1	35	156	47	0	58	139	23	0
13	110	34	1	36	184	68	0	59	180	70	0
14	128	42	1	37	138	43	0	60	165	56	0
15	130	48	1	38	172	68	0	61	172	62	0
16	135	45	1	39	168	57	0	62	160	51	0
17	114	17	1	40	176	65	0	63	157	48	0
18	116	20	1	41	164	57	0	64	170	59	0
19	124	19	1	42	154	61	0	65	153	40	0
20	136	36	1	43	124	36	0	66	148	35	0
21	142	50	1	44	142	44	0	67	140	33	0
22	120	39	1	45	144	50	0	68	132	26	0
23	120	21	1	46	149	47	0	69	169	61	0

Example 5.1 The data set given in Table 5.1, adapted from Kleinbaum *et al.* (1998), describes how systolic blood pressure (Y) changes with age (x_1) for the two gender groups, male ($z = 0$) and female ($z = 1$). The data points, circle for female and plus sign for male, are plotted in Figure 5.1. It is clear from the plots that a simple linear regression of Y on x_1 provides a sensible approximation to how Y changes with x_1 for each of the two gender groups. The fitted straight lines are calculated to be $Y = 97.08 + 0.95x_1$ for female and $Y = 110.04 + 0.96x_1$ for male, $\hat{\sigma}$ is equal to 8.95 with $\nu = 65$ degrees of freedom, $\Delta = \begin{pmatrix} 0.6144, & -0.0121 \\ -0.0121, & 0.0003 \end{pmatrix}$ and R^2 value is given by 0.776. The two fitted lines are also plotted in Figure 5.1. The statistic of the partial F test in (5.6) or (5.7) is 19.114 and so the p-value of the test is equal to 3×10^{-7}. Hence it is very unlikely that the true regression line for female coincides with that for male. But this is all we can say with the partial F test; in particular, no information is provided by the partial F test on how different these two regression lines are.

As another example, the data set in Table 5.2, also adapted from Kleinbaum *et*

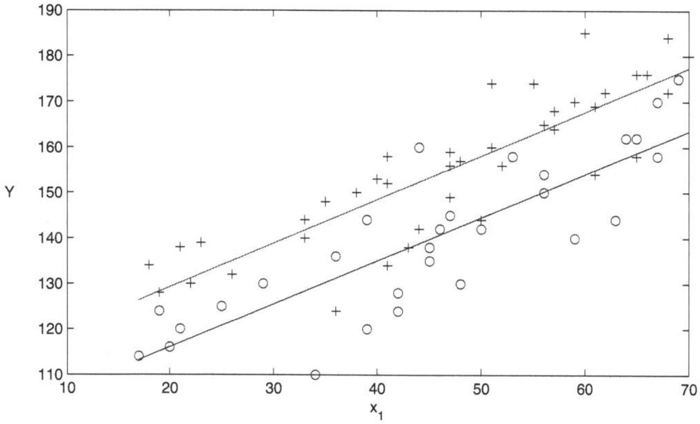

Figure 5.1: *The two fitted regression lines and the data points*

Table 5.2: *Data adapted from Kleinbaum et al. (1998, page 356)*

House	Y	10^{x_1}	10^{x_2}	z	House	Y	10^{x_1}	10^{x_2}	z
1	84.0	7	10	1	16	88.5	7	11	0
2	93.0	7	22	1	17	40.6	5	5	0
3	83.1	7	15	1	18	81.6	7	8	1
4	85.2	7	12	1	19	86.7	7	9	1
5	85.2	7	8	1	20	89.7	7	12	0
6	85.2	7	12	1	21	86.7	7	9	1
7	85.2	7	8	1	22	89.7	7	12	0
8	63.3	6	2	1	23	75.9	6	6	1
9	84.3	7	11	1	24	78.9	6	11	1
10	84.3	7	11	1	25	87.9	7	15	1
11	77.4	7	5	1	26	91.0	7	8	1
12	92.4	7	18	0	27	92.0	8	13	1
13	92.4	7	18	0	28	87.9	7	15	1
14	61.5	5	8	0	29	90.9	7	8	1
15	88.5	7	11	0	30	91.9	8	13	1

al. (1998), indicates how the sale price of a residential house (Y) changes with x_1 and x_2 for either an in-town house ($z = 0$) or a suburb house ($z = 1$). In the table 10^{x_1} is the number of bedrooms in the house and 10^{x_2} is the age (in years) of the house. In Kleinbaum *et al.* (1998), regression of Y is on 10^{x_1} and 10^{x_2}. Here we consider the regression of Y on x_1 and x_2 which results in better fit to the data. The two fitted regression planes are given by $Y = -105.00 + 175.56x_1 + 41.62x_2$ for an in-town house and $Y = -18.70 + 102.95x_1 + 17.04x_2$ for a suburb house,

$\hat{\sigma}$ is equal to 3.526 with $\nu = 24$ d.f.,

$$\Delta = \begin{pmatrix} 69.918, & -101.003, & 13.252 \\ -101.003, & 167.788, & -35.676 \\ 13.252, & -35.676, & 15.188 \end{pmatrix},$$

and the R^2-value of the fit is given by 0.916. The two fitted regression planes are plotted in Figure 5.2 over the observed covariate region which is indicated in the $Y = 0$ plane. The statistic of the partial F test in (5.6) or (5.7) is 6.550 and so the p-value of the test is equal to 0.00216. Hence it is unlikely that the true regression plane for an in-town house coincides with that for a suburb house. Again this is all we can say with the partial F test. All the computations in this example are done using MATLAB® program example0501.m.

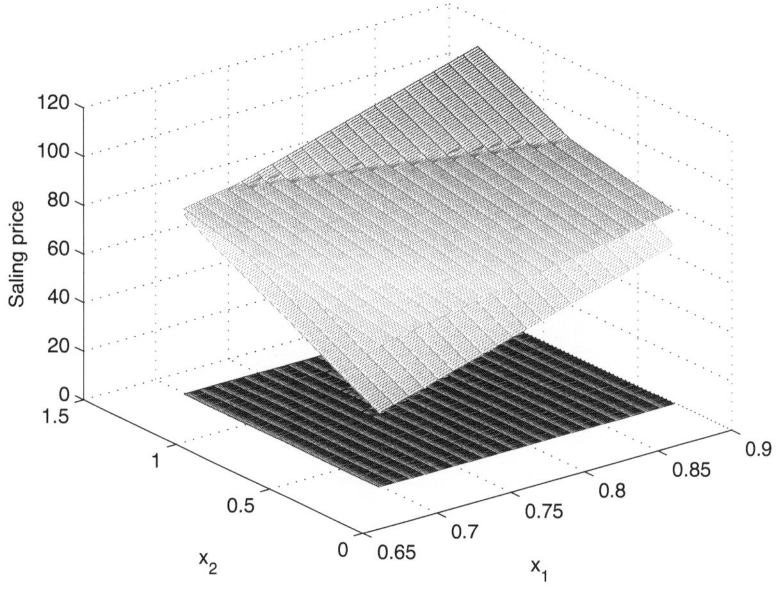

Figure 5.2: *The two fitted regression planes and the zero plane*

5.2 Hyperbolic bands over the whole space

To assess the magnitude of difference between the two models, it is natural to use a simultaneous confidence band to bound the difference between the two models

$$\mathbf{x}'\boldsymbol{\beta}_2 - \mathbf{x}'\boldsymbol{\beta}_1 = (1, x_1, \cdots, x_p)\boldsymbol{\beta}_2 - (1, x_1, \cdots, x_p)\boldsymbol{\beta}_1$$

over a covariate region of interest.

Let $\hat{\beta} = \hat{\beta}_2 - \hat{\beta}_1$ and $\beta = \beta_2 - \beta_1$. Then it is clear that $\hat{\beta} \sim \mathcal{N}_{p+1}(\beta, \sigma^2\Delta)$ where $\Delta = (\mathbf{X}_1'\mathbf{X}_1)^{-1} + (\mathbf{X}_2'\mathbf{X}_2)^{-1}$, $\hat{\sigma}/\sigma \sim \sqrt{\chi_\nu^2/\nu}$ where $\nu = n_1 + n_2 - 2(p+1)$, and $\hat{\beta}$ and $\hat{\sigma}$ are independent random variables. Applying Theorem 3.1 we have the following exact $1-\alpha$ simultaneous confidence band for $\mathbf{x}'\beta_2 - \mathbf{x}'\beta_1$ over the whole covariate space

$$\mathbf{x}'\beta_2 - \mathbf{x}'\beta_1 \in \mathbf{x}'\hat{\beta}_2 - \mathbf{x}'\hat{\beta}_1 \pm \sqrt{(p+1)f_{p+1,\nu}^{\alpha}}\, \hat{\sigma}\sqrt{\mathbf{x}'\Delta\mathbf{x}} \ \ \forall\, \mathbf{x}_{(0)} \in \Re^p. \tag{5.8}$$

Confidence band (5.8) quantifies the magnitude of differences between the two regression models at all the points $\mathbf{x}_{(0)} \in \Re^p$. In particular, if $H_0 : \beta_1 = \beta_2$ is true, that is, the two models are the same, then the difference between the two models is the zero hyper-plane $\mathbf{x}'\mathbf{0}$ which is included inside the confidence band (5.8) with probability $1-\alpha$. So a size α test of the hypotheses in (5.2) by using the confidence band (5.8) is to reject H_0

$$\text{if and only if } \mathbf{x}'\mathbf{0} \text{ is outside the band (5.8) for at least one } \mathbf{x}_{(0)} \in \Re^p. \tag{5.9}$$

Similar to Theorem 3.2, one can easily show that this size α test is just the partial F test (5.6) or (5.7), that is, the two tests (5.9) and (5.6) always reject and accept H_0 at the same time.

So, as long as the confidence band (5.8) contains the zero hyper-plane the null hypothesis H_0 will not be rejected. On the one hand, this is not surprising since the inclusion of the zero hyper-plane in the confidence band (5.8) implies that the difference between the two regression models is possibly zero and so H_0 is true. On the other hand, any $\mathbf{x}'\mathbf{d}_0$ included in the band (5.8) is possibly the true difference between the two regression models which can be very different from the zero hyper-plane if the band (5.8) is wide. So the non-rejection of H_0 from the partial F test (5.6) cannot be taken as H_0 being true. The confidence band (5.8) provides the extra information on the magnitude of the difference between the two models. If the difference between the two models from the confidence band is small then one may draw the inference that the two models are nearly the same, even if H_0 is rejected by the test. This inference is not available from the partial F test, however, for the reasons given above, even if H_0 is not rejected. So the confidence band is more informative than the partial F test.

Confidence band (5.8) is two-sided and so suitable for two-sided comparison of the two regression models. If one is only interested in assessing whether one regression model has a mean response lower than the other regression model, such as whether the females tend to have lower blood pressure than the same-age males, one-sided confidence bands are more pertinent. Following directly from the band (3.5) and Theorem 3.3, an exact $1-\alpha$ level lower confidence band is given by

$$\mathbf{x}'\beta_2 - \mathbf{x}'\beta_1 > \mathbf{x}'\hat{\beta}_2 - \mathbf{x}'\hat{\beta}_1 - c\hat{\sigma}\sqrt{\mathbf{x}'\Delta\mathbf{x}} \ \ \forall\, \mathbf{x}_{(0)} \in \Re^p, \tag{5.10}$$

where the critical constant c is the solution to the equation

$$\frac{1}{2}F_{p+1,\nu}(c^2/(p+1)) + \frac{1}{2}F_{p,\nu}(c^2/p) = 1-\alpha.$$

While the two-sided confidence band (5.8) induces the partial F test (5.6), the one-sided confidence band (5.10) induces a test for testing

$$H_0 : \mathbf{x}'\beta_2 = \mathbf{x}'\beta_1 \ \forall \mathbf{x}_{(0)} \in \Re^p \Longleftrightarrow H_a : \mathbf{x}'\beta_2 > \mathbf{x}'\beta_1 \text{ for at least one } \mathbf{x}_{(0)} \in \Re^p;$$

the test rejects H_0 if and only if

$$\mathbf{x}'\hat{\beta}_2 - \mathbf{x}'\hat{\beta}_1 - c\hat{\sigma}\sqrt{\mathbf{x}'\Delta\mathbf{x}} > 0 \ \text{ for at least one } \mathbf{x}_{(0)} \in \Re^p \quad (5.11)$$

$$\Longleftrightarrow \quad S := \sup_{\mathbf{x}_{(0)} \in \Re^p} \frac{\mathbf{x}'(\hat{\beta}_2 - \hat{\beta}_1)}{\hat{\sigma}\sqrt{\mathbf{x}'\Delta\mathbf{x}}} > c. \quad (5.12)$$

To see that the size of this test is exactly α, note from the band (5.10) that, when H_0 is true (i.e. $\beta_1 = \beta_2$), we have

$$\begin{aligned} 1-\alpha &= \text{P}\{\, 0 > \mathbf{x}'\hat{\beta}_2 - \mathbf{x}'\hat{\beta}_1 - c\hat{\sigma}\sqrt{\mathbf{x}'\Delta\mathbf{x}} \ \forall\, \mathbf{x}_{(0)} \in \Re^p \,\} \\ &= 1 - \text{P}\{\, \mathbf{x}'\hat{\beta}_2 - \mathbf{x}'\hat{\beta}_1 - c\hat{\sigma}\sqrt{\mathbf{x}'\Delta\mathbf{x}} > 0 \ \text{ for at least one } \mathbf{x}_{(0)} \in \Re^p \,\} \\ &= 1 - \text{P}\{ \text{ Reject } H_0 \text{ by the test in (5.11) } \}. \end{aligned}$$

Although this test is easy to implement via (5.11) when p is equal to 1 or 2 by plotting the confidence band, it is best implemented for a general p via its p-value using the test statistic in (5.12) in the following way. First, from the observed data, one calculates the observed value of the test statistic S, denoted as S^*, from the expression in (5.12); see Section 5.4.2 below for details on how to calculate S^*. Then one calculates the p-value from

$$\text{P}\{S > S^*\} = 1 - \frac{1}{2}F_{p+1,\nu}({S^*}^2/(p+1)) - \frac{1}{2}F_{p,\nu}({S^*}^2/p).$$

The null hypothesis H_0 is rejected at significant level α if and only if the p-value is less than α.

Example 5.2 For the blood pressure data set in Example 5.1, the critical values c are given by 2.505 for the two-sided band (5.8) and 2.315 for the one-sided band (5.10) respectively, and the confidence bands are plotted in Figure 5.3. As the two-sided confidence band, given by the two outer curves, does not contain the zero line completely, the two regression lines for female and male do not coincide, i.e., H_0 is rejected. This agrees with the observed very small p-value 3×10^{-7} of the partial F-test. One can make more inferences from this confidence band however. For example, one can infer that females tend to have lower blood pressure than males for the age range [13, 81] and one can also get some information on the

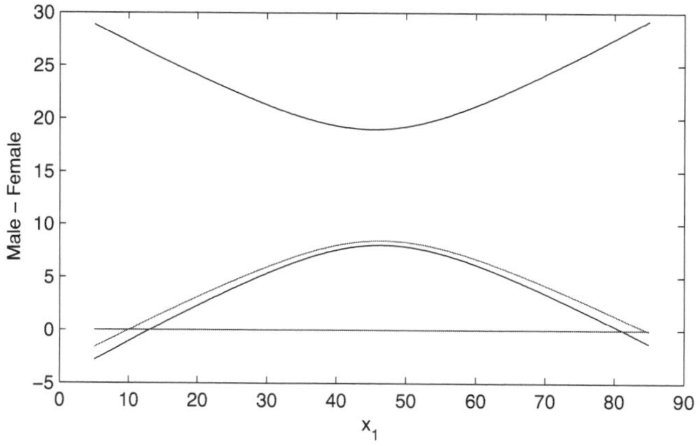

Figure 5.3: *The two-sided and one-sided confidence bands for* $\mathbf{x}'\beta_M - \mathbf{x}'\beta_F$

magnitude of mean blood pressure difference between females and males at any specific age. If one is interested only in assessing whether males tend to have higher blood pressure than females then the one-sided confidence band should be the focus of attention. From the one-sided band, one can infer that males tend to have higher blood pressure than females for the age range [10,85]. The p-value of the one-sided test (5.12) in this example is given by 1.7×10^{-7} which is smaller than the p-value of the two-sided partial F test as expected.

For the house sale price data set, the two-sided and one-sided confidence bands are plotted in Figure 5.4 with the critical constants given by 3.004 and 2.840 respectively. Because the two critical values are quite close to each other, the lower part of the two-sided band almost overlaps with the lower confidence band in the plot. From the plots, it is clear that a suburb area is more expensive than an in-town area for smaller (fewer bedrooms) and newer houses, and the opposite is true for bigger and older houses. The p-value of the two-sided partial F test is 0.00216 while the p-value of the one-sided test is 0.00146. All the computations in this example are done using MATLAB program example0502.m.

5.3 Confidence bands over a rectangular region

Often we are interested in the difference between the two regression models only over a restricted region of the covariates. Furthermore the regression models rarely hold over the whole space of the covariates. So a confidence band over the whole space of covariates is likely to be neither efficient if only the differences over a restricted region are of interest, nor appropriate if the models hold only over certain restricted region. A confidence band for the difference between the models over a restricted region is therefore useful. In this section we consider hyperbolic

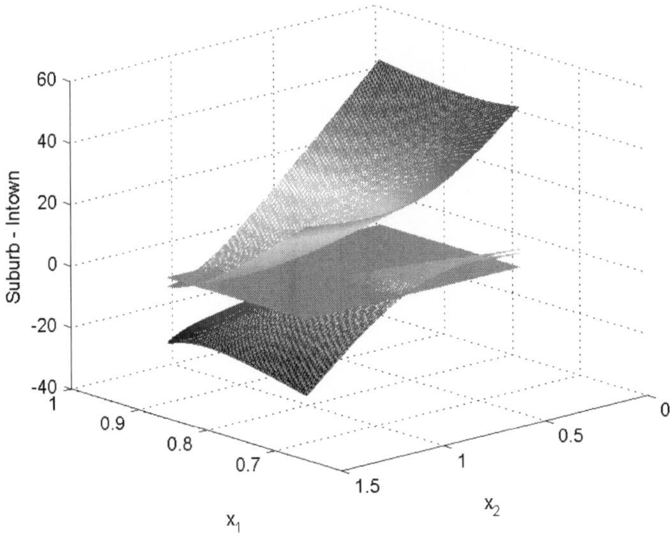

Figure 5.4: *The two-sided and one-sided confidence bands for* $\mathbf{x}'\boldsymbol{\beta}_S - \mathbf{x}'\boldsymbol{\beta}_I$

and constant-width simultaneous confidence bands for $\mathbf{x}'\boldsymbol{\beta}_2 - \mathbf{x}'\boldsymbol{\beta}_1$ over a given rectangular region of the covariates of the form

$$\mathcal{X}_r = \{(x_1, \cdots, x_p)' : a_i \le x_i \le b_i, i = 1, \cdots, p\}, \tag{5.13}$$

where $-\infty \le a_i < b_i \le \infty, i = 1, \cdots, p$ are given constants.

Simulation methods are used to find the critical values and observed p-values, which work for a general $p \ge 1$. Note, however, for the special case of $p = 1$, one can use the methods provided in Chapter 2 to compute the critical values c and observed p-values exactly by using one-dimensional numerical integration. This section is based on the work of Liu *et al.* (2007a).

5.3.1 *Two-sided hyperbolic band*

A two-sided hyperbolic simultaneous confidence band for $\mathbf{x}'\boldsymbol{\beta}_2 - \mathbf{x}'\boldsymbol{\beta}_1$ over the covariate region $\mathcal{X}_r$ in (5.13) has the form

$$\mathbf{x}'\boldsymbol{\beta}_2 - \mathbf{x}'\boldsymbol{\beta}_1 \in \mathbf{x}'\hat{\boldsymbol{\beta}}_2 - \mathbf{x}'\hat{\boldsymbol{\beta}}_1 \pm c\hat{\sigma}\sqrt{\mathbf{x}'\Delta\mathbf{x}} \;\; \forall\, \mathbf{x}_{(0)} \in \mathcal{X}_r, \tag{5.14}$$

where c is a critical constant chosen so that the simultaneous confidence level of the band is equal to $1 - \alpha$.

Let $\mathbf{P}$ be the unique square root matrix of $\Delta = (\mathbf{X}_1'\mathbf{X}_1)^{-1} + (\mathbf{X}_2'\mathbf{X}_2)^{-1}$, and define $\mathbf{T} = \mathbf{P}^{-1}(\hat{\boldsymbol{\beta}}_2 - \boldsymbol{\beta}_2 - \hat{\boldsymbol{\beta}}_1 + \boldsymbol{\beta}_1)/\hat{\sigma}$ which has $\mathcal{T}_{p+1,\nu}$ distribution. It follows from similar lines as in Section 3.2.1 that the simultaneous confidence level of the band (5.14) is given by $P\{S < c\}$ where

$$S = \sup_{\mathbf{x}_{(0)} \in \mathcal{X}_r} \frac{|\mathbf{x}'(\hat{\beta}_2 - \beta_2 - \hat{\beta}_1 + \beta_1)|}{\hat{\sigma}\sqrt{\mathbf{x}'\Delta\mathbf{x}}}$$
$$= \sup_{\mathbf{x}_{(0)} \in \mathcal{X}_r} \frac{\left|(\mathbf{Px})'\left(\mathbf{P}^{-1}(\hat{\beta}_2 - \beta_2 - \hat{\beta}_1 + \beta_1)/\hat{\sigma}\right)\right|}{\sqrt{(\mathbf{Px})'(\mathbf{Px})}}$$
$$= \sup_{\mathbf{x}_{(0)} \in \mathcal{X}_r} \frac{|(\mathbf{Px})'\mathbf{T}|}{\|\mathbf{Px}\|}. \tag{5.15}$$

Hence the critical constant c can be calculated by using simulation as in Section 3.2.1. The only difference is that the matrix $\mathbf{P}$ here is the square root matrix of Δ while the matrix $\mathbf{P}$ in Section 3.2.1 is the square root matrix of $(\mathbf{X}'\mathbf{X})^{-1}$.

Over the covariate region $\mathcal{X}_r$ of interest, confidence band (5.14) is narrower and so provides more accurate information on the magnitude of difference between the two regression models than does confidence band (5.8).

Confidence band (5.14) can also be used to test the hypotheses in (5.2):

$$\text{reject } H_0 \iff \sup_{\mathbf{x}_{(0)} \in \mathcal{X}_r} \frac{|\mathbf{x}'(\hat{\beta}_2 - \hat{\beta}_1)|}{\hat{\sigma}\sqrt{\mathbf{x}'\Delta\mathbf{x}}} > c \tag{5.16}$$
$$\iff \mathbf{x}'\mathbf{0} \text{ is outside the band (5.14) for at least one } \mathbf{x}_{(0)} \in \mathcal{X}_r.$$

It is clear that the size of this test is equal to α if the simultaneous confidence level of the band (5.14) is equal to $1-\alpha$. It is also noteworthy that this test requires only that the two regression models hold over the region $\mathcal{X}_r$. On the other hand, the partial F test (5.6) requires that the two regression models hold over the whole covariate space $\Re^p$ since its rejection of H_0 may be due to the exclusion of the zero plane from the band (5.8) at any $\mathbf{x}_{(0)} \in \Re^p$.

The p-value of the test (5.16) can be calculated in the following way. Let $\hat{\beta}_1$, $\hat{\beta}_2$ and $\hat{\sigma}$ be the estimates of β_1, β_2 and σ respectively calculated based on the observed data, and so the observed value of the test statistic in (5.16) is given by

$$S^* = \sup_{\mathbf{x}_{(0)} \in \mathcal{X}_r} \frac{|\mathbf{x}'(\hat{\beta}_1 - \hat{\beta}_2)|}{\hat{\sigma}\sqrt{\mathbf{x}'\Delta\mathbf{x}}} = \sup_{\mathbf{x}_{(0)} \in \mathcal{X}_r} \frac{|(\mathbf{Px})'\mathbf{T}^*|}{\|\mathbf{Px}\|}$$

where $\mathbf{T}^* = \mathbf{P}^{-1}(\hat{\beta}_2 - \hat{\beta}_1)/\hat{\sigma}$. The quadratic-programming method discussed in Section 3.2.1 can be used to compute S^*. The observed p-value is then given by

$$\mathrm{P}\{S > S^*\} = \mathrm{P}\left\{ \sup_{\mathbf{x}_{(0)} \in \mathcal{X}_r} \frac{|(\mathbf{Px})'\mathbf{T}|}{\|\mathbf{Px}\|} > S^* \right\}$$

where $\mathbf{T} \sim \mathcal{T}_{p+1,\nu}$. This p-value can be approximated by simulation: simulate a large number of independent replicates of the random variable S and use the proportion of the replicates that are larger than S^* as an approximation to the p-value.

5.3.2 *One-sided hyperbolic band*

For one-sided inferences, one requires one-sided confidence bands. A lower hyperbolic confidence band for $\mathbf{x}'\beta_2 - \mathbf{x}'\beta_1$ over the covariate region $\mathcal{X}_r$ is given by

$$\mathbf{x}'\beta_2 - \mathbf{x}'\beta_1 > \mathbf{x}'\hat{\beta}_2 - \mathbf{x}'\hat{\beta}_1 - c\hat{\sigma}\sqrt{\mathbf{x}'\Delta\mathbf{x}} \;\; \forall\, \mathbf{x}_{(0)} \in \mathcal{X}_r \tag{5.17}$$

where the critical constant c is chosen so that the simultaneous confidence level of this band is equal to $1-\alpha$.

Let $\mathbf{P}$ and $\mathbf{T}$ be as defined in Section 5.3.1. Then the simultaneous confidence level of the band (5.17) is given by $P\{S < c\}$ where

$$S = \sup_{\mathbf{x}_{(0)} \in \mathcal{X}_r} \frac{\mathbf{x}'(\hat{\beta}_2 - \beta_2 - \hat{\beta}_1 + \beta_1)}{\hat{\sigma}\sqrt{\mathbf{x}'\Delta\mathbf{x}}} = \sup_{\mathbf{x}_{(0)} \in \mathcal{X}_r} \frac{(\mathbf{P}\mathbf{x})'\mathbf{T}}{\|\mathbf{P}\mathbf{x}\|}.$$

Hence the critical constant $c > 0$ can be calculated by using simulation as in Section 3.2.2; again the only difference is that the matrix $\mathbf{P}$ here is the square root matrix of Δ while the matrix $\mathbf{P}$ in Section 3.2.2 is the square root matrix of $(\mathbf{X}'\mathbf{X})^{-1}$.

From the lower confidence band (5.17), one can infer whether the mean response of model 2, $\mathbf{x}'\beta_2$, is greater than the mean response of model 1, $\mathbf{x}'\beta_1$, in only some parts or over the whole region of $\mathbf{x}_{(0)} \in \mathcal{X}_r$. Over the covariate region $\mathcal{X}_r$ of interest, the lower confidence band (5.17) is higher and so provides more accurate one-sided inference on the magnitude of difference between the two regression models than the lower part of the two-sided confidence band (5.14) and the lower confidence band (5.10).

The lower confidence band (5.17) can also be used to test the hypotheses

$$H_0 : \mathbf{x}'\beta_2 = \mathbf{x}'\beta_1 \;\forall \mathbf{x}_{(0)} \in \mathcal{X}_r \Longleftrightarrow H_a : \mathbf{x}'\beta_2 > \mathbf{x}'\beta_1 \text{ for at least one } \mathbf{x}_{(0)} \in \mathcal{X}_r; \tag{5.18}$$

the test rejects H_0 if and only if

$$\begin{aligned} & \mathbf{x}'\hat{\beta}_2 - \mathbf{x}'\hat{\beta}_1 - c\hat{\sigma}\sqrt{\mathbf{x}'\Delta\mathbf{x}} > 0 \text{ for at least one } \mathbf{x}_{(0)} \in \mathcal{X}_r \\ \Longleftrightarrow \quad & \sup_{\mathbf{x}_{(0)} \in \mathcal{X}_r} \frac{\mathbf{x}'(\hat{\beta}_2 - \hat{\beta}_1)}{\hat{\sigma}\sqrt{\mathbf{x}'\Delta\mathbf{x}}} > c. \end{aligned} \tag{5.19}$$

It is again straightforward to show that the size of this test is equal to α if the simultaneous confidence level of the band (5.17) is equal to $1-\alpha$. It is noteworthy that the null hypothesis H_0 in (5.18) is just $H_0 : \beta_2 = \beta_1$.

The p-value of the test (5.19) can be calculated in the following way. Let $\hat{\beta}_1$,

$\hat{\beta}_2$ and $\hat{\sigma}$ be the estimates of β_1, β_2 and σ, respectively, based on the observed data, and so the observed value of the test statistic in (5.19) is given by

$$S^* = \sup_{\mathbf{x}_{(0)} \in \mathcal{X}_r} \frac{\mathbf{x}'(\hat{\beta}_2 - \hat{\beta}_1)}{\hat{\sigma}\sqrt{\mathbf{x}'\Delta\mathbf{x}}} = \sup_{\mathbf{x}_{(0)} \in \mathcal{X}_r} \frac{(\mathbf{Px})'\mathbf{T}^*}{\|\mathbf{Px}\|}$$

where $\mathbf{T}^* = \mathbf{P}^{-1}(\hat{\beta}_2 - \hat{\beta}_1)/\hat{\sigma}$.

It is pointed out in Section 3.2.2 that S^* may not be equal to $\|\pi(\mathbf{T}^*, \mathbf{P}, \mathcal{X}_r)\|$. In fact it can be shown that

$$S^* = \begin{cases} \|\pi(\mathbf{T}^*, \mathbf{P}, \mathcal{X}_r)\| & \text{if } \|\pi(\mathbf{T}^*, \mathbf{P}, \mathcal{X}_r)\| > 0 \\ \text{a non}-\text{positive number} & \text{if } \|\pi(\mathbf{T}^*, \mathbf{P}, \mathcal{X}_r)\| = 0. \end{cases}$$

When $\|\pi(\mathbf{T}^*, \mathbf{P}, \mathcal{X}_r)\| > 0$, S^* can therefore be computed using the quadratic-programming method discussed in Section 3.2.1. When $\|\pi(\mathbf{T}^*, \mathbf{P}, \mathcal{X}_r)\| = 0$, however, no method is available for computing S^* for a general $p \geq 1$ (though methods can be developed for the special case of $p = 1$ by using the results of Chapter 2).

On the other hand it can be shown that, when $\|\pi(\mathbf{T}^*, \mathbf{P}, \mathcal{X}_r)\| = 0$, the observed p-value $\mathrm{P}\{S > S^*\}$ is at least 0.5 since $S^* \leq 0$. When $\|\pi(\mathbf{T}^*, \mathbf{P}, \mathcal{X}_r)\| > 0$, we have $S^* > 0$ and so the p-value is given by

$$\mathrm{P}\{S > S^*\} = 1 - \mathrm{P}\{S \leq S^*\} = 1 - \mathrm{P}\{\|\pi(\mathbf{T}, \mathbf{P}, \mathcal{X}_r)\| \leq S^*\}$$

from the discussion in Section 3.2.2. This p-value can therefore be computed by using simulation as in the two-sided case.

Example 5.3 For the blood pressure example, if one is interested in confidence bands over the observed range $17 \leq x_1 = age \leq 70$ then the two-sided confidence band (5.14) uses critical value $c = 2.485$ (s.e. 0.0060) and the one-sided band (5.17) uses critical constant $c = 2.1904$ (s.e. 0.0062) for $\alpha = 0.05$. These critical constants are smaller than the corresponding critical constants for the confidence bands over the whole covariate space given in the last example as expected. Hence the restricted two-sided band (5.14) is narrower than the unrestricted two-sided confidence band (5.8) and the one-sided restricted confidence band (5.17) is higher than the un-restricted one-sided confidence band (5.11) over the observed range of 'age'. In particular, from both the two-sided and one-sided confidence bands, one can conclude that males tend to have significantly higher blood pressure than females $17 \leq age \leq 70$. The p-values of the two-sided test (5.16) and one-sided test (5.19) are both given by 0.00000.

For the house sale price example, the critical constants of the two-sided band (5.14) and one-sided band (5.17) over the observed range $5 \leq 10^{x_1} \leq 8$ (i.e., $0.699 \leq x_1 \leq 0.903$) and $2 \leq 10^{x_2} \leq 22$ (i.e. $0.301 \leq x_2 \leq 1.342$) are given by 2.978 (s.e. 0.0068) and 2.706 (s.e. 0.0070) respectively. The p-values of the corresponding tests (5.16) and (5.19) are given by 0.0051 and 0.0034 respectively. All the computations in this example are done using MATLAB program example0503.m.

5.3.3 *Two-sided constant width band*

In addition to the hyperbolic confidence bands above, one may also construct constant width confidence bands. A two-sided constant width simultaneous confidence band over the covariate region $\mathcal{X}_r$ is given by

$$\mathbf{x}'\beta_2 - \mathbf{x}'\beta_1 \in \mathbf{x}'\hat{\beta}_2 - \mathbf{x}'\hat{\beta}_1 \pm c\hat{\sigma} \ \ \forall\, \mathbf{x}_{(0)} \in \mathcal{X}_r \tag{5.20}$$

where c is a critical constant chosen so that the simultaneous confidence level of the band is equal to $1-\alpha$. This two-sided constant width band for the special case of $p=1$ and over a particular covariate interval was considered by Aitkin (1973) and Hayter *et al.* (2007).

For a general $p \geq 1$, we use simulation to approximate the value of c as before. Note that c satisfies $P\{S < c\} = 1-\alpha$ where

$$S = \sup_{\mathbf{x}_{(0)} \in \mathcal{X}_r} \frac{|\mathbf{x}'(\hat{\beta}_2 - \beta_2 - \hat{\beta}_1 + \beta_1)|}{\hat{\sigma}} = \sup_{\mathbf{x}_{(0)} \in \mathcal{X}_r} |\mathbf{x}'\mathbf{T}| \tag{5.21}$$

where $\mathbf{T} = (\hat{\beta}_2 - \beta_2 - \hat{\beta}_1 + \beta_1)/\hat{\sigma}$ has the multivariate t distribution $\mathcal{T}_{p+1,v}(\mathbf{0}, \Delta)$. So the critical constant c can be approximated by simulation as in Section 3.3.1.

This constant width confidence band can also be used to test the hypotheses in (5.2):

$$\begin{aligned} \text{reject } H_0 \quad &\Longleftrightarrow \quad \mathbf{x}'\mathbf{0} \text{ is outside the band (5.20) for at least one } \mathbf{x}_{(0)} \in \mathcal{X}_r \\ &\Longleftrightarrow \quad \sup_{\mathbf{x}_{(0)} \in \mathcal{X}_r} \frac{|\mathbf{x}'(\hat{\beta}_2 - \hat{\beta}_1)|}{\hat{\sigma}} > c. \end{aligned} \tag{5.22}$$

The size of this test is equal to α if the confidence level of the band (5.20) is equal to $1-\alpha$. Similar to the test (5.16), this test requires only that the two regression models hold over the region $\mathcal{X}_r$.

The p-value of this test can be approximated by simulation as before. Let $\hat{\beta}_1$, $\hat{\beta}_2$ and $\hat{\sigma}$ be the estimates of β_1, β_2 and σ, respectively, based on the observed data, and so the observed statistic of the test (5.22) is

$$S^* = \sup_{\mathbf{x}_{(0)} \in \mathcal{X}_r} \frac{|\mathbf{x}'(\hat{\beta}_2 - \hat{\beta}_1)|}{\hat{\sigma}}.$$

The observed p-value is then given by

$$\mathrm{P}\left\{ \sup_{\mathbf{x}_{(0)} \in \mathcal{X}_r} |\mathbf{x}'\mathbf{T}| > S^* \right\}$$

where $\mathbf{T}$ is a $\mathcal{T}_{p+1,v}(\mathbf{0}, \Delta)$ random vector. This p-value can therefore be approximated by simulation as before: simulate a large number of independent replicates of the random variable $\sup_{\mathbf{x}_{(0)} \in \mathcal{X}_r} |\mathbf{x}'\mathbf{T}|$; use the proportion of the replicates that are larger than S^* as an approximation to the p-value.

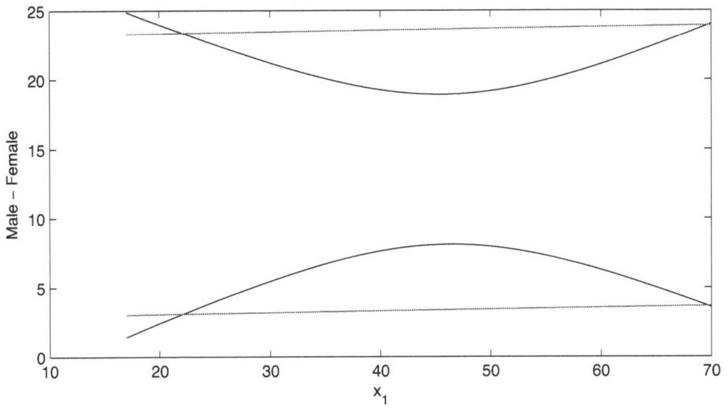

Figure 5.5: *Two-sided hyperbolic and constant width bands for blood pressure example*

Example 5.4 For the blood pressure example, the critical value c of the two-sided constant width band (5.20) over the observed range of $17 \le x_1 = age \le 70$ is given by 1.1315 (s.e. 0.00288). The p-value of the two-sided test (5.22) is given by 0.0058. The constant width band (5.20) and the corresponding hyperbolic band (5.14) are plotted in Figure 5.5. It is clear from the plots that the hyperbolic band tends to be narrower near the mean of the observed x_1-values but wider near the two ends of the observed x_1-values than the constant width band.

For the house sale price example, the critical value c of the two-sided constant width band (5.20) over the observed range of x_1 and x_2 is given by 7.8646 (s.e. 0.02543). The p-value of the two-sided test (5.22) is given by 0.04691. The constant width band (5.20) and the corresponding hyperbolic band (5.14) are plotted in Figure 5.6. It can be seen from the figure that the constant width band only just crosses the zero plane near $10^{x_1} = 5$ and $10^{x_2} = 2$. This agrees with the observed p-value 0.04691 which is very close to $\alpha = 0.05$.

5.3.4 *One-sided constant width band*

A lower constant width simultaneous confidence band over the covariate region $\mathcal{X}_r$ is given by

$$\mathbf{x}'\beta_2 - \mathbf{x}'\beta_1 > \mathbf{x}'\hat{\beta}_2 - \mathbf{x}'\hat{\beta}_1 - c\hat{\sigma} \ \ \forall\, \mathbf{x}_{(0)} \in \mathcal{X}_r \tag{5.23}$$

where c is a critical constant chosen so that the simultaneous confidence level of the band is equal to $1-\alpha$.

We again use simulation to approximate the value of c for a general $p \ge 1$ by noting that c satisfies $\mathrm{P}\{S < c\} = 1 - \alpha$ where

$$S = \sup_{\mathbf{x}_{(0)} \in \mathcal{X}_r} \frac{\mathbf{x}'(\hat{\beta}_2 - \beta_2 - \hat{\beta}_1 + \beta_1)}{\hat{\sigma}} = \sup_{\mathbf{x}_{(0)} \in \mathcal{X}_r} \mathbf{x}'\mathbf{T} \tag{5.24}$$

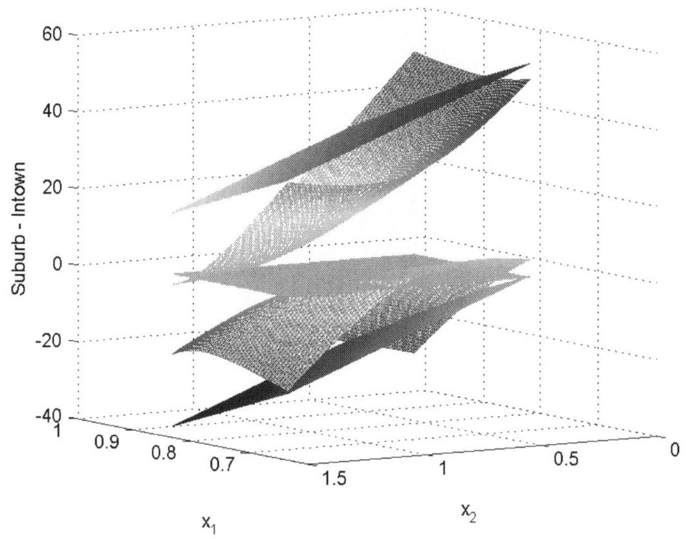

Figure 5.6: *Two-sided hyperbolic and constant width bands for house sale price example*

and $\mathbf{T}$ is defined in the last subsection.

This constant width confidence band can be used to test the hypotheses in (5.18):

$$\begin{aligned} \text{reject } H_0 &\iff \mathbf{x}'\hat{\beta}_2 - \mathbf{x}'\hat{\beta}_1 - c\hat{\sigma} > 0 \text{ for at least one } \mathbf{x}_{(0)} \in \mathcal{X}_r \\ &\iff \sup_{\mathbf{x}_{(0)} \in \mathcal{X}_r} \frac{\mathbf{x}'(\hat{\beta}_2 - \hat{\beta}_1)}{\hat{\sigma}} > c. \end{aligned} \tag{5.25}$$

The size of this test is equal to α if the confidence level of the band (5.23) is equal to $1 - \alpha$.

To calculate the p-value of this test, let $\hat{\beta}_1$, $\hat{\beta}_2$ and $\hat{\sigma}$ be the estimates of β_1, β_2 and σ, respectively, based on the observed data. So the observed statistic of the test (5.25) is

$$S^* = \sup_{\mathbf{x}_{(0)} \in \mathcal{X}_r} \frac{\mathbf{x}'(\hat{\beta}_2 - \hat{\beta}_1)}{\hat{\sigma}}.$$

The observed p-value is then given by

$$\mathrm{P}\left\{ \sup_{\mathbf{x}_{(0)} \in \mathcal{X}_r} \mathbf{x}'\mathbf{T} > S^* \right\}$$

where $\mathbf{T}$ is a $\mathcal{T}_{p+1,v}(\mathbf{0}, \Delta)$ random vector. This p-value can therefore be approximated by simulation as before.

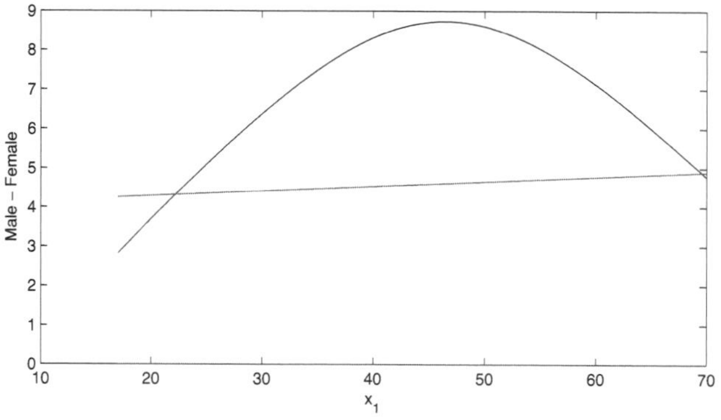

Figure 5.7: *One-sided hyperbolic and constant width bands for blood pressure example*

Example 5.5 For the blood pressure example, the critical value c of the one-sided constant width band (5.23) over the observed range of $17 \leq x_1 = age \leq 70$ is given by 0.9957 (s.e. 0.00306). The p-value of the one-sided test (5.25) is given by 0.0032. The constant width band (5.23) and the corresponding one-sided hyperbolic band (5.17) are plotted in Figure 5.7. It is clear from the plots that the hyperbolic band tends to be higher near the mean of the observed x_1-values but lower near the two ends of the observed x_1-values than the constant width band.

For the house sale price example, the critical value c of the one-sided constant width band (5.23) over the observed range of x_1 and x_2 is given by 6.7913 (s.e. 0.0238). The p-value of the one-sided test (5.25) is given by 0.02572. The constant width band (5.23) and the corresponding one-sided hyperbolic band (5.17) are plotted in Figure 5.8. It can be seen from the figure that the two bands cross over each other over certain parts of the covariate region and so no one dominates the other.

5.4 Confidence bands over an ellipsoidal region

In this section we consider simultaneous confidence bands over an ellipsoidal region of the covariates. These confidence bands can be constructed to have exact $1-\alpha$ confidence level by using one dimensional numerical integration as in Chapter 3. We focus on the case of $p \geq 2$ since for $p = 1$ the problem becomes the construction of confidence bands over an interval of the only covariate, which is covered by the methods provided in the last section.

First, let us define the ellipsoidal region of the covariates. Following the notation of Section 3.4, let $\mathbf{X}_{i(0)}$ be the $n_i \times p$ matrix produced from the design matrix $\mathbf{X}_i$ by deleting the first column of 1's from $\mathbf{X}_i$, and $\bar{\mathbf{x}}_{i(0)} = (\bar{x}_{i\cdot 1}, \cdots, \bar{x}_{i\cdot p})'$ where $\bar{x}_{i\cdot j} = \frac{1}{n_i}\sum_{l=1}^{n_i} x_{ilj}$ is the mean of the observations from the i-th model (or group)

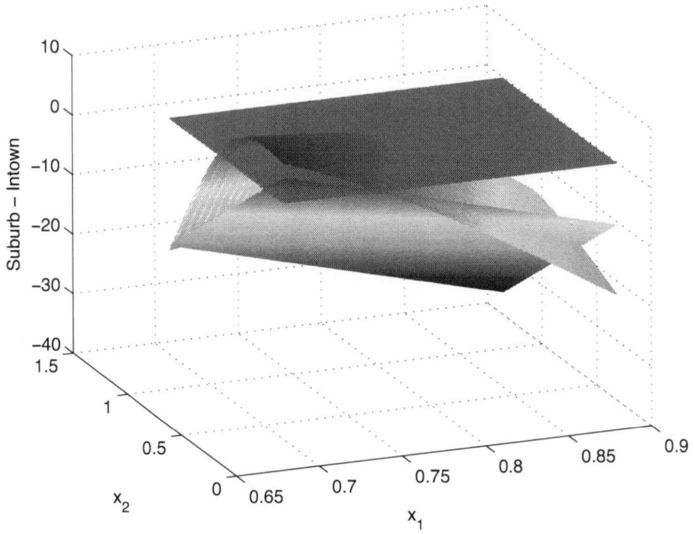

Figure 5.8: *One-sided hyperbolic and constant width bands for house sale price example*

on the jth predictor variable ($1 \leq j \leq p$, $i = 1, 2$). Define $p \times p$ matrix

$$\mathbf{V}_i = \frac{1}{n_i}\left(\mathbf{X}_{i(0)} - \mathbf{1}\bar{\mathbf{x}}'_{i(0)}\right)'\left(\mathbf{X}_{i(0)} - \mathbf{1}\bar{\mathbf{x}}'_{i(0)}\right) = \frac{1}{n_i}\left(\mathbf{X}'_{i(0)}\mathbf{X}_{i(0)} - n_i\bar{\mathbf{x}}_{i(0)}\bar{\mathbf{x}}'_{i(0)}\right)$$

where $\mathbf{1}$ is the n_i-vector of 1's ($i = 1, 2$), define number

$$\begin{aligned}\kappa &= \left(\frac{1}{n_1}+\frac{1}{n_2}\right) + \bar{\mathbf{x}}'_{1(0)}\frac{\mathbf{V}_1^{-1}}{n_1}\bar{\mathbf{x}}_{1(0)} + \bar{\mathbf{x}}'_{2(0)}\frac{\mathbf{V}_2^{-1}}{n_2}\bar{\mathbf{x}}_{2(0)} - \left(\bar{\mathbf{x}}'_{1(0)}\frac{\mathbf{V}_1^{-1}}{n_1} + \bar{\mathbf{x}}'_{2(0)}\frac{\mathbf{V}_2^{-1}}{n_2}\right) \\ &\quad \times \left(\frac{\mathbf{V}_1^{-1}}{n_1}+\frac{\mathbf{V}_2^{-1}}{n_2}\right)^{-1}\left(\frac{\mathbf{V}_1^{-1}}{n_1}\bar{\mathbf{x}}_{1(0)} + \frac{\mathbf{V}_2^{-1}}{n_2}\bar{\mathbf{x}}_{2(0)}\right), \end{aligned} \tag{5.26}$$

and define p-vector

$$\zeta = \left(\frac{\mathbf{V}_1^{-1}}{n_1}+\frac{\mathbf{V}_2^{-1}}{n_2}\right)^{-1}\left(\frac{\mathbf{V}_1^{-1}}{n_1}\bar{\mathbf{x}}_{1(0)} + \frac{\mathbf{V}_2^{-1}}{n_2}\bar{\mathbf{x}}_{2(0)}\right). \tag{5.27}$$

Note that κ and ζ depend on $\mathbf{X}_1$ and $\mathbf{X}_2$ only. Now the ellipsoidal region is defined as

$$\mathcal{X}_e = \left\{\mathbf{x}_{(0)} : \left(\mathbf{x}_{(0)} - \zeta\right)'\left(\frac{\mathbf{V}_1^{-1}}{n_1}+\frac{\mathbf{V}_2^{-1}}{n_2}\right)\left(\mathbf{x}_{(0)} - \zeta\right) \leq a^2\kappa\right\} \tag{5.28}$$

where $a \geq 0$ is a given constant. It is clear that this region is centered at ζ and has an ellipsoidal shape. The value of a determines the size of $\mathcal{X}_e$.

Figure 5.9 gives several examples of $\mathcal{X}_e$ corresponding to various values of a for the house sale prices example in which $p = 2$. The center of all these $\mathcal{X}_e$ is denoted by the star. Note in particular that the centers of the two ellipsoidal regions corresponding to the two individual groups, in-town and suburb houses, defined in (3.17) of Chapter 3, are denoted by the two '+' signs, and the star is not on the straight-line segment that connects the two '+' signs. The smaller rectangle in the figure depicts the observed region of the covariates.

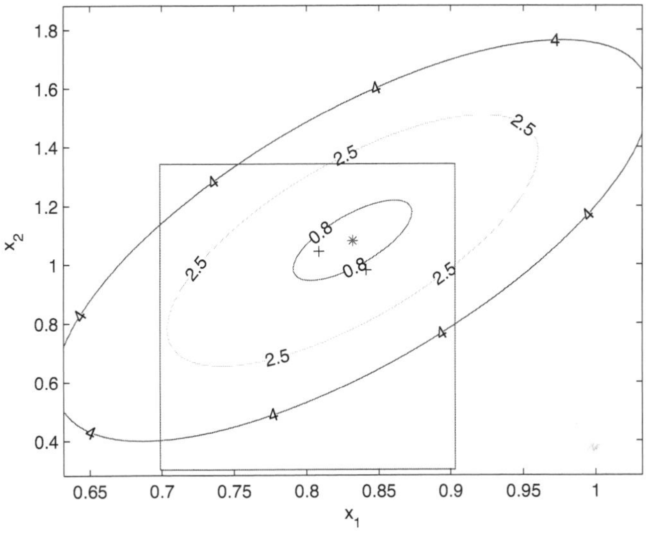

Figure 5.9: *Several examples of $\mathcal{X}_e$ for the house sale price example*

Note that

$$\mathbf{x}'\Delta\mathbf{x} = \left(\mathbf{x}_{(0)} - \zeta\right)' \left(\mathbf{V}_1^{-1}/n_1 + \mathbf{V}_2^{-1}/n_2\right) \left(\mathbf{x}_{(0)} - \zeta\right) + \kappa.$$

Hence $\kappa = (1, \zeta')\Delta(1, \zeta')' > 0$ and

$$\mathcal{X}_e = \{\mathbf{x}_{(0)} : \mathbf{x}'\Delta\mathbf{x} \leq (a^2 + 1)\kappa\}. \tag{5.29}$$

Next we define some new p-vector $\mathbf{w}_{(0)}$ which has a one-to-one correspondence with $\mathbf{x}_{(0)}$ and then express $\mathcal{X}_e$ in terms of $\mathbf{w}_{(0)}$. Let $\mathbf{z}^* = (z_0^*, \cdots, z_p^*)'$ be the first column of Δ^{-1}, that is, $\mathbf{z}^* = \Delta^{-1}\mathbf{e}_1$ where $(p+1)$-vector $\mathbf{e}_1$ has the first element equal to one and the other elements equal to zero. Note that $z_0^* > 0$ since Δ is positive definite. Furthermore, it can be shown that $\kappa z_0^* = 1$, which is required below. Define $\mathbf{z} = \mathbf{z}^* / \sqrt{z_0^*}$. Then it is clear that $\mathbf{z}'\Delta\mathbf{z} = 1$.

Let $\mathbf{Q}$ be the square root matrix of Δ. Let $\{\mathbf{V}_1,\cdots,\mathbf{V}_p\}$ be an orthonormal basis of the null space of $\mathbf{Qz}$; i.e., the $(p+1)\times p$-matrix $\mathbf{V}=(\mathbf{V}_1,\cdots,\mathbf{V}_p)$ satisfies

$$\mathbf{V}'\mathbf{V}=\mathbf{I}_p \text{ and } \mathbf{V}'(\mathbf{Qz})=\mathbf{0}.$$

Define $(p+1)\times p$-matrix $\mathbf{Z}=\mathbf{Q}^{-1}\mathbf{V}$. Then it is clear that

$$(\mathbf{z},\mathbf{Z})'\Delta(\mathbf{z},\mathbf{Z})=\mathbf{I}_{p+1}. \tag{5.30}$$

Define $\mathbf{N}=(\mathbf{z},\mathbf{Z})^{-1}\Delta^{-1}(\hat{\beta}_2-\beta_2-\hat{\beta}_1+\beta_1)/\sigma$. It follows directly from (5.30) that $\mathbf{N}\sim\mathcal{N}_{p+1}(\mathbf{0},\mathbf{I})$ and so $\mathbf{T}=\mathbf{N}/(\hat{\sigma}/\sigma)\sim\mathcal{T}_{p+1,\nu}$.

Now define

$$\mathbf{w}=(w_0,w_1,\cdots,w_p)'=(\mathbf{z},\mathbf{Z})'\Delta\mathbf{x}. \tag{5.31}$$

Then $w_0=\mathbf{z}'\Delta\mathbf{x}=1/\sqrt{z_0^*}$ from the definition of $\mathbf{z}$. From (5.31), it is clear that there is a one-to-one correspondence between $\mathbf{w}_{(0)}=(w_1,\cdots,w_p)'$ and $\mathbf{x}_{(0)}=(x_1,\cdots,x_p)'$. Furthermore, when $\mathbf{x}_{(0)}$ varies over $\mathcal{X}_e$ given by (5.29), the corresponding $\mathbf{w}_{(0)}$ forms the set

$$\mathcal{W}_e=\left\{\mathbf{w}_{(0)}:\ \|\mathbf{w}\|\le\sqrt{\kappa(1+a^2)}\right\} \tag{5.32}$$

due to (5.30).

Throughout Section 5.4, $\theta^*=\arccos(1/\sqrt{1+a^2})$ and $\mathbf{T}^*=(\mathbf{z},\mathbf{Z})^{-1}\Delta^{-1}(\hat{\beta}_2-\hat{\beta}_1)/\hat{\sigma}=(T_0^*,\mathbf{T}_{(0)}^{*\prime})'$. The preparation is now complete for the construction of confidence bands over $\mathcal{X}_e$.

5.4.1 *Two-sided hyperbolic band*

A two-sided hyperbolic simultaneous confidence band for $\mathbf{x}'\beta_2-\mathbf{x}'\beta_1$ over the covariate region $\mathcal{X}_e$ in (5.28) or the equivalent (5.29) has the form

$$\mathbf{x}'\beta_2-\mathbf{x}'\beta_1\in\mathbf{x}'\hat{\beta}_2-\mathbf{x}'\hat{\beta}_1\pm c\hat{\sigma}\sqrt{\mathbf{x}'\Delta\mathbf{x}}\ \ \forall\,\mathbf{x}_{(0)}\in\mathcal{X}_e, \tag{5.33}$$

where c is a critical constant chosen so that the simultaneous confidence level of the band is equal to $1-\alpha$.

The simultaneous confidence level of this band is given by $P\{S<c\}$ where

$$\begin{aligned}
S &= \sup_{\mathbf{x}_{(0)}\in\mathcal{X}_e}\frac{|\mathbf{x}'(\hat{\beta}_2-\beta_2-\hat{\beta}_1+\beta_1)|}{\hat{\sigma}\sqrt{\mathbf{x}'\Delta\mathbf{x}}}\\
&= \frac{\sup_{\mathbf{x}_{(0)}\in\mathcal{X}_e}\left|((\mathbf{z},\mathbf{Z})'\Delta\mathbf{x})'\left((\mathbf{z},\mathbf{Z})^{-1}\Delta^{-1}(\hat{\beta}_2-\beta_2-\hat{\beta}_1+\beta_1)/\hat{\sigma}\right)\right|}{\sqrt{((\mathbf{z},\mathbf{Z})'\Delta\mathbf{x})'\left((\mathbf{z},\mathbf{Z})'\Delta\mathbf{x}\right)}}\\
&= \sup_{\mathbf{w}_{(0)}\in\mathcal{W}_e}\frac{|\mathbf{w}'\mathbf{T}|}{\|\mathbf{w}\|} \qquad (5.34)\\
&= \sup_{\mathbf{v}\in\mathcal{V}_e}\frac{|\mathbf{v}'\mathbf{T}|}{\|\mathbf{v}\|} \qquad (5.35)
\end{aligned}$$

where the second equality above follows directly from (5.30), $\mathcal{W}_e$ is given in (5.32), and

$$\begin{aligned}
\mathcal{V}_e &= \left\{\mathbf{v} = u\mathbf{w} = (v_0, \cdots, v_p)' : u > 0, \mathbf{w}_{(0)} \in \mathcal{W}_e\right\} \\
&= \left\{\mathbf{v} : \|\mathbf{v}\| \le v_0 \sqrt{\kappa(1+a^2)/w_0}\right\} \\
&= \left\{\mathbf{v} : \|\mathbf{v}\| \le v_0 \sqrt{1+a^2}\sqrt{\kappa z_0^*}\right\} \\
&= \left\{\mathbf{v} : \|\mathbf{v}\| \le v_0 \sqrt{1+a^2}\right\} \qquad (5.36)
\end{aligned}$$

by noting that $w_0 = 1/\sqrt{z_0^*}$ and $\kappa z_0^* = 1$.

From the expression of S in (5.35) and the expression of $\mathcal{V}_e$ in (5.36), the results of Section 3.4.1 can be used to provide an expression for $P\{S < c\}$. Specifically, it follows from (3.28) and (3.34) that

$$\begin{aligned}
P\{S < c\} &= \int_0^{\theta^*} 2k \sin^{p-1}\theta d\theta \cdot F_{p+1,\nu}\left(\frac{c^2}{p+1}\right) \\
&+ \int_0^{\pi/2-\theta^*} 2k \sin^{p-1}(\theta+\theta^*) \cdot F_{p+1,\nu}\left(\frac{c^2}{(p+1)\cos^2\theta}\right) d\theta \qquad (5.37)
\end{aligned}$$

where $\theta^* = \arccos(1/\sqrt{1+a^2})$.

The confidence band (5.33) can also be used to test the hypotheses in (5.2):

$$\begin{aligned}
\text{reject } H_0 &\iff \sup_{\mathbf{x}_{(0)} \in \mathcal{X}_e} \frac{|\mathbf{x}'(\hat{\beta}_2 - \hat{\beta}_1)|}{\hat{\sigma}\sqrt{\mathbf{x}'\Delta\mathbf{x}}} > c \qquad (5.38) \\
&\iff \mathbf{x}'\mathbf{0} \text{ is outside the band (5.33) for at least one } \mathbf{x}_{(0)} \in \mathcal{X}_e.
\end{aligned}$$

It is clear that the size of this test is equal to α if the simultaneous confidence level of the band (5.33) is equal to $1-\alpha$. It is also noteworthy that this test requires only that the two regression models hold over the region $\mathcal{X}_e$.

The p-value of the test (5.38) can be calculated in the following way. Let $\hat{\beta}_1$, $\hat{\beta}_2$ and $\hat{\sigma}$ be the estimates of β_1, β_2 and σ respectively calculated based on the observed data, and so the observed value of the test statistic in (5.38) is given by

$$\begin{aligned}
S^* &= \sup_{\mathbf{x}_{(0)} \in \mathcal{X}_e} \frac{|\mathbf{x}'(\hat{\beta}_1 - \hat{\beta}_2)|}{\hat{\sigma}\sqrt{\mathbf{x}'\Delta\mathbf{x}}} \\
&= \sup_{\mathbf{v} \in \mathcal{V}_e} \frac{|\mathbf{v}'\mathbf{T}^*|}{\|\mathbf{v}\|} \\
&= \begin{cases} \|\mathbf{T}^*\| & if |T_0^*| \ge \frac{1}{a}\|\mathbf{T}_{(0)}^*\| \\ \frac{|T_0^*| + a\|\mathbf{T}_{(0)}^*\|}{\sqrt{1+a^2}} & \text{otherwise} \end{cases} \qquad (5.39)
\end{aligned}$$

where $\mathbf{T}^* = (\mathbf{z},\mathbf{Z})^{-1}\Delta^{-1}(\hat{\beta}_2 - \hat{\beta}_1)/\hat{\sigma} = (T_0^*, \mathbf{T}_{(0)}^{*\prime})'$ from the derivation of (5.35), and the expressions in (5.39) follow from Lemma 3.5. The observed p-value is then given by

$$\mathrm{P}\{S > S^*\} = 1 - \mathrm{P}\{S < S^*\}$$

where the probability $\mathrm{P}\{S < S^*\}$ can be computed by using expression (5.37) but with c replaced with S^*. For the special case of $a = \infty$, the test (5.38) is just the partial F test (5.6) or (5.7).

5.4.2 *One-sided hyperbolic band*

A lower hyperbolic simultaneous confidence band for $\mathbf{x}'\beta_2 - \mathbf{x}'\beta_1$ over the covariate region $\mathcal{X}_e$ in (5.28) has the form

$$\mathbf{x}'\beta_2 - \mathbf{x}'\beta_1 > \mathbf{x}'\hat{\beta}_2 - \mathbf{x}'\hat{\beta}_1 - c\hat{\sigma}\sqrt{\mathbf{x}'\Delta\mathbf{x}} \ \ \forall\, \mathbf{x}_{(0)} \in \mathcal{X}_e, \tag{5.40}$$

where c is a critical constant chosen so that the simultaneous confidence level of the band is equal to $1-\alpha$.

A similar argument as in the two-sided case of the last section shows that the simultaneous confidence level of this band is given by $P\{S < c\}$ where

$$\begin{aligned} S &= \sup_{\mathbf{x}_{(0)} \in \mathcal{X}_e} \frac{\mathbf{x}'(\hat{\beta}_2 - \beta_2 - \hat{\beta}_1 + \beta_1)}{\hat{\sigma}\sqrt{\mathbf{x}'\Delta\mathbf{x}}} \\ &= \sup_{\mathbf{v} \in \mathcal{V}_e} \frac{\mathbf{v}'\mathbf{T}}{\|\mathbf{v}\|} \end{aligned} \tag{5.41}$$

where $\mathbf{T}$ and $\mathcal{V}_e$ are the same as in the last section.

From the expressions (3.26) and (3.42), the confidence level $\mathrm{P}\{S < c\}$, assuming $c > 0$, is given by

$$\begin{aligned} \mathrm{P}\{S < c\} &= \int_0^{\frac{\pi}{2}} k\sin^{p-1}(\theta + \theta^*) \cdot F_{p+1,v}\left(\frac{c^2}{(p+1)\cos^2\theta}\right) d\theta \\ &+ \int_0^{\theta^*} k\sin^{p-1}\theta d\theta \cdot F_{p+1,v}\left(c^2/(p+1)\right) + \int_0^{\frac{\pi}{2}-\theta^*} k\sin^{p-1}\theta d\theta. \end{aligned} \tag{5.42}$$

The confidence band (5.40) can also be used to test the hypotheses in (5.18):

$$\begin{aligned} \text{reject } H_0 &\iff \sup_{\mathbf{x}_{(0)} \in \mathcal{X}_e} \frac{\mathbf{x}'(\hat{\beta}_2 - \hat{\beta}_1)}{\hat{\sigma}\sqrt{\mathbf{x}'\Delta\mathbf{x}}} > c \qquad (5.43) \\ &\iff \mathbf{x}'\mathbf{0} \text{ is outside the band (5.40) for at least one } \mathbf{x}_{(0)} \in \mathcal{X}_e. \end{aligned}$$

It is clear that the size of this test is equal to α if the simultaneous confidence level of the band (5.40) is equal to $1-\alpha$.

The p-value of the test (5.43) can be calculated in the following way. Let $\hat{\beta}_1$,

$\hat{\beta}_2$ and $\hat{\sigma}$ be the estimates of β_1, β_2 and σ, respectively, and so the observed value of the test statistic in (5.43) is given by

$$\begin{aligned} S^* &= \sup_{\mathbf{x}_{(0)} \in \mathcal{X}_e} \frac{\mathbf{x}'(\hat{\beta}_1 - \hat{\beta}_2)}{\hat{\sigma}\sqrt{\mathbf{x}'\Delta\mathbf{x}}} \\ &= \sup_{\mathbf{v} \in \mathcal{V}_e} \frac{\mathbf{v}'\mathbf{T}^*}{\|\mathbf{v}\|} \\ &= \begin{cases} \|\mathbf{T}^*\| & if T_0^* \geq \frac{1}{a}\|\mathbf{T}^*_{(0)}\| \\ \frac{T_0^* + a\|\mathbf{T}^*_{(0)}\|}{\sqrt{1+a^2}} & \text{otherwise} \end{cases} \end{aligned} \tag{5.44}$$

where the espressions in (5.44) follow from Lemma 3.8.

The observed p-value is then given by

$$\mathrm{P}\{S > S^*\} = 1 - \mathrm{P}\{S < S^*\}.$$

When $S^* \geq 0$, the probability $\mathrm{P}\{S < S^*\}$ can be computed by using expression (5.42) but with c replaced with S^*. However, when $c < 0$, expression (5.42) is no longer valid since all the derivations in Section 3.4.2 assume $c > 0$. Next we derive an expression for $\mathrm{P}\{S < S^*\}$ for $S^* < 0$. It follows from (5.41) and Lemma 3.8 that

$$\begin{aligned} & \mathrm{P}\{S < S^*\} = \mathrm{P}\left\{ \sup_{\mathbf{v} \in \mathcal{V}_e} \frac{\mathbf{v}'\mathbf{T}}{\|\mathbf{v}\|} \leq S^* \right\} \\ = \;& \mathrm{P}\{\mathbf{T} \in \mathcal{V}_e,\ \|\mathbf{T}\| \leq S^*\} + \mathrm{P}\left\{ \mathbf{T} \notin \mathcal{V}_e,\ \frac{T_0 + a\|\mathbf{T}_{(0)}\|}{\sqrt{1+a^2}} \leq S^* \right\}. \end{aligned} \tag{5.45}$$

Now the first probability in (5.45) is clearly equal to zero since $S^* < 0$, and the second probability in (5.45) is equal to

$$\mathrm{P}\left\{ \theta_{\mathbf{T}1} \in (\theta^*, \theta^* + \pi/2),\ \frac{T_0 + a\|\mathbf{T}_{(0)}\|}{\sqrt{1+a^2}} \leq S^* \right\} \tag{5.46}$$

$$+ \quad \mathrm{P}\left\{ \theta_{\mathbf{T}1} \in [\theta^* + \pi/2, \pi],\ \frac{T_0 + a\|\mathbf{T}_{(0)}\|}{\sqrt{1+a^2}} \leq S^* \right\}. \tag{5.47}$$

The probability in (5.46) is equal to

$$\begin{aligned} & \mathrm{P}\left\{ \theta_{\mathbf{T}1} \in (\theta^*, \theta^* + \pi/2),\ \frac{R_{\mathbf{T}} \cos\theta_{\mathbf{T}1} + aR_{\mathbf{T}} \sin\theta_{\mathbf{T}1}}{\sqrt{1+a^2}} \leq S^* \right\} \\ = \;& \mathrm{P}\{\theta_{\mathbf{T}1} \in (\theta^*, \theta^* + \pi/2),\ R_{\mathbf{T}} \cos(\theta_{\mathbf{T}1} - \theta^*) \leq S^*\} \\ = \;& 0 \end{aligned}$$

since $S^* < 0$ and $\cos(\theta_{\mathbf{T}1} - \theta^*) > 0$ for $\theta_{\mathbf{T}1} \in (\theta^*, \theta^* + \pi/2)$. The probability in

(5.47) is equal to

$$
\begin{aligned}
& \mathrm{P}\{\theta_{\mathbf{T}1} \in [\theta^* + \pi/2, \pi],\ R_{\mathbf{T}} \cos(\theta_{\mathbf{T}1} - \theta^*) \le S^*\} \\
= & \int_{\theta^*+\pi/2}^{\pi} k \sin^{p-1}\theta \cdot \mathrm{P}\left\{ R_{\mathbf{T}} \ge \frac{S^*}{\cos(\theta - \theta^*)} \right\} d\theta \\
= & \int_{\pi/2}^{\pi-\theta^*} k \sin^{p-1}(\theta + \theta^*) \left(1 - F_{p+1,\nu}\left(\frac{S^{*2}}{(p+1)\cos^2\theta} \right) \right) d\theta \quad (5.48)
\end{aligned}
$$

by using the density of $\theta_{\mathbf{T}}$ in (3.30) and the cdf of $R_{\mathbf{T}}$ in (3.31). Expression (5.48) is used for computing $\mathrm{P}\{S < S^*\}$ when $S^* < 0$.

For the special case of $a = +\infty$ (and so $\theta^* = \pi/2$), the confidence band (5.40) becomes the lower hyperbolic band over the whole covariate space in (5.10). From (5.44), the observed statistic is given by

$$
S^* = \begin{cases} \|\mathbf{T}^*\| & \text{if } T_0^* \ge 0 \\ \|\mathbf{T}^*_{(0)}\| & \text{if } T_0^* < 0 \end{cases}
$$

which is always non-negative. So expression (5.42) can be used to compute the p-value, and this results in the p-value formula given in Section 5.2.

Example 5.6 For the house sale price example, if we are interested in the comparison of sale price between the in-town and suburb houses over the covariate region $\mathcal{X}_e$ with $a = 4.0$, which is depicted in Figure 5.9, then one-sided and two-sided confidence bands can be constructed for this purpose. For $\alpha = 0.05$, the critical values are given by 2.997 and 2.755 for the two-sided band (5.33) and one-sided band (5.40), respectively. The p-values are given by 0.0024 and 0.0022 respectively for the two-sided test (5.38) and the one-sided test (5.43). All the computations in this example are done using MATLAB program example0506.m.

5.4.3 *Two-sided constant width band*

Let a two-sided constant width simultaneous confidence band for $\mathbf{x}'\beta_2 - \mathbf{x}'\beta_1$ over the covariate region $\mathcal{X}_e$ in (5.28) be of the form

$$
\mathbf{x}'\beta_2 - \mathbf{x}'\beta_1 \in \mathbf{x}'\hat{\beta}_2 - \mathbf{x}'\hat{\beta}_1 \pm c\hat{\sigma}\sqrt{\kappa(1+a^2)} \ \ \forall\, \mathbf{x}_{(0)} \in \mathcal{X}_e, \tag{5.49}
$$

where c is a critical constant chosen so that the simultaneous confidence level of the band is equal to $1 - \alpha$.

The simultaneous confidence level of this band is given by $P\{S < c\}$ where

$$
\begin{aligned}
S &= \sup_{\mathbf{x}_{(0)} \in \mathcal{X}_e} \frac{|\mathbf{x}'(\hat{\beta}_2 - \beta_2 - \hat{\beta}_1 + \beta_1)|}{\hat{\sigma}\sqrt{\kappa(1+a^2)}} \\
&= \sup_{\mathbf{w}_{(0)} \in \mathcal{W}_e} \frac{|\mathbf{w}'\mathbf{T}|}{\sqrt{\kappa(1+a^2)}} \qquad (5.50)
\end{aligned}
$$

where $\mathbf{T}$ is the same as in expression (5.34), and expression (5.50) follows from an argument similar to the derivation of expression (5.34). Now note that

$$\sup_{\mathbf{w}_{(0)}\in\mathcal{W}_e} |\mathbf{w}'\mathbf{T}| = \sup_{\mathbf{w}_{(0)}\in\mathcal{W}_e} \left|\sqrt{\kappa}T_0 + \mathbf{w}_{(0)}'\mathbf{T}_{(0)}\right|$$
$$\leq \sqrt{\kappa}|T_0| + \sqrt{\kappa a^2}\|\mathbf{T}_{(0)}\| \quad (5.51)$$

since $w_0 = 1/\sqrt{z_0^*} = \sqrt{\kappa}$ and $\|\mathbf{w}_{(0)}\| \leq \sqrt{\kappa a^2}$ for $\mathbf{w}_{(0)} \in \mathcal{W}_e$. Furthermore, the inequality in (5.51) becomes equality for $\mathbf{w}_{(0)} = sign(T_0)\sqrt{\kappa a^2}\ \mathbf{T}_{(0)}/\|\mathbf{T}_{(0)}\| \in \mathcal{W}_e$. It follows therefore from (5.50) that

$$S = \frac{|T_0| + a\|\mathbf{T}_{(0)}\|}{\sqrt{1+a^2}}. \quad (5.52)$$

The result of Section 3.5.1 can now be applied to this expression of S to derive an expression for $\mathrm{P}\{S < c\}$ and, specifically, we have

$$\mathrm{P}\{S < c\} = \int_{-\theta^*}^{\pi/2-\theta^*} 2k\sin^{p-1}(\theta+\theta^*)F_{p+1,v}\left(\frac{c^2}{(p+1)\cos^2\theta}\right)d\theta. \quad (5.53)$$

Again the confidence band (5.49) can be used to test the hypotheses in (5.2):

$$\text{reject } H_0 \iff \sup_{\mathbf{x}_{(0)}\in\mathcal{X}_e} \frac{|\mathbf{x}'(\hat{\beta}_2 - \hat{\beta}_1)|}{\hat{\sigma}\sqrt{\kappa(1+a^2)}} > c \quad (5.54)$$
$$\iff \mathbf{x}'\mathbf{0} \text{ is outside the band (5.49) for at least one } \mathbf{x}_{(0)} \in \mathcal{X}_e.$$

It is clear that the size of this test is equal to α if the simultaneous confidence level of the band (5.49) is equal to $1-\alpha$.

The p-value of the test (5.54) can be calculated in the following way. Let $\hat{\beta}_1$, $\hat{\beta}_2$ and $\hat{\sigma}$ be the estimates of β_1, β_2 and σ, respectively, and so the observed value of the test statistic in (5.54) is given by

$$S^* = \sup_{\mathbf{x}_{(0)}\in\mathcal{X}_e} \frac{|\mathbf{x}'(\hat{\beta}_1 - \hat{\beta}_2)|}{\hat{\sigma}\sqrt{\kappa(1+a^2)}}$$
$$= \frac{|T_0^*| + a\|\mathbf{T}_{(0)}^*\|}{\sqrt{1+a^2}}. \quad (5.55)$$

The observed p-value is then given by

$$\mathrm{P}\{S > S^*\} = 1 - \mathrm{P}\{S < S^*\}$$

where the probability $\mathrm{P}\{S < S^*\}$ can be computed by using expression (5.53) but with c replaced with S^*.

5.4.4 *One-sided constant width band*

Let a lower constant width simultaneous confidence band for $\mathbf{x}'\beta_2 - \mathbf{x}'\beta_1$ over the covariate region $\mathcal{X}_e$ in (5.28) be of the form

$$\mathbf{x}'\beta_2 - \mathbf{x}'\beta_1 > \mathbf{x}'\hat{\beta}_2 - \mathbf{x}'\hat{\beta}_1 - c\hat{\sigma}\sqrt{\kappa(1+a^2)} \;\; \forall\, \mathbf{x}_{(0)} \in \mathcal{X}_e, \tag{5.56}$$

where c is a critical constant chosen so that the simultaneous confidence level of the band is equal to $1-\alpha$.

Similar to the two-sided case, the simultaneous confidence level of this band is given by $P\{S < c\}$ where

$$\begin{aligned} S &= \sup_{\mathbf{x}_{(0)} \in \mathcal{X}_e} \frac{\mathbf{x}'(\hat{\beta}_2 - \beta_2 - \hat{\beta}_1 + \beta_1)}{\hat{\sigma}\sqrt{\kappa(1+a^2)}} \\ &= \sup_{\mathbf{w}_{(0)} \in \mathcal{W}_e} \frac{\mathbf{w}'\mathbf{T}}{\sqrt{\kappa(1+a^2)}} \end{aligned} \tag{5.57}$$

where $\mathbf{T}$ is the same as in the last section. Now note that

$$\begin{aligned} \sup_{\mathbf{w}_{(0)} \in \mathcal{W}_e} \mathbf{w}'\mathbf{T} &= \sup_{\mathbf{w}_{(0)} \in \mathcal{W}_e} \left(\sqrt{\kappa}T_0 + \mathbf{w}_{(0)}'\mathbf{T}_{(0)}\right) \\ &\leq \sqrt{\kappa}T_0 + \sqrt{\kappa a^2}\|\mathbf{T}_{(0)}\| \end{aligned} \tag{5.58}$$

since $w_0 = 1/\sqrt{z_0^*} = \sqrt{\kappa}$ and $\|\mathbf{w}_{(0)}\| \leq \sqrt{\kappa a^2}$ for $\mathbf{w}_{(0)} \in \mathcal{W}_e$. Furthermore, the inequality in (5.58) becomes equality for $\mathbf{w}_{(0)} = \sqrt{\kappa a^2}\,\mathbf{T}_{(0)}/\|\mathbf{T}_{(0)}\| \in \mathcal{W}_e$. It follows therefore from (5.57) that

$$S = \frac{T_0 + a\|\mathbf{T}_{(0)}\|}{\sqrt{1+a^2}}. \tag{5.59}$$

The result of Section 3.5.2 can now be applied to this expression of S to provide an expression for $\mathrm{P}\{S < c\}$. Specifically, for $c > 0$, we have

$$\begin{aligned} \mathrm{P}\{S < c\} &= \int_{-\theta^*}^{\pi/2} k\sin^{p-1}(\theta+\theta^*) F_{p+1,\nu}\left(\frac{c^2}{(p+1)\cos^2\theta}\right) d\theta \\ &\quad + \int_{\theta^*+\pi/2}^{\pi} k\sin^{p-1}\theta\, d\theta. \end{aligned} \tag{5.60}$$

Again the confidence band (5.56) can be used to test the hypotheses in (5.18):

$$\begin{aligned} \text{reject } H_0 &\iff \sup_{\mathbf{x}_{(0)} \in \mathcal{X}_e} \frac{\mathbf{x}'(\hat{\beta}_2 - \hat{\beta}_1)}{\hat{\sigma}\sqrt{\kappa(1+a^2)}} > c \qquad (5.61) \\ &\iff \mathbf{x}'\mathbf{0} \text{ is outside the band (5.56) for at least one } \mathbf{x}_{(0)} \in \mathcal{X}_e. \end{aligned}$$

It is clear that the size of this test is equal to α if the simultaneous confidence level of the band (5.56) is equal to $1-\alpha$.

The p-value of the test (5.61) can be calculated in the following way. Let $\hat{\beta}_1$, $\hat{\beta}_2$ and $\hat{\sigma}$ be the estimates of β_1, β_2 and σ respectively, and so the observed value of the test statistic in (5.61) is given by

$$S^* = \sup_{\mathbf{x}_{(0)} \in \mathcal{X}_e} \frac{\mathbf{x}'(\hat{\beta}_1 - \hat{\beta}_2)}{\hat{\sigma}\sqrt{\kappa(1+a^2)}} = \frac{T_0^* + a\|\mathbf{T}_{(0)}^*\|}{\sqrt{1+a^2}}. \quad (5.62)$$

The observed p-value is then given by

$$P\{S > S^*\} = 1 - P\{S < S^*\}.$$

When $S^* \geq 0$, the probability $P\{S < S^*\}$ can be computed by using expression (5.60) but with c replaced with S^*. However, when $c < 0$, expression (5.60) is no longer valid and next we derive an expression for $P\{S < S^*\}$ when $S^* < 0$. It follows from expression (5.59) that

$$\begin{aligned} P\{S < S^*\} = P\left\{\frac{T_0 + a\|\mathbf{T}_{(0)}\|}{\sqrt{1+a^2}} \leq S^*\right\} & \\ = \; P\{R_\mathbf{T}\cos(\theta_{\mathbf{T}1} - \theta^*) \leq S^*\} & \\ = \; P\{\theta_{\mathbf{T}1} \in [0, \theta^* + \pi/2), R_\mathbf{T}\cos(\theta_{\mathbf{T}1} - \theta^*) \leq S^*\} & \quad (5.63) \\ + P\{\theta_{\mathbf{T}1} \in [\theta^* + \pi/2, \pi], R_\mathbf{T}\cos(\theta_{\mathbf{T}1} - \theta^*) \leq S^*\}. & \quad (5.64) \end{aligned}$$

Now the probability in (5.63) is clearly equal to zero since $S^* < 0$ and $\cos(\theta_{\mathbf{T}1} - \theta^*) > 0$ for $\theta_{\mathbf{T}1} \in [0, \theta^* + \pi/2)$. The probability in (5.64) is equal to

$$\begin{aligned} & \int_{\theta^*+\pi/2}^{\pi} k\sin^{p-1}\theta \cdot P\left\{R_\mathbf{T} \geq \frac{S^*}{\cos(\theta - \theta^*)}\right\} d\theta \\ = & \int_{\pi/2}^{\pi-\theta^*} k\sin^{p-1}(\theta + \theta^*)\left(1 - F_{p+1,\nu}\left(\frac{S^{*2}}{(p+1)\cos^2\theta}\right)\right) d\theta \quad (5.65) \end{aligned}$$

by using the density function of $\theta_\mathbf{T}$ in (3.30) and the cdf of $R_\mathbf{T}$ in (3.31). Expression (5.65) is used for computing $P\{S < S^*\}$ when $S^* < 0$.

Example 5.7 For the house sale price example with $a = 4.0$ and $\alpha = 0.05$, the critical values are given by 2.773 and 2.594 respectively for the two-sided band (5.49) and one-sided band (5.56). The p-values are given by 0.0013 and 0.0014 respectively for the two-sided test (5.54) and the one-sided test (5.61). All the computations in this example are done using MATLAB program example0507.m.

5.5 Assessing the equivalence of two models

There are usually two very different purposes in the comparison of two regression models. One purpose is to show, hopefully at least, that the two models are different. This can be achieved by testing the hypotheses in (5.2): if the null hypothesis H_0 is rejected then the two regression models can be claimed to be statistically significantly different. The confidence band corresponding to the test provides further information on where and by how much the two models differ. The second purpose, opposite to the first one, is to establish that the two models have only negligible differences and so are equivalent for practical purposes. Testing the hypotheses in (5.2) is not suitable for this second purpose, since the non-rejection of H_0 does not imply that the two models are close to each other. The confidence bands discussed so far in this chapter do provide information on the magnitude of the difference between the two models and therefore can be used for the second purpose. But more efficient approaches are provided in this section for establishing the equivalence of two regression models, while the confidence bands in the last three sections are more suitable for detecting the differences between the two models.

5.5.1 Non-superiority

Regression model one $m_1(\mathbf{x},\beta_1) = \mathbf{x}'\beta_1$ is defined to be non-superior to regression model two $m_2(\mathbf{x},\beta_2) = \mathbf{x}'\beta_2$ over a given covariate region $\mathbf{x}_{(0)} \in \mathcal{X}$ if

$$m_1(\mathbf{x},\beta_1) < m_2(\mathbf{x},\beta_2) + \delta \;\; \forall \mathbf{x}_{(0)} \in \mathcal{X}, \text{ i.e., } \max_{\mathbf{x}_{(0)} \in \mathcal{X}} (m_1(\mathbf{x},\beta_1) - m_2(\mathbf{x},\beta_2)) < \delta$$

where $\delta > 0$ is a pre-specified threshold value. Therefore model one is non-superior to model two if and only if the amount by which model one is higher than model two is strictly less than the threshold δ at any point $\mathbf{x}_{(0)} \in \mathcal{X}$. Although one purpose of introducing non-superiority (and non-inferiority in the next subsection) is to establish the equivalence of the two models, non-superiority and non-inferiority may be of interest in their own right. For example, QT/QTc studies are increasingly used by drug developers and regulatory agencies to assess the safety of a new drug; see e.g., Li *et al.* (2004) and Ferber (2005). Liu, Bretz, Hayter and Wynn (2009) argued that the essence of the problem involved is to establish that the response curve over time corresponding to the new treatment is non-superior to that corresponding to a control treatment.

It is clear that non-superiority of model one to model two can be deduced from an upper simultaneous confidence band on $m_1(\mathbf{x},\beta_1) - m_2(\mathbf{x},\beta_2)$ over $\mathbf{x}_{(0)} \in \mathcal{X}$. Now we construct a special upper simultaneous confidence band for this purpose. Let $U_P(\mathbf{Y}_1,\mathbf{Y}_2,\mathbf{x})$ be a $1-\alpha$ level pointwise upper confidence limit on $m_1(\mathbf{x},\beta_1) - m_2(\mathbf{x},\beta_2)$ for each $\mathbf{x}_{(0)} \in \mathcal{X}$, i.e.,

$$P\{m_1(\mathbf{x},\hat{\beta}_1) - m_2(\mathbf{x},\hat{\beta}_2) \leq U_P(\mathbf{Y}_1,\mathbf{Y}_2,\mathbf{x})\} \geq 1-\alpha \;\; \forall\, \mathbf{x}_{(0)} \in \mathcal{X}. \qquad (5.66)$$

Then the following theorem provides an upper simultaneous confidence band on $m_1(\mathbf{x}, \beta_1) - m_2(\mathbf{x}, \beta_2)$ over the region $\mathbf{x}_{(0)} \in \mathcal{X}$.

Theorem 5.1 Under the notations above, we have

$$P\left\{ m_1(\mathbf{x}, \beta_1) - m_2(\mathbf{x}, \beta_2) \leq \max_{\mathbf{x}_{(0)} \in \mathcal{X}} U_P(\mathbf{Y}_1, \mathbf{Y}_2, \mathbf{x}) \ \forall\, \mathbf{x}_{(0)} \in \mathcal{X} \right\} \geq 1 - \alpha \quad (5.67)$$

irrespective of the values of the unknown parameters β_1, β_2 and σ.

Proof 5.1 *For arbitrary given β_1, β_2 and σ, let $\mathbf{x}^*_{(0)} \in \mathcal{X}$ be such that*

$$m_1(\mathbf{x}^*, \beta_1) - m_2(\mathbf{x}^*, \beta_2) = \max_{\mathbf{x}_{(0)} \in \mathcal{X}} \{m_1(\mathbf{x}, \beta_1) - m_2(\mathbf{x}, \beta_2)\}$$

where $\mathbf{x}^* = (1, \mathbf{x}^{*\prime}_{(0)})'$. We therefore have

$$\begin{aligned} 1 - \alpha &\leq P\{m_1(\mathbf{x}^*, \beta_1) - m_2(\mathbf{x}^*, \beta_2) \leq U_P(\mathbf{Y}_1, \mathbf{Y}_2, \mathbf{x}^*)\} \\ &= P\left\{ \max_{\mathbf{x}_{(0)} \in \mathcal{X}} (m_1(\mathbf{x}, \beta_1) - m_2(\mathbf{x}, \beta_2)) \leq U_P(\mathbf{Y}_1, \mathbf{Y}_2, \mathbf{x}^*) \right\} \\ &\leq P\left\{ \max_{\mathbf{x}_{(0)} \in \mathcal{X}} (m_1(\mathbf{x}, \beta_1) - m_2(\mathbf{x}, \beta_2)) \leq \max_{\mathbf{x}_{(0)} \in \mathcal{X}} U_P(\mathbf{Y}_1, \mathbf{Y}_2, \mathbf{x}) \right\} \\ &= P\left\{ m_1(\mathbf{x}, \beta_1) - m_2(\mathbf{x}, \beta_2) \leq \max_{\mathbf{x}_{(0)} \in \mathcal{X}} U_P(\mathbf{Y}_1, \mathbf{Y}_2, \mathbf{x}) \ \forall\, \mathbf{x}_{(0)} \in \mathcal{X} \right\} \end{aligned}$$

as required. The proof is thus complete. ∎

It is clear from the proof that the specific form $m_i(\mathbf{x}, \beta_i) = \mathbf{x}'\beta_i$ is not necessary for the theorem to hold. Furthermore, the two model functions $m_i(\mathbf{x}, \beta_i)$ may have different forms. All that is required is the pointwise upper confidence limit $U_P(\mathbf{Y}_1, \mathbf{Y}_2, \mathbf{x})$ specified in (5.66). See LBHW (2009) for how to set $U_P(\mathbf{Y}_1, \mathbf{Y}_2, \mathbf{x})$ for general model functions $m_i(\mathbf{x}, \beta_i)$.

For the special case of $m_i(\mathbf{x}, \beta_i) = \mathbf{x}'\beta_i$ considered here, it is clear that the pointwise upper limit can be set as

$$U_P(\mathbf{Y}_1, \mathbf{Y}_2, \mathbf{x}) = \mathbf{x}'\hat{\beta}_1 - \mathbf{x}'\hat{\beta}_2 + t_\nu^\alpha \sqrt{\mathbf{x}'\Delta\mathbf{x}} \quad (5.68)$$

where t_ν^α is the upper α point of a t distribution with ν degrees of freedom. In this case, we point out that the minimum coverage probability in (5.67) over all possible values of the unknown parameters β_1, β_2 and σ is actually equal to $1 - \alpha$, without giving further details. So the confidence band in (5.67) has an exact confidence level $1 - \alpha$. This confidence band is different from the upper confidence bands seen in the last three sections in that the band is a hyper-plane of constant level $\max_{\mathbf{x}_{(0)} \in \mathcal{X}} U_P(\mathbf{Y}_1, \mathbf{Y}_2, \mathbf{x})$ over $\mathbf{x}_{(0)} \in \mathcal{X}$. Since this confidence band requires only the pointwise upper confidence limit $U_P(\mathbf{Y}_1, \mathbf{Y}_2, \mathbf{x})$, its construction is straightforward.

This confidence band can be used to assess the non-superiority of regression model one to regression model two:

$$\text{Claim non-superiority if and only if } \max_{\mathbf{x}_{(0)} \in \mathcal{X}} U_P(\mathbf{Y}_1, \mathbf{Y}_2, \mathbf{x}) < \delta.$$

Alternatively, one may set the hypotheses

$$\begin{aligned} & H_0^S: \max_{\mathbf{x}_{(0)} \in \mathcal{X}} (m_1(\mathbf{x}, \beta_1) - m_2(\mathbf{x}, \beta_2)) \geq \delta \\ \Longleftrightarrow \quad & H_a^S: \max_{\mathbf{x}_{(0)} \in \mathcal{X}} (m_1(\mathbf{x}, \beta_1) - m_2(\mathbf{x}, \beta_2)) < \delta \end{aligned} \tag{5.69}$$

and test these hypotheses by

$$\text{rejecting } H_0^S \text{ if and only if } \max_{\mathbf{x}_{(0)} \in \mathcal{X}} U_P(\mathbf{Y}_1, \mathbf{Y}_2, \mathbf{x}) < \delta. \tag{5.70}$$

The size of this test is α since the test is implied by the $1-\alpha$ confidence band (5.67). For the special case of $\delta = 0$, test (5.70) was proposed by Tsutakawa and Hewett (1978) for the special case of two regression straight lines and by Hewett and Lababidi (1980) and Berger (1984) for multiple linear regression models. But the confidence statement (5.67) is more useful than the dichotomous inferences of the test (5.70) since the upper confidence bound $\max_{\mathbf{x}_{(0)} \in \mathcal{X}} U_P(\mathbf{Y}_1, \mathbf{Y}_2, \mathbf{x})$ provides useful information on the magnitude of $\max_{\mathbf{x}_{(0)} \in \mathcal{X}} (m_1(\mathbf{x}, \beta_1) - m_2(\mathbf{x}, \beta_2))$ regardless whether or not test (5.70) rejects H_0^S.

It is tempting to assess the non-superiority of $m_1(\mathbf{x}, \beta_1)$ to $m_2(\mathbf{x}, \beta_2)$ by using a usual simultaneous upper confidence band on $m_1(\mathbf{x}, \beta_1) - m_2(\mathbf{x}, \beta_2)$ over the region $\mathbf{x}_{(0)} \in \mathcal{X}$ given in the last two sections. Suppose one has the following simultaneous upper confidence band

$$P\{m_1(\mathbf{x}, \beta_1) - m_2(\mathbf{x}, \beta_2) \leq U_S(\mathbf{Y}_1, \mathbf{Y}_2, \mathbf{x}) \ \forall\, \mathbf{x}_{(0)} \in \mathcal{X}\} \geq 1 - \alpha.$$

From this simultaneous confidence band, the non-superiority of $m_1(\mathbf{x}, \beta_1)$ to $m_2(\mathbf{x}, \beta_2)$ can be claimed

$$\text{if and only if } \max_{\mathbf{x}_{(0)} \in \mathcal{X}} U_S(\mathbf{Y}_1, \mathbf{Y}_2, \mathbf{x}) < \delta.$$

Note, however, $U_S(\mathbf{Y}_1, \mathbf{Y}_2, \mathbf{x}) \geq U_P(\mathbf{Y}_1, \mathbf{Y}_2, \mathbf{x})$ for each $\mathbf{x}_{(0)} \in \mathcal{X}$ since $U_S(\mathbf{Y}_1, \mathbf{Y}_2, \mathbf{x})$ is a simultaneous upper band while $U_P(\mathbf{Y}_1, \mathbf{Y}_2, \mathbf{x})$ is a pointwise upper limit. So

$$\max_{\mathbf{x}_{(0)} \in \mathcal{X}} U_S(\mathbf{Y}_1, \mathbf{Y}_2, \mathbf{x}) \geq \max_{\mathbf{x}_{(0)} \in \mathcal{X}} U_P(\mathbf{Y}_1, \mathbf{Y}_2, \mathbf{x})$$

and often the inequality holds strictly. Hence the non-superiority of $m_1(\mathbf{x}, \beta_1)$

to $m_2(\mathbf{x}, \beta_2)$ that can be established from the simultaneous upper band U_S can always be established from the pointwise upper limit U_P, but not vice versa. Furthermore, a usual simultaneous upper confidence band is often hard to construct, involving numerical integration or statistical simulation to find the critical constants as seen in the last two sections, while the construction of the pointwise upper confidence limit is often straightforward. So for the purpose of establishing the non-superiority of $m_1(\mathbf{x}, \beta_1)$ to $m_2(\mathbf{x}, \beta_2)$ one should use the confidence band (5.67).

Example 5.8 For the blood pressure example, if one hopes to show that male blood pressure is not excessively higher than female blood pressure then confidence band (5.67) for assessing non-superiority is suitable for this purpose. For the observed range $\mathcal{X} = [17, 70]$, it is straightforward to compute

$$\max_{\mathbf{x}_{(0)} \in \mathcal{X}} U_P(\mathbf{Y}_1, \mathbf{Y}_2, \mathbf{x}) = 21.058.$$

So 'non-superiority' can be claimed if and only if the threshold δ is set to be larger than 21.058. On the other hand, if $U_S(\mathbf{Y}_1, \mathbf{Y}_2, \mathbf{x})$ is the upper simultaneous confidence band of hyperbolic shape (corresponding to the lower band (5.17)) then

$$\max_{\mathbf{x}_{(0)} \in \mathcal{X}} U_S(\mathbf{Y}_1, \mathbf{Y}_2, \mathbf{x}) = 23.525.$$

Hence 'non-superiority' can be claimed from this simultaneous upper band if and only if the threshold δ is set to be larger than 23.525, though it is very unlikely that δ will be set as large as 23.525 considering the range of the observed Y values. All the computations in this example are done using MATLAB program example0508.m.

5.5.2 *Non-inferiority*

Regression model one $m_1(\mathbf{x}, \beta_1)$ is defined to be non-inferior to regression model two $m_2(\mathbf{x}, \beta_2)$ over a given covariate region $\mathbf{x}_{(0)} \in \mathcal{X}$ if

$$m_1(\mathbf{x}, \beta_1) > m_2(\mathbf{x}, \beta_2) - \delta \;\; \forall \mathbf{x}_{(0)} \in \mathcal{X}, \text{ i.e., } \min_{\mathbf{x}_{(0)} \in \mathcal{X}} (m_1(\mathbf{x}, \beta_1) - m_2(\mathbf{x}, \beta_2)) > -\delta$$

where $\delta > 0$ is a pre-specified threshold value. Therefore model one is non-inferior to model two if and only if the amount by which model one is lower than model two is strictly less than the threshold δ at any point $\mathbf{x}_{(0)} \in \mathcal{X}$. It is clear from the definitions of non-superiority and non-inferiority that, over the region $\mathbf{x}_{(0)} \in \mathcal{X}$, the non-inferiority of $m_1(\mathbf{x}, \beta_1)$ to $m_2(\mathbf{x}, \beta_2)$ is just the non-superiority of $m_2(\mathbf{x}, \beta_2)$ to $m_1(\mathbf{x}, \beta_1)$. So by interchanging the positions of the two models, we immediately have the following simultaneous lower confidence band for $m_1(\mathbf{x}, \beta_1) - m_2(\mathbf{x}, \beta_2)$ over $\mathbf{x}_{(0)} \in \mathcal{X}$ from Theorem 5.1.

Theorem 5.2 Suppose $L_P(\mathbf{Y}_1, \mathbf{Y}_2, \mathbf{x})$ be a $1-\alpha$ level pointwise lower confidence limit on $m_1(\mathbf{x}, \beta_1) - m_2(\mathbf{x}, \beta_2)$ for each $\mathbf{x}_{(0)} \in \mathcal{X}$, i.e.,

$$\mathrm{P}\{m_1(\mathbf{x}, \beta_1) - m_2(\mathbf{x}, \beta_2) \geq L_P(\mathbf{Y}_1, \mathbf{Y}_2, \mathbf{x})\} \geq 1-\alpha \ \forall\, \mathbf{x}_{(0)} \in \mathcal{X}. \tag{5.71}$$

Then we have

$$\mathrm{P}\left\{ m_1(\mathbf{x}, \beta_1) - m_2(\mathbf{x}, \beta_2) \geq \min_{\mathbf{x}_{(0)} \in \mathcal{X}} L_P(\mathbf{Y}_1, \mathbf{Y}_2, \mathbf{x}) \ \forall\, \mathbf{x}_{(0)} \in \mathcal{X} \right\} \geq 1-\alpha \tag{5.72}$$

regardless the values of the unknown parameters β_1, β_2 and σ.

For the case of $m_i(\mathbf{x}, \beta_i) = \mathbf{x}'\beta_i$ considered here, it is clear that the pointwise lower limit can be set as

$$L_P(\mathbf{Y}_1, \mathbf{Y}_2, \mathbf{x}) = \mathbf{x}'\hat{\beta}_1 - \mathbf{x}'\hat{\beta}_2 - t_\nu^\alpha \sqrt{\mathbf{x}'\Delta\mathbf{x}} \tag{5.73}$$

where t_ν^α is the upper α point of a t distribution with ν degrees of freedom.

This confidence band can be used to assess the non-inferiority of regression model one to regression model two:

$$\text{Claim non-inferiority if and only if } \min_{\mathbf{x}_{(0)} \in \mathcal{X}} L_P(\mathbf{Y}_1, \mathbf{Y}_2, \mathbf{x}) > -\delta.$$

Alternatively, one may set the hypotheses

$$H_0^I: \min_{\mathbf{x}_{(0)} \in \mathcal{X}} (m_1(\mathbf{x}, \beta_1) - m_2(\mathbf{x}, \beta_2)) \leq -\delta$$

$$\Longleftrightarrow \ H_a^I: \min_{\mathbf{x}_{(0)} \in \mathcal{X}} (m_1(\mathbf{x}, \beta_1) - m_2(\mathbf{x}, \beta_2)) > -\delta \tag{5.74}$$

and test these hypotheses by

$$\text{rejecting } H_0^I \text{ if and only if } \min_{\mathbf{x}_{(0)} \in \mathcal{X}} L_P(\mathbf{Y}_1, \mathbf{Y}_2, \mathbf{x}) > -\delta. \tag{5.75}$$

The size of this test is α since the test is implied by the $1-\alpha$ confidence band (5.72). But the lower confidence band (5.72) is more informative than the test (5.75).

For reasons similar to those pointed out in the non-superiority case, the lower confidence band (5.72) is easier to construct and has a greater chance to establish non-inferiority than a usual simultaneous lower confidence band on $m_1(\mathbf{x}, \beta_1) - m_2(\mathbf{x}, \beta_2)$ over the region $\mathcal{X}$ given in the last two sections.

5.5.3 Equivalence of two models

Two regression models $m_1(\mathbf{x}, \beta_1)$ and $m_2(\mathbf{x}, \beta_2)$ are defined to be equivalent over the region $\mathbf{x}_{(0)} \in \mathcal{X}$ if

$$\max_{\mathbf{x}_{(0)} \in \mathcal{X}} |m_1(\mathbf{x}, \beta_1) - m_2(\mathbf{x}, \beta_2)| < \delta$$

where $\delta > 0$ is a pre-specified threshold value. Therefore the two models are equivalent if and only if the difference between the two models is strictly less than the threshold δ at any point $\mathbf{x}_{(0)} \in \mathcal{X}$. It is clear from the definitions of non-superiority, non-inferiority and equivalence that, over the region $\mathbf{x}_{(0)} \in \mathcal{X}$, the two models are equivalent if and only if one model is neither superior nor inferior to the other model. So the equivalence of the two models can be claimed by establishing both the non-superiority and non-inferiority of one model to the other model.

We set up the equivalence hypotheses

$$\begin{aligned} & H_0^E \ : \ \max_{\mathbf{x}_{(0)} \in \mathcal{X}} |m_1(\mathbf{x}, \beta_1) - m_2(\mathbf{x}, \beta_2)| \geq \delta \\ \iff \ & H_a^E \ : \ \max_{\mathbf{x}_{(0)} \in \mathcal{X}} |m_1(\mathbf{x}, \beta_1) - m_2(\mathbf{x}, \beta_2)| < \delta. \end{aligned} \tag{5.76}$$

So the equivalence of the two models can be claimed if and only if the null hypothesis H_0^E is rejected. From (5.70) the non-superiority null hypothesis H_0^S is rejected if and only if

$$\max_{\mathbf{x}_{(0)} \in \mathcal{X}} U_P(\mathbf{Y}_1, \mathbf{Y}_2, \mathbf{x}) < \delta,$$

and from (5.75) the non-inferiority null hypothesis H_0^I is rejected if and only if

$$\min_{\mathbf{x}_{(0)} \in \mathcal{X}} L_P(\mathbf{Y}_1, \mathbf{Y}_2, \mathbf{x}) > -\delta.$$

The aggregation of these two individual tests gives a test of equivalence: rejecting H_0^E if and only if

$$-\delta < \min_{\mathbf{x}_{(0)} \in \mathcal{X}} L_P(\mathbf{Y}_1, \mathbf{Y}_2, \mathbf{x}) \ \text{ and } \ \max_{\mathbf{x}_{(0)} \in \mathcal{X}} U_P(\mathbf{Y}_1, \mathbf{Y}_2, \mathbf{x}) < \delta. \tag{5.77}$$

This test is also an intersection-union test of Berger (1982, see Appendix D) derived from the two size α tests (5.70) and (5.75), and so its size is still α in fact. To see this, notice that

$$H_0^E = H_0^S \cup H_0^I \ \text{ and } \ H_a^E = H_a^S \cap H_a^I$$

and that the rejection region of H_0^E in (5.77) is the intersection of the rejection region of H_0^S in (5.70) and the rejection region of H_0^I in (5.75).

It is clear from the discussions of this section that a usual simultaneous two-sided confidence band for $\mathbf{x}'\beta_1 - \mathbf{x}'\beta_2$ over $\mathbf{x}_{(0)} \in \mathcal{X}$, such as the band in (5.14), has a much smaller chance than the test (5.77) for establishing the equivalence of the two models since the critical constant of a usual two-sided simultaneous confidence band is even larger than that of the one-sided simultaneous confidence band of a similar shape.

Table 5.3: *Data from Ruberg and Stegeman (1991)*

Batch 1		Batch 2	
Y (% drug content)	x (years)	Y(% drug content)	x (years)
100.4	0.014	100.7	0.022
100.3	0.280	100.6	0.118
99.7	0.514	100.3	0.272
99.2	0.769	99.9	0.566
98.9	1.074	98.6	1.165
98.2	1.533	97.6	2.022
97.3	2.030	96.4	3.077
95.7	3.071		
94.5	4.049		

Example 5.9 The purpose of a drug stability study is to study how the content of a drug degrades over time in order to produce the expiry date of the drug. Because a same drug is usually manufactured in several batches, one needs to assess whether different batches have similar degradation profiles; different batches that have similar degradation profiles can be pooled to produce one expiry date.

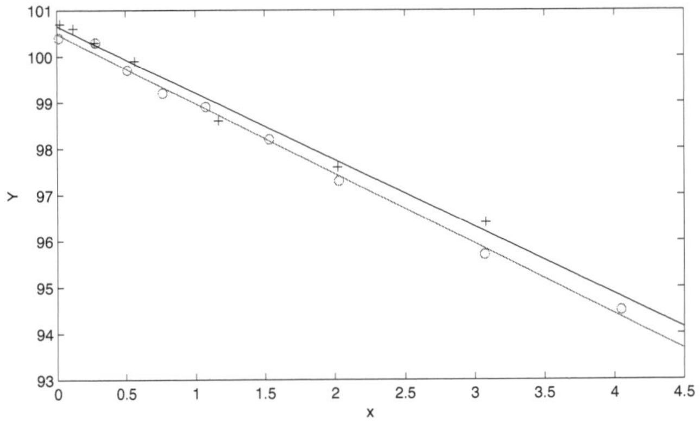

Figure 5.10: *The two samples and fitted regression lines for the drug stability example*

Two samples of observations on drug content from two batches of the same drug are given in Table 5.3 and plotted in Figure 5.10; these are taken from Ruberg and Stegeman (1991). It is clear that the degradation profiles of the two batches are well described by two regression straight lines. The fitted regression lines are calculated to be $Y_1 = 100.656 - 1.449x_1$ and $Y_2 = 100.489 - 1.515x_1$, the pooled variance estimate is $\hat{\sigma}^2 = 0.029$ with $\nu = 12$ degrees of freedom, and the matrix $\Delta = (X_1^T X_1)^{-1} + (X_2^T X_2)^{-1}$ is given by $\Delta_{11} = 0.5407$, $\Delta_{12} = \Delta_{21} = -0.2334$ and

$\Delta_{22} = 0.1961$. We want to assess whether the two regression lines $\mathbf{x}'\beta_1$ and $\mathbf{x}'\beta_2$ are equivalent.

Let $\mathcal{X} = [0,2]$. It is straightforward to compute

$$\max_{\mathbf{x}_{(0)} \in \mathcal{X}} U_P(\mathbf{Y}_1, \mathbf{Y}_2, \mathbf{x}) = 0.489 \text{ and } \min_{\mathbf{x}_{(0)} \in \mathcal{X}} L_P(\mathbf{Y}_1, \mathbf{Y}_2, \mathbf{x}) = -0.056.$$

Therefore the equivalence of the two regression lines over $x \in [0,2]$ can be claimed if and only if the threshold δ is larger than 0.489; in drug stability studies δ is sometimes set as large as 5.

Finally it is noteworthy that the test statistic of the partial F-test in (5.6) for this example is equal to 4.354, with p-value 0.038. So the null hypothesis in (5.2) of the coincidence of the two regression lines is rejected at $\alpha = 5\%$. But this has little relevance to the assessment of equivalence of the two regression lines.

For assessing the equivalence of the two linear regression models given in (5.1) over a given region $\mathcal{X}$ of the covariates $\mathbf{x}_{(0)}$, Liu, Hayter and Wynn (2007) provided an exact $1-\alpha$ level upper confidence bound on

$$\max_{\mathbf{x}_{(0)} \in \mathcal{X}} |\mathbf{x}'\beta_2 - \mathbf{x}'\beta_1| \Big/ \sigma,$$

and this upper confidence bound is based on the F statistic in (5.6) or (5.7). Note, however, the maximum difference between the two models over $\mathbf{x} \in \mathcal{X}$ is measured in units of σ and so not directly comparable with the measurement given in (5.76) above when σ is unknown. Furthermore, when σ is assumed to be known, the upper bound in Liu, Hayter and Wynn (2007) is equivalent to a $1-\alpha$ two-sided simultaneous confidence band on $\mathbf{x}'\beta_2 - \mathbf{x}'\beta_1$ over $\mathbf{x}_{(0)} \in \mathcal{X}$ and therefore, as pointed out above, has a smaller chance to establish equivalence than the test given above; on the other hand the upper confidence bound provides useful information on the magnitude of $\max_{\mathbf{x}_{(0)} \in \mathcal{X}} |\mathbf{x}'\beta_2 - \mathbf{x}'\beta_1|$ while the test here provides only inferences on whether or not H_0^E is rejected. Bofinger (1999) gave an upper confidence bound on $\max_{\mathbf{x}_{(0)} \in \mathcal{X}} |\mathbf{x}'\beta_2 - \mathbf{x}'\beta_1|$ but only for two regression straight lines.

Another interesting problem is the construction of simultaneous confidence bands when the error variances σ^2 of the two models are different. Hayter *et al.* (2007) considered this problem but only for simple regression lines.

6

Comparison of More Than Two Regression Models

The comparison of more than two regression models is characteristically different from the comparison of two regression models considered in the last chapter in many aspects as we shall see in this chapter. Simultaneous confidence bands can be more advantageous than the usual partial F test approach in providing directly quantitative information on the differences among the models of interest.

Suppose k linear regression models describe the relationship between the same response variable Y and the same set of covariates $x_1, \cdots, x_p$ corresponding to k treatments or groups. Specifically let the ith linear regression model be given by

$$\mathbf{Y}_i = \mathbf{X}_i \boldsymbol{\beta}_i + \mathbf{e}_i, \quad i = 1, \cdots, k \tag{6.1}$$

where $\mathbf{Y}_i = (y_{i,1}, \cdots, y_{i,n_i})'$ is a vector of random observations, $\mathbf{X}_i$ is a $n_i \times (p+1)$ full column-rank design matrix with first column $(1, \cdots, 1)'$ and $l (\geq 2)$th column $(x_{i,1,l-1}, \cdots, x_{i,n_i,l-1})'$, $\boldsymbol{\beta}_i = (\beta_{i,0}, \cdots, \beta_{i,p})'$, and $\mathbf{e}_i = (e_{i,1}, \cdots, e_{i,n_i})'$ with all the $\{e_{i,j}, j = 1, \cdots, n_i, i = 1, \cdots, k\}$ being iid $N(0, \sigma^2)$ random variables, where σ^2 is an unknown parameter. Since $\mathbf{X}_i'\mathbf{X}_i$ is non-singular, the least squares estimator of $\boldsymbol{\beta}_i$ is given by $\hat{\boldsymbol{\beta}}_i = (\mathbf{X}_i'\mathbf{X}_i)^{-1}\mathbf{X}_i'\mathbf{Y}_i, i = 1, \cdots, k$. Let

$$\hat{\sigma}^2 = \sum_{i=1}^{k} \|\mathbf{Y}_i - \mathbf{X}_i\hat{\boldsymbol{\beta}}_i\|^2 / \nu$$

be the pooled error mean square with $\nu = \sum_{i=1}^{k}(n_i - p - 1)$ degrees of freedom. Then $\hat{\sigma}^2$ is independent of the $\hat{\boldsymbol{\beta}}_i$'s.

6.1 Partial F test approach

One frequently encountered problem is the comparison of the k regression models and, in particular, whether all the k regression models are the same or not. One common approach to this problem is to test

$$H_0: \boldsymbol{\beta}_1 = \cdots = \boldsymbol{\beta}_k \quad \text{against} \quad H_a: \text{not } H_0 \tag{6.2}$$

and the partial F test (1.7) for this case can be implemented in the following way. Define $k-1$ dummy variables

$$z_i = \begin{cases} 1 & \text{if } Y \text{ is from the model } i \\ 0 & \text{if } Y \text{ is not from the model } i \end{cases}, \quad i = 1, \cdots, k-1$$

and set up an overall model for the aggregated data from the k individual models as

$$Y = \mathbf{x}'\mathbf{d}_1 + z_1\mathbf{x}'\mathbf{d}_2 + \cdots + z_{k-1}\mathbf{x}'\mathbf{d}_k + e = (\mathbf{x}', z_1\mathbf{x}', \cdots, z_{k-1}\mathbf{x}') \begin{pmatrix} \mathbf{d}_1 \\ \vdots \\ \mathbf{d}_k \end{pmatrix} + e \quad (6.3)$$

where $\mathbf{x} = (1, x_1, \cdots, x_p)' \in \Re^{p+1}$ as before, $\mathbf{d}_1 = \beta_k$ and $\mathbf{d}_i = \beta_{i-1} - \beta_k$ for $i = 2, \cdots, k$. This overall model implies the k individual models specified in (6.1):

$$Y = \mathbf{x}'(\mathbf{d}_1 + \mathbf{d}_2) + e = \mathbf{x}'\beta_1 + e \;\; \text{for a } Y \text{ from the individual model 1,}$$

$$\cdots\cdots$$

$$Y = \mathbf{x}'\mathbf{d}_1 + e = \mathbf{x}'\beta_k + e \;\; \text{for a } Y \text{ from the individual model k.}$$

In terms of the overall model (6.3), the hypotheses in (6.2) of the coincidence of the k individual models become

$$H_0 : \mathbf{d}_2 = \cdots = \mathbf{d}_k = \mathbf{0} \quad \text{against} \quad H_a : \text{not } H_0$$

Under H_0 the overall model (6.3) is reduced to the simpler model

$$Y = \mathbf{x}'\mathbf{d}_1 + e. \quad (6.4)$$

So the partial F test (1.7) rejects H_0 if and only if

$$\frac{(SS_R \text{ of model } (6.3) - SS_R \text{ of model } (6.4))\,/q}{\text{MS residual of model } (6.3)} > f^{\alpha}_{q,v} \quad (6.5)$$

where $f^{\alpha}_{q,v}$ is the upper α point of an F distribution with $q = (k-1)(p+1)$ and $v = \sum_{i=1}^{k}(n_i - (p+1))$ degrees of freedom.

The equivalent form (1.11) of this partial F test can in this case be calculated in the following way. Represent all the observations from the k individual models $(\mathbf{Y}_i, \mathbf{X}_i), i = 1, \cdots, k$ by

$$\mathbf{Y} = \mathbf{X}\beta + \mathbf{e} \quad (6.6)$$

where

$$\mathbf{Y} = \begin{pmatrix} \mathbf{Y}_1 \\ \vdots \\ \mathbf{Y}_k \end{pmatrix}, \; \mathbf{X} = \begin{pmatrix} \mathbf{X}_1 & & \\ & \ddots & \\ & & \mathbf{X}_k \end{pmatrix}, \; \beta = \begin{pmatrix} \beta_1 \\ \vdots \\ \beta_k \end{pmatrix}, \; \mathbf{e} = \begin{pmatrix} \mathbf{e}_1 \\ \vdots \\ \mathbf{e}_k \end{pmatrix}.$$

From this overall model, the least squares estimators of β and σ are clearly given by $\hat{\beta} = (\hat{\beta}_1', \cdots, \hat{\beta}_k')'$ and $\hat{\sigma}$, respectively. Under this model, the hypotheses in (6.2) become

$$H_0 : \; \mathbf{A}\beta = \mathbf{0} \quad \text{against} \quad H_a : \; \text{not } H_0$$

where the $q \times k(p+1)$ partition matrix $\mathbf{A}$ is given by

$$\mathbf{A} = \begin{pmatrix} \mathbf{I}_{p+1} & -\mathbf{I}_{p+1} & & & \\ & \mathbf{I}_{p+1} & -\mathbf{I}_{p+1} & & \\ & & \ddots & \ddots & \\ & & & \mathbf{I}_{p+1} & -\mathbf{I}_{p+1} \end{pmatrix}.$$

The partial F test (1.11) therefore rejects H_0 if and only if

$$\frac{(\mathbf{A}\hat{\beta})'\{\mathbf{A}(\mathbf{X}'\mathbf{X})^{-1}\mathbf{A}'\}^{-1}(\mathbf{A}\hat{\beta})}{\hat{\sigma}^2} > c^2 \tag{6.7}$$

where $c^2 = qf^{\alpha}_{q,\nu}$.

The inferences that can be drawn from the partial F test (6.5) or the equivalent form (6.7) are that if H_0 is rejected then the k regression models are deemed to be not the same, and that if H_0 is not rejected then there is not sufficient statistical evidence to conclude that some regression models are different. Even though we can claim that not all the k models are the same when H_0 is rejected, no information is available from the partial F test on which models may be different and by how much. When H_0 is not rejected, little can be said since non-rejection of H_0 cannot be taken as the k regression models are the same. So, whether H_0 is rejected or not, little useful information is available from this approach of hypotheses testing.

Example 6.1 As discussed in Example 5.9, the purpose of a drug stability study is to study how a drug degrades over time in order to produce the expiry date of the drug. Because the same drug is usually manufactured in several batches, one needs to assess whether different batches have similar degradation profiles; different batches that have similar degradation profiles can be pooled to produce one expiry date.

Random observations on drug content from six batches of the same drug are given in Table 6.1 and taken from Ruberg and Stegeman (1991). One can plot the data points and see clearly that the degradation profile of drug content (Y–% drug content) over time (x_1 – years) for each of the six batches is well described by a regression straight line. The fitted regression straight lines for the six batches are given respectively by $Y_1 = 100.49 - 1.515x_1$, $Y_2 = 100.66 - 1.449x_1$, $Y_3 = 100.26 - 1.687x_1$, $Y_4 = 100.45 - 1.393x_1$, $Y_5 = 100.45 - 1.999x_1$, $Y_6 = 99.98 - 1.701x_1$, and the pooled variance estimate is $\hat{\sigma}^2 = 0.0424$ with $\nu = 25$ degrees of freedom. The R^2-value of the fit is equal to 0.987.

If one applies the partial F test (6.5) or (6.7) to this problem in order to assess whether the $k = 6$ regression straight lines $\mathbf{x}'\beta_i$ $(i = 1, \cdots, 6)$ are similar, then the F statistic is calculated to be equal to 7.469 with p-value 2.3×10^{-5}. Hence H_0 is highly significant, that is, it is highly unlikely that the six regression straight lines are all equal. But the test does not provide information on which regression straight lines are significantly different and which are not significantly different, and on the magnitude of differences between different regression lines.

Table 6.1: *Data from Ruberg and Stegeman (1991)*

Batch 1		Batch 2		Batch 3		Batch 4		Batch 5		Batch 6	
Y	x_1	Y	x_1	Y	x_1	Y	x_1	Y	x_1	Y	x_1
100.4	0.014	100.7	0.022	100.2	0.025	100.4	0.066	100.5	0.011	100.1	0.011
100.3	0.280	100.6	0.118	99.7	0.275	100.0	0.343	99.8	0.310	99.5	0.310
99.7	0.514	100.3	0.272	99.2	0.574	99.5	0.533	99.1	0.624	98.5	0.624
99.2	0.769	99.9	0.566	99.0	0.797	99.3	0.802	98.4	1.063	98.4	1.063
98.9	1.074	98.6	1.165	98.8	1.041	99.3	1.033				
98.2	1.533	97.6	2.022	96.4	2.058	98.2	1.538				
97.3	2.030	96.4	3.077	96.2	2.519						
95.7	3.071										
94.5	4.049										

As another example, Dielman (2001, pp. 399) is interested in the comparison of Meddicorp Company's sales in three regions, south ($i = 1$), west ($i = 2$) and midwest ($i = 3$) of a country. For territories in each region, the sales value in \$1000 (Y) is modelled by linear regression on two explanatory variables: x_1, the advertising cost (in \$1000) Meddicorp spent in a territory, and x_2, the total amount of bonus (in \$1000) paid out in a territory. The data set given in Dielman (2001, Table 7.2) is adapted in Table 6.2 here.

Following the usual model fitting procedure, it is clear that a linear regression model of sales Y on advertising cost x_1 and bonus x_2 fits the observed data well for each of the three regions. The fitted regression models for the three regions are given respectively by $Y_1 = 281.55 + 1.1039x_1 + 1.025x_2$, $Y_2 = -28.14 + 0.9620x_1 + 2.647x_2$ and $Y_3 = 660.59 + 1.5115x_1 - 0.080x_2$, and the pooled variance estimate is $\hat{\sigma}^2 = 2905.4481$ with $\nu = 16$ degrees of freedom. The R^2-value of the fits is equal to 0.963.

If one applies the partial F test (6.5) or (6.7) to this problem in order to assess whether the $k = 3$ regression models $\mathbf{x}'\beta_i$ $(i = 1, \cdots, k)$ are similar, then the F statistic is calculated to be equal to 7.726 with p-value 5×10^{-4}. Hence it is highly unlikely that the three regression models are all equal. But again that is all that can be said from the test.

6.2 Hyperbolic confidence bands for all contrasts

In this section we consider simultaneous confidence bands that bound all the contrasts of the k regression models. From these confidence bands, information on the magnitude of any contrast of the regression models is available. These confidence bands can be regarded as extensions of simultaneous confidence intervals for all the contrasts among the population means studied by Scheffé (1953, 1959). We focus in this section on hyperbolic bands since constant-width or other shape bands are not available in the literature.

Table 6.2: *Data adapted from Dielman (2001, Table 7.2, page 401)*

Obs	Y	x_1	x_2	Region
1	963.50	374.27	230.98	1
2	893.00	408.50	236.28	1
3	1057.25	414.31	271.57	1
4	1071.50	446.86	305.69	1
5	1078.25	489.59	238.41	1
6	1040.00	453.39	235.63	1
7	1159.25	524.56	292.87	1
8	1183.25	448.42	291.20	2
9	1122.50	500.56	271.38	2
10	1045.25	440.86	249.68	2
11	1102.25	487.79	232.99	2
12	1225.25	537.67	272.20	2
13	1202.75	535.17	268.27	2
14	1294.25	486.03	309.85	2
15	1124.75	499.15	272.55	2
16	1419.50	517.88	282.17	3
17	1547.75	637.60	321.16	3
18	1580.00	635.72	294.32	3
19	1304.75	484.18	332.64	3
20	1552.25	618.07	261.80	3
21	1508.00	612.21	266.64	3
22	1564.25	601.46	277.44	3
23	1634.75	585.10	312.35	3
24	1467.50	540.17	291.03	3
25	1583.75	583.85	289.29	3

6.2.1 *Conservative bands over the whole space*

First we derive, from the partial F test given in Section 6.1, a set of conservative simultaneous confidence bands for all the contrasts of the k regression models

$$\sum_{i=1}^{k} d_i \mathbf{x}' \beta_i \ \text{ for all } \mathbf{d} = (d_1, \cdots, d_k)' \in \mathcal{D}$$

where $\mathcal{D}$ is the set of all contrasts

$$\mathcal{D} = \left\{ \mathbf{d} = (d_1, \cdots, d_k)' \in \Re^k : \sum_{i=1}^{k} d_i = 0 \right\}.$$

Note that for $\mathbf{A}$ given above the expression (6.7), $\mathbf{A}\hat{\beta} \sim \mathcal{N}_q(\mathbf{A}\beta, \sigma^2 \mathbf{A}(\mathbf{X}'\mathbf{X})^{-1}\mathbf{A}')$ and so

$$P\left\{ (\mathbf{A}(\hat{\beta} - \beta))' \left(\mathbf{A}(\mathbf{X}'\mathbf{X})^{-1}\mathbf{A}'\right)^{-1} (\mathbf{A}(\hat{\beta} - \beta)) \ < \ c^2 \hat{\sigma}^2 \right\} = 1 - \alpha \qquad (6.8)$$

where $c = \sqrt{qf^{\alpha}_{q,\nu}}$. Of course this last probability statement can be regarded as a confidence region statement for β, which underlies the partial F test (6.7).

Let $q \times q$ matrix $\mathbf{P}$ be the unique square root matrix of $(\mathbf{A}(\mathbf{X}'\mathbf{X})^{-1}\mathbf{A}')^{-1}$. Then we have

$$\begin{aligned}
1-\alpha &= \mathrm{P}\left\{ (\mathbf{A}(\hat{\beta}-\beta))' \left(\mathbf{A}(\mathbf{X}'\mathbf{X})^{-1}\mathbf{A}'\right)^{-1} (\mathbf{A}(\hat{\beta}-\beta)) < c^2\hat{\sigma}^2 \right\} \\
&= \mathrm{P}\left\{ (\mathbf{PA}(\hat{\beta}-\beta))' (\mathbf{PA}(\hat{\beta}-\beta)) < c^2\hat{\sigma}^2 \right\} \\
&= \mathrm{P}\left\{ -c\hat{\sigma} < \frac{\mathbf{v}'}{\|\mathbf{v}\|}\mathbf{PA}(\hat{\beta}-\beta) < c\hat{\sigma} \ \forall\, \mathbf{v} \in \Re^q \right\} \qquad (6.9) \\
&= \mathrm{P}\left\{ -c\hat{\sigma} < \frac{\mathbf{w}'\mathbf{A}(\hat{\beta}-\beta)}{\sqrt{\mathbf{w}'\mathbf{A}(\mathbf{X}'\mathbf{X})^{-1}\mathbf{A}'\mathbf{w}}} < c\hat{\sigma} \ \forall\, \mathbf{w} \in R^q \right\} \qquad (6.10)
\end{aligned}$$

where the equality in (6.9) follows directly from the Cauchy-Schwarz inequality as used in the proof of Theorem 3.1, and the equality in (6.10) follows directly from the definition $\mathbf{w} = \mathbf{Pv} \in \Re^q$.

For a $\mathbf{w} \in \Re^q$ of the special form $\mathbf{w} = (d_1\mathbf{x}', (d_1+d_2)\mathbf{x}', \cdots, (d_1+\cdots+d_{k-1})\mathbf{x}')'$ with $\mathbf{d} = (d_1, \cdots, d_k) \in \mathcal{D}$, we have

$$\begin{aligned}
\mathbf{w}'\mathbf{A} &= (d_1\mathbf{x}', \cdots, d_k\mathbf{x}') = \mathbf{x}'(d_1 I_{p+1}, \cdots, d_k I_{p+1}), \\
\mathbf{w}'\mathbf{A}(\hat{\beta}-\beta) &= \sum_{i=1}^{k} d_i(\mathbf{x}'\hat{\beta}_i - \mathbf{x}'\beta_i), \\
\mathbf{w}'\mathbf{A}(\mathbf{X}'\mathbf{X})^{-1}\mathbf{A}'\mathbf{w} &= \sum_{i=1}^{k} d_i^2\mathbf{x}'(\mathbf{X}_i'\mathbf{X}_i)^{-1}\mathbf{x}.
\end{aligned}$$

Therefore the probability statement in (6.10) implies that, with probability at least $1-\alpha$,

$$\sum_{i=1}^{k} d_i\mathbf{x}'\beta_i \in \sum_{i=1}^{k} d_i\mathbf{x}'\hat{\beta}_i \pm c\hat{\sigma}\sqrt{\sum_{i=1}^{k} d_i^2\mathbf{x}'(\mathbf{X}_i'\mathbf{X}_i)^{-1}\mathbf{x}} \quad \forall \mathbf{x}_{(0)} \in \Re^p, \ \forall \mathbf{d} \in \mathcal{D} \qquad (6.11)$$

where $\mathbf{x}_{(0)} = (x_1, \cdots, x_p)'$ as before. This provides a set of simultaneous confidence bands for all the contrasts of the k regression models over $\mathbf{x}_{(0)} \in \Re^p$ with confidence level no less than $1-\alpha$.

For $k = 2$ there is only one confidence band for $\mathbf{x}'\beta_1 - \mathbf{x}'\beta_2$ in (6.11), and it is just the confidence band given in (5.8). So the confidence level is exactly $1-\alpha$. But for $k \geq 3$, which is the focus of this chapter, the confidence bands in (6.11) are conservative since they are derived from (6.10) for only a special class of $\mathbf{w} \in \Re^q$.

6.2.2 *Bands for regression straight lines under the same design*

It is not known to date what critical constant should be in the place of c in (6.11) so that the simultaneous confidence level of the bands in (6.11) is equal to $1-\alpha$

for the general situation of $k \geq 3$ and $p \geq 1$. For the special situation of $p = 1$ and $\mathbf{X}_1 = \cdots = \mathbf{X}_k$, that is, for the comparison of k regression straight lines with the same design matrices, Spurrier (1999) provided the answer to this question, which is given next.

For notational simplicity, denote the only covariate of the k regression lines by x. Since the k regression lines have the same design points, they are denoted by $x_1, \cdots, x_n$ and so the second column of all the design matrices $\mathbf{X}_i$ is given by $(x_1, \cdots, x_n)'$. Let

$$\bar{x} = \frac{1}{n}\sum_{j=1}^{n} x_j, \quad s_{xx}^2 = \sum_{j=1}^{n}(x_j - \bar{x})^2.$$

Then the simultaneous confidence level of the set of confidence bands for all the contrasts of the k regression lines over the whole range of the covariate $x \in (-\infty, \infty)$ in (6.11) is clearly given by

$$P\left\{ \sup_{x \in (-\infty,\infty) \text{ and } \mathbf{d} \in \mathcal{D}} (T_{x,\mathbf{d}}^2) \leq c^2 \right\} \tag{6.12}$$

where

$$T_{x,\mathbf{d}} = \frac{\sum_{i=1}^{k} d_i(\mathbf{x}'\hat{\beta}_i - \mathbf{x}'\beta_i)}{\hat{\sigma}\sqrt{\sum_{i=1}^{k} d_i^2 \mathbf{x}'(X_i'X_i)^{-1}\mathbf{x}}} = \frac{\sum_{i=1}^{k} d_i(\mathbf{x}'\hat{\beta}_i - \mathbf{x}'\beta_i)}{\hat{\sigma}\sqrt{[(1/n) + (x-\bar{x})^2/s_{xx}^2]\sum_{i=1}^{k} d_i^2}}.$$

The next theorem is due to Spurrier (1999).

Theorem 6.1 Let F_{ν_1,ν_2} denote the cumulative distribution function of the F distribution with ν_1 and ν_2 degrees of freedom and let $p_i = \Pi_{j=1}^{i}(2j-1)$ for positive integer i. For odd $k \geq 3$, the probability in (6.12) is equal to

$$\frac{k-1}{2^{(k-1)/2}} F_{k-1,\nu}\left(2c^2/(k-1)\right)$$
$$- \frac{\pi^{1/2} c^{(k-2)} \Gamma((\nu-k-2)/2)}{\Gamma((k-1)/2)\Gamma(\nu/2)\nu^{(k-2)/2}(1+(c^2/\nu))^{(\nu+k-2)/2}} F_{1,\nu+k-2}\left(\frac{c^2(\nu+k-2)}{(c^2+\nu)}\right)$$
$$+ \frac{1}{\Gamma((k-1)/2)2^{(k-1)/2}} \sum_{i=1}^{(k-5)/2} \frac{(k-3-2i)\Gamma((k-1+2i)/2)}{p_{i+1}} F_{k-1+2i,\nu}\left(\frac{2c^2}{(k-1+2i)}\right)$$

where the summation is defined to be zero for $k = 3$ or 5. For even $k \geq 4$, the probability in (6.12) is equal to

$$\left(\Gamma(k/2)2^{(k-2)/2}\left(F_{k,\nu}\left(\frac{c^2}{k}\right) - F_{k-2,\nu}\left(\frac{c^2}{k-2}\right)\right)\right)$$
$$+ \left(\sum_{i=1}^{(k-2)/2} \frac{\Gamma(k-2-i)i}{2^{(k-2)/2-i}\Gamma(k/2-i)} F_{2(k-i-2),\nu}\left(\frac{c^2}{k-i-2}\right)\right) \frac{1}{p_{(k-2)/2}}.$$

So the simultaneous confidence level is a linear combination of F distribution functions, which is easy to compute. The reader is referred to Spurrier (1999) for a proof. Recently Lu and Chen (2009) extended the result of Spurrier (1999) to the situation that the covariate is restricted to a finite interval. Note, however, the assumption of same design points for all the regression lines is crucial for the proofs of both Spurrier (1999) and Lu and Chen (2009).

6.2.3 Bands for regression straight lines under different designs

Note that the confidence bands in (6.11) are over the entire range of the covariates $\mathbf{x}_{(0)} \in \Re^p$. As argued in the previous chapters, a regression model is often a reasonable approximation only over certain range of the covariates. So it is interesting to construct a set of confidence bands for all the contrasts of the k regression models over a given covariate region $\mathbf{x}_{(0)} \in \mathcal{X}$ of the form

$$\sum_{i=1}^{k} d_i \mathbf{x}'\beta_i \in \sum_{i=1}^{k} d_i \mathbf{x}'\hat{\beta}_i \pm c\hat{\sigma}\sqrt{\sum_{i=1}^{k} d_i^2 \mathbf{x}'(\mathbf{X}_i'\mathbf{X}_i)^{-1}\mathbf{x}} \ \forall \mathbf{x}_{(0)} \in \mathcal{X}, \ \forall \mathbf{d} \in \mathcal{D}. \quad (6.13)$$

In this section, we continue to consider the special case of $p = 1$, that is, there is only one covariate x_1 and $\mathcal{X}$ is a given interval (a,b) $(\infty \le a < b \le \infty)$ for covariate x_1. A simulation method is provided in Jamshidian *et al.* (2010) to approximate the exact value of c so that the simultaneous confidence level of the bands in (6.13) is equal to $1-\alpha$. Note that the design matrices of the k regression lines are assumed to be of full column-rank but can be different.

The simultaneous confidence level of the confidence bands in (6.13) is given by $\mathrm{P}\{S \le c\}$, where

$$S = \frac{W}{\hat{\sigma}/\sigma} \quad \text{and} \quad W = \sup_{x_1 \in (a,b),\ \mathbf{d} \in \mathcal{D}} \frac{\left|\sum_{i=1}^{k} d_i \mathbf{x}'(\hat{\beta}_i - \beta_i)\right|}{\sqrt{\mathrm{Var}\left(\sum_{i=1}^{k} d_i \mathbf{x}'(\hat{\beta}_i - \beta_i)\right)}}. \quad (6.14)$$

Here W is independent of the random variable $\hat{\sigma}/\sigma \sim \sqrt{\chi_\nu^2/\nu}$. As before, our approach is to simulate R replicates of S and use the $(1-\alpha)$-quantile of the R simulated values as an approximation to c.

The difficulty in simulating S is the maximization over $\mathbf{d} \in \mathcal{D}$ and $x_1 \in (a,b)$ in (6.14). We first show that the maximum over $\mathbf{d} \in \mathcal{D}$ can be carried out analytically to result in a function of x_1. We then provide an efficient method to maximize this function of x_1 over $x_1 \in (a,b)$. More details are given in Jamshidian *et al.* (2010), to which the reader is referred.

For any $\mathbf{d} \in \mathcal{D}$, we have

$$\sum_{i=1}^{k} d_i \mathbf{x}'(\hat{\beta}_i - \beta_i) = (d_1\mathbf{x}', (d_1+d_2)\mathbf{x}', \cdots, (d_1+\cdots+d_{k-1})\mathbf{x}')\,\mathbf{U},$$

where the random vector $\mathbf{U}$ is given by

$$\mathbf{U} = \begin{pmatrix} \hat{\beta}_1 - \beta_1 - \hat{\beta}_2 + \beta_2 \\ \hat{\beta}_2 - \beta_2 - \hat{\beta}_3 + \beta_3 \\ \vdots \\ \hat{\beta}_{k-1} - \beta_{k-1} - \hat{\beta}_k + \beta_k \end{pmatrix} \sim \mathcal{N}_q(\mathbf{0}, \sigma^2 \Sigma),$$

with Σ being the tri-block-diagonal matrix

$$\Sigma = \begin{pmatrix} \mathbf{V}_1 + \mathbf{V}_2 & -\mathbf{V}_2 & \mathbf{0} & \cdots & \mathbf{0} & \mathbf{0} \\ -\mathbf{V}_2 & \mathbf{V}_2 + \mathbf{V}_3 & -\mathbf{V}_3 & \cdots & \mathbf{0} & \mathbf{0} \\ \mathbf{0} & -\mathbf{V}_3 & \ddots & \ddots & \vdots & \mathbf{0} \\ \vdots & \vdots & \ddots & \ddots & \ddots & \vdots \\ \mathbf{0} & \mathbf{0} & \cdots & -\mathbf{V}_{k-2} & \mathbf{V}_{k-2} + \mathbf{V}_{k-1} & -\mathbf{V}_{k-1} \\ \mathbf{0} & \mathbf{0} & \cdots & \mathbf{0} & -\mathbf{V}_{k-1} & \mathbf{V}_{k-1} + \mathbf{V}_k \end{pmatrix},$$

where $\mathbf{V}_i = (\mathbf{X}_i'\mathbf{X}_i)^{-1}$. Let $\mathbf{Q}$ be the unique square root matrix of Σ. Then

$$\mathbf{Z} = \mathbf{Q}^{-1}\mathbf{U}/\sigma \sim \mathcal{N}_q(\mathbf{0}, \mathbf{I}). \tag{6.15}$$

Let $\mathbf{Q} = (\mathbf{Q}_1, \cdots, \mathbf{Q}_{k-1})$ where each $\mathbf{Q}_i$ is a $q \times (p+1))$ matrix. Since Q is nonsingular, $\mathbf{Q}_1\mathbf{x}, \cdots, \mathbf{Q}_{k-1}\mathbf{x}$ must be $k-1$ linearly independent q-vectors and so the matrix $L_{\mathbf{x}} = (\mathbf{Q}_1\mathbf{x}, \cdots, \mathbf{Q}_{k-1}\mathbf{x})$ is of full column-rank $k-1$ for any $x_1 \in \Re^1$. It is shown in Jamshidian *et al.* (2010) that the maximization over $\mathbf{d} \in \mathcal{D}$ in (6.14) is given by $\sqrt{\mathbf{Z}^T L_x (L_x^T L_x)^{-1} L_x^T \mathbf{Z}}$ and so

$$W = \sup_{x_1 \in (a,b)} \sqrt{\mathbf{Z}' L_x (L_x' L_x)^{-1} L_x' \mathbf{Z}}. \tag{6.16}$$

This completes the maximization over $\mathbf{d} \in \mathcal{D}$ analytically, which works for a general $p \geq 1$ in fact.

Next we solve the maximization over $x_1 \in (a,b)$ in (6.16). For this, it is shown in Jamshidian *et al.* (2010) that $\mathbf{Z}' L_x (L_x' L_x)^{-1} L_x' \mathbf{Z}$ can be written as a rational function $p(x_1)/q(x_1)$, where explicit formulas for the polynomials $p(x_1)$ and $q(x_1)$ are provided. The computation of W reduces therefore to maximization of the rational function $p(x_1)/q(x_1)$ over the interval $x_1 \in (a,b)$ since

$$W = \sup_{x_1 \in (a,b)} \sqrt{p(x_1)/q(x_1)} = \sqrt{\sup_{x_1 \in (a,b)} p(x_1)/q(x_1)}.$$

The global maximum of $p(x_1)/q(x_1)$ over the interval $x_1 \in (a,b)$ is attained at one of the boundary points a and b, or one of the the stationary points of $p(x_1)/q(x_1)$ in the interval (a,b). The stationary points of $p(x_1)/q(x_1)$ are the real roots of the polynomial equation

$$p'(x_1)q(x_1) - p(x_1)q'(x_1) = 0 \tag{6.17}$$

where $p'(x_1)$ and $q'(x_1)$ denote respectively the derivatives of $p(x)$ and $q(x)$. The real roots of this polynomial equation can be easily computed numerically by using, for example, the MATLAB® routine `roots`.

Hence we apply the following algorithm to obtain copies of $S = W/(\hat{\sigma}/\sigma)$, and subsequently an approximation to the critical constant c:

Step 1. Generate one $\mathbf{Z} \sim \mathcal{N}_q(\mathbf{0}, \mathbf{I})$, and obtain the polynomials $p(x_1)$ and $q(x_1)$.

Step 2. Obtain the real roots of the polynomial on the left hand side of (6.17). Denote the real roots in the interval (a,b) by $r_1, \cdots, r_m$.

Step 3. Compute $W = \sqrt{\max_{x_1 \in \{a,b,r_1,\cdots,r_m\}} p(x_1)/q(x_1)}$.

Step 4. Generate a value χ^2 from χ^2_ν and form a realization of S via $W/\sqrt{\chi^2/\nu}$.

Step 5. Repeat Steps 1-4 R times to obtain R replications of S.

Step 6. Obtain the $(1-\alpha)$-quantile of the R replications and use this value as an approximation to c.

The accuracy of this approximation of c can be assessed as before by using the methods provided in Appendix A. A MATLAB program is written to implement this algorithm, which is illustrated using the drug stability example next.

Example 6.2 For the drug stability data set given in Example 6.1, one can construct the simultaneous confidence bands in (6.13) for assessing any contrast among the six regression lines. For $\alpha = 0.05$, (a,b) is equal to the observed range $(0.011, 4.049)$ and the number of simulations is equal to 200,000, the critical value c is calculated to be 3.834. If $(a,b) = (-\infty, \infty)$ then c is computed to be 4.1758. On the other, it is rare nowadays for a pharmaceutical company to seek an expiry date longer than two years. It is therefore sensible to set $(a,b) = (0,2)$ in order to use the simultaneous bands in (6.13) to assess the similarity of the six regression lines over $x_1 \in (0,2)$. In this case, the critical value c is calculated to be 3.805, which is marginally smaller than the critical value c for $(a,b) = (0.011, 4.049)$. The conservative critical value given in (6.11) is equal to 4.729, which is substantially larger than all the three critical values given above.

After the critical value c is computed, one can plot the confidence band in (6.13) for any given contrast among the k regression lines in order to assess the magnitude of that particular contrast. For example, for contrast $\mathbf{d} = (0, 2, -1, 0, 0, -1)'$ with $\alpha = 0.05$ and $(a,b) = (0.0, 2.0)$, the confidence band in (6.13) for this given contrast is plotted in Figure 6.1. From this plot, one can infer that

$$\sum_{i=1}^{k} d_i \mathbf{x}'\beta_i = 2\mathbf{x}'\beta_2 - (\mathbf{x}'\beta_3 + \mathbf{x}'\beta_6)$$

is almost always positive for all $x_1 \in (0,2)$ since the lower part of the band is almost always above zero. One can also get useful information about the magnitude of this contrast from the band.

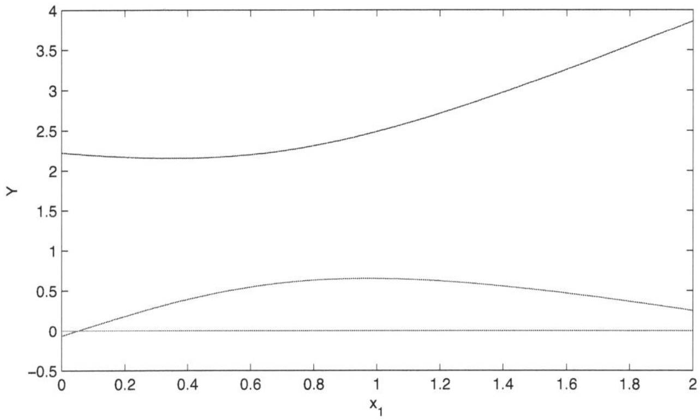

Figure 6.1: *The confidence band for one particular contrast* **d**

One can make an inference about any contrasts of interest from the corresponding confidence bands, and the probability that all the inferences are correct simultaneously is at least $1-\alpha$.

The set of simultaneous confidence bands for all the contrasts in (6.13) can be used to test the hypotheses in (6.2). If the null hypothesis H_0 is true then $\sum_{i=1}^{k} d_i \mathbf{x}'\beta_i = 0$ for all $\mathbf{x}_{(0)} \in \mathcal{X}$ and all $\mathbf{d} \in \mathcal{D}$. So, with probability $1-\alpha$, we have

$$0 \in \sum_{i=1}^{k} d_i \mathbf{x}'\hat{\beta}_i \pm c\hat{\sigma}\sqrt{\sum_{i=1}^{k} d_i^2 \mathbf{x}'(X_i'X_i)^{-1}\mathbf{x}} \;\; \forall \mathbf{x}_{(0)} \in \mathcal{X},\ \forall \mathbf{d} \in \mathcal{D}$$

$$\Longleftrightarrow \quad \sup_{\mathbf{x}_{(0)} \in \mathcal{X}} \sup_{\mathbf{d} \in \mathcal{D}} \frac{|\sum_{i=1}^{k} d_i \mathbf{x}'\hat{\beta}_i|}{\hat{\sigma}\sqrt{\sum_{i=1}^{k} d_i^2 \mathbf{x}'(X_i'X_i)^{-1}\mathbf{x}}} \le c.$$

Therefore the test

$$\text{reject } H_0 \text{ if and only if } \sup_{\mathbf{x}_{(0)} \in \mathcal{X}} \sup_{\mathbf{d} \in \mathcal{D}} \frac{|\sum_{i=1}^{k} d_i \mathbf{x}'\hat{\beta}_i|}{\hat{\sigma}\sqrt{\sum_{i=1}^{k} d_i^2 \mathbf{x}'(X_i'X_i)^{-1}\mathbf{x}}} > c$$

is of size α. The test statistic may be computed using the ideas outlined above for simulating S. As before this test is not as informative as the confidence bands in (6.13). Furthermore, when this test rejects H_0, it may be difficult to identify the contrasts that cause the rejection of H_0 since $\mathcal{D}$ contains infinite contrasts.

6.3 Bands for finite contrasts over rectangular region

In many real problems, one is often interested only in the direct comparison between the k regression models

$$\mathbf{x}'\beta_i - \mathbf{x}'\beta_j \quad \forall\ (i,j) \in \Lambda \tag{6.18}$$

over a chosen covariate region $\mathbf{x}_{(0)} \in \mathcal{X}$. In this section we focus on the rectangular covariate region

$$\mathcal{X}_r = \{\mathbf{x}_{(0)} = (x_1, \cdots, x_p)' : \ a_i \leq x_i \leq b_i \text{ for } i = 1, \cdots, p\}$$

which is one of the most useful covariate regions. The set Λ is a given index set that determines the comparisons of interest. For example, if all pairwise comparisons of the k models are of interest then $\Lambda = \{(i,j) : 1 \leq i \neq j \leq k\}$. If (many-one) comparisons of one particular model, the first model say, with the other models is of interest, then $\Lambda = \{(i,j) : 2 \leq i \leq k, j = 1\}$. If the k regression models involve a certain (e.g. temporal or spatial) ordering then the successive comparisons of the k models may be of interest for which $\Lambda = \{(i, i+1) : 1 \leq i \leq k-1\}$).

There is a rich literature concerning the pairwise, many-one and successive comparisons of several population means. For instance, Tukey (1953) and Hayter (1984) considered two-sided pairwise comparison of k population means while Hayter (1990) and Hayter and Liu (1996) investigated the one-sided pairwise comparison of k population means. Dunnett (1955) discussed the comparison of several means with a control mean, and Hochberg and Marcus (1978), Lee and Spurrier (1995) and Liu *et al.* (2000) studied the successive comparisons of several population means. Miller (1981), Hochberg and Tamhane (1987) and Hsu (1996) provided excellent overview of the work in this area. Wilcox (1987) considered the pairwise comparison of several regression straight lines by constructing a set of $1-\alpha$ simultaneous confidence intervals for all the pairwise differences of the intercepts and all the pairwise differences of the slopes of the regression lines.

The important features of the comparisons in (6.18) are that there are only a finite number of contrasts involved and that the contrasts can be very flexibly chosen to suit the need. In particular, the k regression models may play roles of different importance by suitable choice of Λ. For example, for the many-one comparisons of the second to kth regression models with the first regression model, the first model plays a more important role than the other $k-1$ models. Next we discuss the construction and use of simultaneous confidence bands for the comparisons in (6.18).

6.3.1 *Hyperbolic bands*

A set of hyperbolic simultaneous confidence bands for the comparisons in (6.18) is given in Liu *et al.* (2004) by

$$\mathbf{x}'\beta_i - \mathbf{x}'\beta_j \in \mathbf{x}'\hat{\beta}_i - \mathbf{x}'\hat{\beta}_j \pm c\hat{\sigma}\sqrt{\mathbf{x}'\Delta_{ij}\mathbf{x}} \quad \forall \mathbf{x}_{(0)} \in \mathcal{X}_r \text{ and } \forall (i,j) \in \Lambda \tag{6.19}$$

where c is the critical constant chosen so that the confidence level of this set of simultaneous confidence bands is equal to $1-\alpha$, and $\Delta_{ij} = (\mathbf{X}_i'\mathbf{X}_i)^{-1} + (\mathbf{X}_j'\mathbf{X}_j)^{-1}$.

Note that the confidence level of the bands in (6.19) is given by $P\{S < c\}$ where

$$S = \sup_{(i,j)\in\Lambda} \sup_{\mathbf{x}_{(0)}\in\mathcal{X}_r} \frac{|\mathbf{x}'\left((\hat{\beta}_i - \beta_i) - (\hat{\beta}_j - \beta_j)\right)|}{\hat{\sigma}\sqrt{\mathbf{x}'\Delta_{ij}\mathbf{x}}}.$$

As before we use simulation to calculate the critical value c by simulating a large number of independent replications of S. Spurrier (2002) provided simultaneous confidence bands for all pairwise comparisons and for comparisons with a control of several regression straight lines over the whole covariate range $(-\infty,\infty)$ under certain design matrices restrictions, while Bhargava and Spurrier (2004) constructed simultaneous confidence bands for comparing two regression straight lines with one regression straight line over a finite covariate interval.

Let $\mathbf{P}_{ij}$ be the unique square root matrix of $\Delta_{ij} = (\mathbf{X}_i'\mathbf{X}_i)^{-1} + (\mathbf{X}_j'\mathbf{X}_j)^{-1}$. Let $\mathbf{Z}_i = (\hat{\beta}_i - \beta_i)/\sigma$ $(i = 1,\cdots,k)$ which are independent normal random vectors independent of $\hat{\sigma}$ and have distribution $\mathbf{Z}_i \sim \mathcal{N}_{p+1}(\mathbf{0}, (\mathbf{X}_i'\mathbf{X}_i)^{-1})$. Denote $\mathbf{T}_{ij} = (\mathbf{P}_{ij})^{-1}(\mathbf{Z}_i - \mathbf{Z}_j)/(\hat{\sigma}/\sigma)$, $1 \le i \ne j \le k$. Then S can be expressed as

$$\begin{aligned}
S &= \sup_{(i,j)\in\Lambda} \sup_{\mathbf{x}_{(0)}\in\mathcal{X}_r} \frac{|(\mathbf{P}_{ij}\mathbf{x})'\mathbf{T}_{ij}|}{\sqrt{(\mathbf{P}_{ij}\mathbf{x})'(\mathbf{P}_{ij}\mathbf{x})}} \\
&= \sup_{(i,j)\in\Lambda} \sup_{\mathbf{x}_{(0)}\in\mathcal{X}_r} \frac{|(\mathbf{P}_{ij}\mathbf{x})'\mathbf{T}_{ij}|}{\|\mathbf{P}_{ij}\mathbf{x}\|} \\
&= \sup_{(i,j)\in\Lambda} \max\left\{\|\pi(\mathbf{T}_{ij},\mathbf{P}_{ij},\mathcal{X}_r)\|,\ \|\pi(-\mathbf{T}_{ij},\mathbf{P}_{ij},\mathcal{X}_r)\|\right\} \qquad (6.20)
\end{aligned}$$

where the notation $\pi(\mathbf{T}_{ij},\mathbf{P}_{ij},\mathcal{X}_r)$ is defined in Section 3.2.1 and the equality in (6.20) follows directly from the results in Section 3.2.1. It is clear from (6.20) that the distribution of S does not depend on the unknown parameters β_i and σ.

Simulation of a random realization of the variable S can be implemented in the following way.

Step 0. Determine $\mathbf{P}_{ij}$ for $(i,j) \in \Lambda$.

Step 1. Simulate independent $\mathbf{Z}_i \sim \mathcal{N}_{p+1}(\mathbf{0}, (\mathbf{X}_i'\mathbf{X}_i)^{-1})$ $(i = 1,\cdots,k)$ and $\hat{\sigma}/\sigma \sim \sqrt{\chi^2_\nu/\nu}$.

Step 2. Compute $\mathbf{T}_{ij} = (\mathbf{P}_{ij})^{-1}(\mathbf{Z}_i - \mathbf{Z}_j)/(\hat{\sigma}/\sigma)$ for $(i,j) \in \Lambda$.

Step 3. Compute $\pi(\mathbf{T}_{ij},\mathbf{P}_{ij},\mathcal{X}_r)$ and $\pi(-\mathbf{T}_{ij},\mathbf{P}_{ij},\mathcal{X}_r)$ by using the algorithm given in Appendix B for each $(i,j) \in \Lambda$.

Step 4. Find S from (6.20)

Repeat Steps 1-4 R times to simulate R independent replicates of the random variable S, and set the $(1-\alpha)$-quantile $\hat{c}$ as the estimator of the critical constant c. The accuracy of $\hat{c}$ as a approximation of c can be assessed using the methods provided in Appendix A as before. By simulating a sufficiently large number R of replicates of S, $\hat{c}$ can be as accurate as one requires. Our experiences indicate that $R = 200,000$ provides a $\hat{c}$ that is accurate enough for most practical purposes.

For the special case of $\mathcal{X}_r = \Re^p$, i.e., the simultaneous confidence bands in (6.19) are over the entire space of the covariates, the random variable S has a much simpler form. Note that

$$\sup_{\mathbf{x}_{(0)} \in \Re^p} |(\mathbf{P}_{ij}\mathbf{x})'\mathbf{T}_{ij}| / \parallel \mathbf{P}_{ij}\mathbf{x} \parallel = \parallel \mathbf{T}_{ij} \parallel$$

from the Cauchy-Schwarz inequality. Hence, for $\mathcal{X}_r = \Re^p$,

$$S = \sup_{(i,j) \in \Lambda} \parallel \mathbf{T}_{ij} \parallel$$

which can be easily simulated. Note that $\mathbf{T}_{ij} \sim \mathcal{T}_{p+1,\nu}$ and so $\parallel \mathbf{T}_{ij} \parallel^2$ has a $(p+1)F_{p+1,\nu}$ distribution. But $\mathbf{T}_{ij}$ for different $(i,j) \in \Lambda$ may be correlated, and the distribution of $\{\parallel \mathbf{T}_{ij} \parallel^2, \ (i,j) \in \Lambda\}$ is related to the so-called multivariate chi-square distribution (see Royen, 1995, and the references therein). Even in this simple case, the distribution function of S is difficult to find analytically. So simulation seems a reasonable way to find the distribution or the percentile of S.

Example 6.3 For the drug stability data set given in Example 6.1, in order to assess the similarity of the six regression lines the simultaneous confidence bands for all pairwise comparisons of the six regression lines in (6.19) can be used. As pointed out previously, pharmaceutical companies usually do not seek approval for an expiry date beyond two years nowadays. This leads us to set the range for the covariate 'time' as $x_1 \in [0.0, 2.0]$. We first construct a set of 95% simultaneous confidence bands for all pairwise comparisons of the six regression lines

$$\beta_{i,0} + \beta_{i,1}x_1 - \beta_{j,0} - \beta_{j,1}x_1 \ \in \ \hat{\beta}_{i,0} + \hat{\beta}_{i,1}x_1 - \hat{\beta}_{j,0} - \hat{\beta}_{j,1}x_1 \ \pm \ c\hat{\sigma}\sqrt{\mathbf{x}'\Delta_{ij}\mathbf{x}}$$
$$\text{for } x_1 \in [0.0, 2.0] \text{ and for all } 1 \le i < j \le 6. \qquad (6.21)$$

Based on 200,000 simulations, the critical value c is calculated to be $c = 3.535$.

It is noteworthy that if the all-contrast confidence bands in (6.13) are used to derive the confidence bands for all pairwise comparisons then critical value $c = 3.805$ is used according to Example 6.2. As expected, this critical value is larger than the critical value $c = 3.535$ given above for the simultaneous confidence bands in (6.21). Hence if one is interested only in the pairwise comparison then the confidence bands for pairwise comparison, instead of the confidence bands for all contrasts, should be used.

The comparison between the ith and jth regression lines can now be carried out by using the confidence band

$$\hat{\beta}_{i,0} + \hat{\beta}_{i,1}x_1 - \hat{\beta}_{j,0} - \hat{\beta}_{j,1}x_1 \ \pm \ c\hat{\sigma}\sqrt{\mathbf{x}'\Delta_{ij}\mathbf{x}} \ \text{ for } x_1 \in [0, 2].$$

For example, Figure 6.2 plots this band for $i = 2$ and $j = 1$. Since the confidence

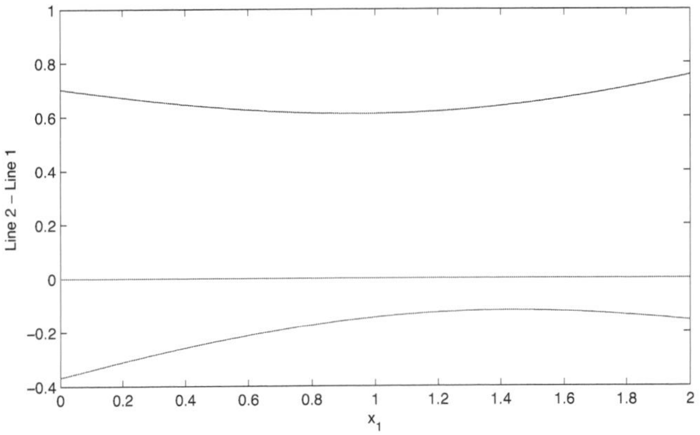

Figure 6.2: *The confidence band for* $\mathbf{x}'\boldsymbol{\beta}_2 - \mathbf{x}'\boldsymbol{\beta}_1$

band contains the zero line for all $x_1 \in (0,2)$, the two regression lines are not significantly different. Furthermore, we can calculate the maximum distance to zero of each band

$$D_{ij} = \max_{x_1 \in [0.0, 2.0]} \left| \hat{\beta}_{i,0} + \hat{\beta}_{i,1}x_1 - \hat{\beta}_{j,0} - \hat{\beta}_{j,1}x_1 \pm c\hat{\sigma}\sqrt{\mathbf{x}'\Delta_{ij}\mathbf{x}} \right|$$

to bound the largest difference between the ith and jth regression lines. For the band plotted in Figure 6.2, it is clear that $D_{2,1}$ is about 0.755 and so the first two regression lines can be declared to be equivalent if the pre-specified threshold of equivalence, δ, is larger than 0.755.

Figure 6.3 plots the confidence band for $i = 3$ and $j = 2$. Since the confidence band does not contain the zero line for all $x_1 \in (0,2)$, the two regression lines are significantly differently. It is clear however that $D_{3,2}$ is about 1.432. Hence the two regression lines can still be declared to be equivalent if the threshold δ is larger than 1.432, even though they are significantly different. If one chooses $\delta = 4.0$, the value used in Ruberg and Hsu (1992), then one can declare that all six regression lines are equivalent by looking at all the pairwise confidence bands. Hence all six batches can be pooled to calculate one expiry date.

Example 6.4 For the sales value data set given in Example 6.1, one may also compute all pairwise comparison simultaneous confidence bands in (6.19) to compare the sales value among the three regions: south, midwest and west. Now suppose that south is Meddicorp's traditional sales base, while midwest and west are two newly expanded sales regions. Hence comparisons of midwest and west with south are of interest. For this purpose, the simultaneous confidence bands for many-one comparison in (6.19) are required. Specifically, we need simultaneous

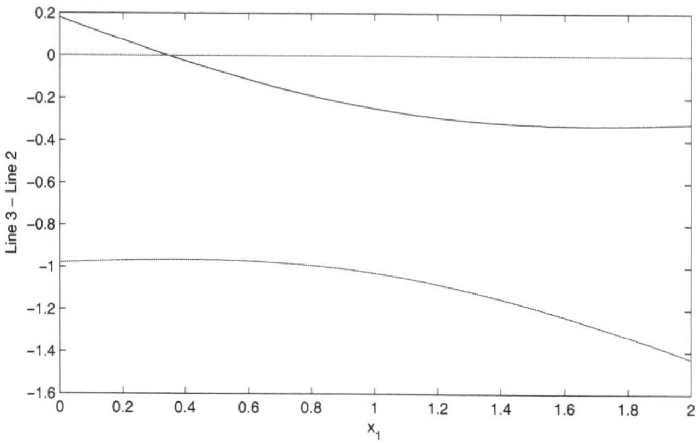

Figure 6.3: *The confidence band for* $\mathbf{x}'\beta_3 - \mathbf{x}'\beta_2$

confidence bands

$$\mathbf{x}'\beta_i - \mathbf{x}'\beta_j \in \mathbf{x}'\hat{\beta}_i - \mathbf{x}'\hat{\beta}_j \pm c\hat{\sigma}\sqrt{\mathbf{x}'\Delta_{ij}\mathbf{x}} \quad \forall \mathbf{x}_{(0)} \in \mathcal{X}_r \text{ and } \forall (i,j) \in \Lambda \qquad (6.22)$$

where $\Lambda = \{(i,j) : i = 2,3 \text{ and } j = 1\}$ and $\mathcal{X}_r$ can be chosen as the observed region of the covariates $x_1 \in [374.27, 637.60]$ and $x_2 \in [230.98, 332.64]$. For $\alpha = 0.05$, the critical constant is calculated to be $c = 3.359$ based on $R = 200,000$ simulations, while the conservative critical value in (6.11) is equal to $c = 4.056$.

Now one can plot the two confidence bands in (6.22), which are given in Figure 6.4 and Figure 6.5, respectively, for $i = 2$ and $i = 3$. From Figure 6.4, the mean sales in midwest and south are not significantly different since the confidence band contains the zero plane over the covariate region. On the other hand, Figure 6.5 indicates that the mean sales in west and south are significantly different since the confidence band excludes the zero plane for a substantial part of the covariate region $\mathcal{X}_r$. More specifically, one can infer that the mean sales in the west are significantly higher than the mean sales in the south over the part of the covariate region $\mathcal{X}_r$ that the confidence band is above the zero plane.

It is clear from the two examples above that one can make more relevant and informative inferences from the simultaneous confidence bands than from the partial F test in Example 6.1. All the simultaneous confidence bands above are two-sided. If one is interested only in one-sided inferences then one-sided simultaneous confidence bands can be constructed accordingly.

Example 6.5 We continue to study Meddicorp's sales value data set given in Example 6.1. Now suppose that one is interested in whether the mean sales in the midwest and west tend to be higher than that in the south. For this purpose,

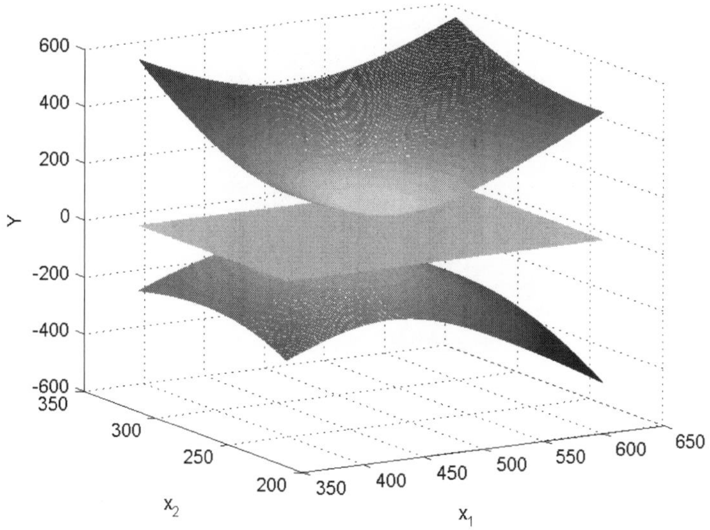

Figure 6.4: *The confidence band for* $\mathbf{x}'\boldsymbol{\beta}_2 - \mathbf{x}'\boldsymbol{\beta}_1$

one-sided simultaneous confidence bands for many-one comparisons are required. Specifically, we need simultaneous confidence bands

$$\mathbf{x}'\boldsymbol{\beta}_i - \mathbf{x}'\boldsymbol{\beta}_j > \mathbf{x}'\hat{\boldsymbol{\beta}}_i - \mathbf{x}'\hat{\boldsymbol{\beta}}_j - c\hat{\sigma}\sqrt{\mathbf{x}'\Delta_{ij}\mathbf{x}} \quad \forall \mathbf{x}_{(0)} \in \mathcal{X}_r \text{ and } \forall (i,j) \in \Lambda \tag{6.23}$$

where $\Lambda = \{(i,j) : i = 2,3 \text{ and } j = 1\}$ and $\mathcal{X}_r$ is the observed region of the covariates $x_1 \in [374.27, 637.60]$ and $x_2 \in [230.98, 332.64]$ as before.

Note that the confidence level of the bands in (6.23) is given by $P\{S < c\}$ where

$$S = \sup_{(i,j)\in\Lambda} \sup_{\mathbf{x}_{(0)}\in\mathcal{X}_r} \frac{\mathbf{x}'\left((\hat{\boldsymbol{\beta}}_i - \boldsymbol{\beta}_i) - (\hat{\boldsymbol{\beta}}_j - \boldsymbol{\beta}_j)\right)}{\hat{\sigma}\sqrt{\mathbf{x}'\Delta_{ij}\mathbf{x}}}.$$

By using the results of Section 3.2.2 and similar manipulation as in the two-sided case of this section, we have

$$\mathrm{P}\{S < c\} = \mathrm{P}\left\{ \sup_{(i,j)\in\Lambda} \|\pi(\mathbf{T}_{ij}, \mathbf{P}_{ij}, \mathcal{X}_r)\| < c \right\}$$

for $c > 0$, where the notations $\mathbf{T}_{ij}$ and $\mathbf{P}_{ij}$ are the same as in (6.20). Hence the critical constant c can be approximated by simulation in a similar way as in the two-sided case. Based on $R = 200,000$ simulations, c is calculated to be 3.040 for $\alpha = 0.05$. As expected, this critical value is smaller than the two-sided critical value $c = 3.359$ given in Example 6.4.

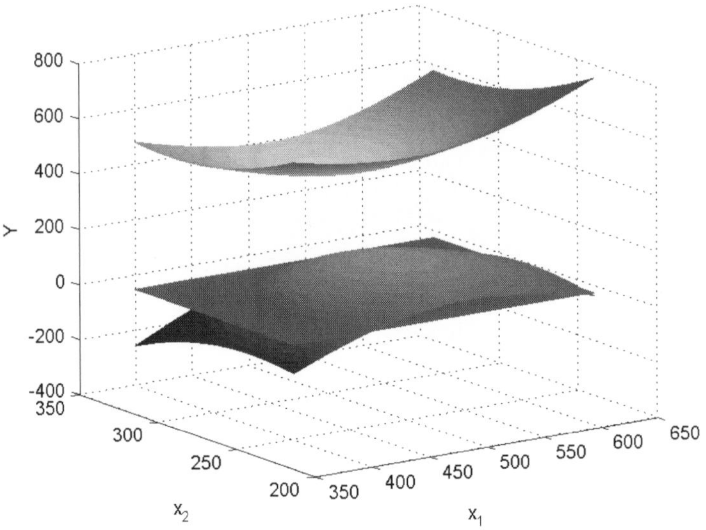

Figure 6.5: *The confidence band for* $\mathbf{x}'\boldsymbol{\beta}_3 - \mathbf{x}'\boldsymbol{\beta}_1$

Again one can plot the two one-sided confidence bands to make required one-sided inferences. For example, Figure 6.6 plots the one-sided band for $\mathbf{x}'\boldsymbol{\beta}_3 - \mathbf{x}'\boldsymbol{\beta}_1$. From this figure, one can infer that the mean sales in the west are higher than the mean sales in the south over the covariate region that the lower band is above zero. It is clear that one cannot claim that the mean sales in the west are higher than the mean sales in the south over the whole covariate region $\mathbf{x}_{(0)} \in \mathcal{X}_r$.

In the examples above, there are only $p = 1$ or $p = 2$ covariates for which it is most intuitive to simply plot the bands and then make inference by inspecting the bands. When there are more than $p = 2$ covariates, one may plot slices of a confidence band by fixing the values of certain covariates and these slices can provide useful information for comparing the regression models. On the other hand, if one is interested only in whether a confidence band contains the zero regression plane over the covariate region $\mathbf{x}_{(0)} \in \mathcal{X}_r$ (in order to assess whether the two corresponding regression models are significantly different) one can simply compute the p-values in the following way.

For the simultaneous confidence bands in (6.19) for the comparison of the ith and jth regression models for all $(i, j) \in \Lambda$, one first calculates the observed value

$$s_{ij} = \sup_{\mathbf{x}_{(0)} \in \mathcal{X}_r} \frac{|\mathbf{x}'(\hat{\boldsymbol{\beta}}_i - \hat{\boldsymbol{\beta}}_j)|}{\hat{\sigma}\sqrt{\mathbf{x}'\Delta_{ij}\mathbf{x}}}$$

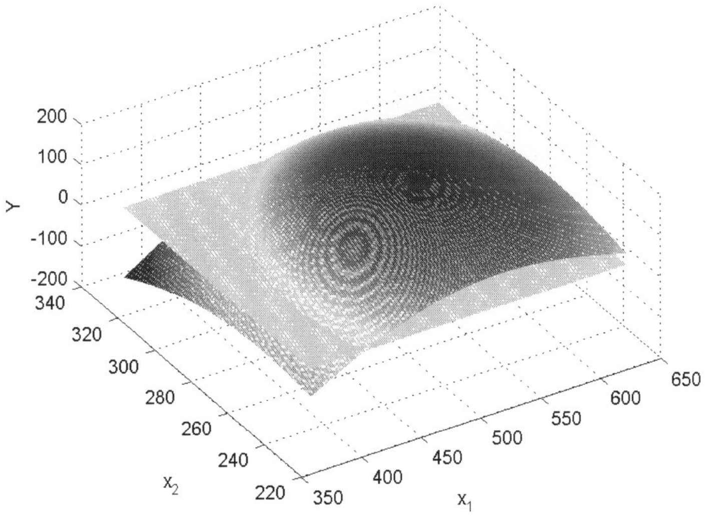

Figure 6.6: *The lower confidence band for* $\mathbf{x}'\beta_3 - \mathbf{x}'\beta_1$

by using the formula

$$s_{ij} = \max\{\|\pi(\mathbf{t}_{ij}, \mathbf{P}_{ij}, \mathcal{X}_r)\|, \|\pi(-\mathbf{t}_{ij}, \mathbf{P}_{ij}, \mathcal{X}_r)\|\}$$

where $\mathbf{P}_{ij}$ is the same as given in (6.20), and $\mathbf{t}_{ij} = (\mathbf{P}_{ij})^{-1}(\hat{\beta}_i - \hat{\beta}_j)/\hat{\sigma}$ with $\hat{\beta}_l$ and $\hat{\sigma}$ being the respective estimates of β_l and σ based on the observed data set. Next one computes

$$p_{ij} = \mathrm{P}\{S > s_{ij}\} \ \forall\, (i,j) \in \Lambda \tag{6.24}$$

where the random variable S is given in (6.20); p_{ij} can be approximated by simulating a large number replications of S using the algorithm given below expression (6.20). Now one can use p_{ij} to judge whether the confidence band in (6.19) for $\mathbf{x}'\beta_i - \mathbf{x}'\beta_j$ contains the zero regression hyper-plane over $\mathbf{x}_{(0)} \in \mathcal{X}_r$ (and so whether the two regression models are significantly different):

$$\mathbf{x}'\hat{\beta}_i - \mathbf{x}'\hat{\beta}_j \pm c\hat{\sigma}\sqrt{\mathbf{x}'\Delta_{ij}\mathbf{x}} \ \text{contains the zero hyper-plane} \iff p_{ij} > \alpha$$

for each $(i,j) \in \Lambda$.

A size α test of the hypotheses in (6.2) based on the simultaneous confidence bands in (6.19) is to

$$\text{reject } H_0 \Leftrightarrow \sup_{(i,j)\in\Lambda} \sup_{\mathbf{x}_{(0)}\in\mathcal{X}_r} \frac{|\mathbf{x}'[\hat{\beta}_i - \hat{\beta}_j]|}{\hat{\sigma}\sqrt{\mathbf{x}'\Delta_{ij}\mathbf{x}}} > c.$$

This test can be represented alternatively by using p-value. One first computes the observed value of the test statistic

$$s = \sup_{(i,j)\in\Lambda} \sup_{\mathbf{x}_{(0)}\in\mathcal{X}_r} \frac{|\mathbf{x}'[\hat{\beta}_i - \hat{\beta}_j]|}{\hat{\sigma}\sqrt{\mathbf{x}'\Delta_{ij}\mathbf{x}}}$$

from the data available. One next calculates the p-value

$$p = \mathrm{P}\{S > s\}$$

by using simulation as before, where the random variable S is specified in (6.20). Finally, one rejects H_0 in (6.2) if and only if $p < \alpha$.

Note that $s = \sup_{(i,j)\in\Lambda} s_{ij}$ and so $p = \min_{(i,j)\in\Lambda} p_{ij}$. Hence $p < \alpha$ (i.e., H_0 is rejected) if and only if $p_{ij} < \alpha$ (i.e., $\mathbf{x}'\beta_i$ and $\mathbf{x}'\beta_j$ are significantly different) for at least one pair $(i,j) \in \Lambda$.

Example 6.6 For the sales value example considered in Example 6.4, it is computed that $p_{2,1} = 0.18939$ and $p_{3,1} = 0.00036$ based on 200,000 simulations. One can therefore conclude that at $\alpha = 0.05$, the second (i.e., west region) and first (i.e., south region) regression models are not significantly different since $p_{2,1} > \alpha$, while the third (i.e., midwest region) and first (i.e., south region) regression models are significantly different since $p_{3,1} < \alpha$. These conclusions agree with the confidence bands given in Figures 6.4 and 6.5 of course. Furthermore $p = \min\{p_{2,1}, p_{3,1}\} = 0.00036$ and so H_0 in (6.2), the coincidence of the three regression models, is rejected at any $\alpha > 0.00036$.

The p-value approach to the two-sided simultaneous confidence bands in (6.19) can also be applied to the one-sided confidence bands in (6.23), and details are omitted here.

6.3.2 *Constant width bands*

Of course one may also use constant width simultaneous confidence bands to assess the differences among the k regression models and, due to the constant width shape, these bands may sometimes be more advantageous than the hyperbolic simultaneous confidence bands in (6.19). A set of constant width simultaneous confidence bands corresponding to the set of hyperbolic bands in (6.19) is given in Liu *et al.* (2007b) by

$$\mathbf{x}'\beta_i - \mathbf{x}'\beta_j \in \mathbf{x}'\hat{\beta}_i - \mathbf{x}'\hat{\beta}_j \pm c\hat{\sigma} \quad \forall\, \mathbf{x}_{(0)} \in \mathcal{X}_r \text{ and } \forall\, (i,j) \in \Lambda \tag{6.25}$$

where c is the critical constant chosen so that the confidence level of this set of simultaneous confidence bands is equal to $1 - \alpha$.

The confidence level of the bands is given by $\mathrm{P}\{S < c\}$ where

$$S = \sup_{(i,j)\in\Lambda} \sup_{\mathbf{x}_{(0)}\in\mathcal{X}_r} \frac{|\mathbf{x}'[(\hat{\beta}_i - \beta_i) - (\hat{\beta}_j - \beta_j)]|}{\hat{\sigma}}. \tag{6.26}$$

Let $\mathbf{Z}_i = (\hat{\beta}_i - \beta_i)/\sigma \sim \mathcal{N}_{p+1}(\mathbf{0}, (\mathbf{X}_i'\mathbf{X}_i)^{-1})$ and denote $\mathbf{T}_{ij} = (\mathbf{Z}_i - \mathbf{Z}_j)/(\hat{\sigma}/\sigma)$ which has a multivariate t distribution. Then S can be written as

$$S = \sup_{(i,j)\in\Lambda} \sup_{\mathbf{x}_{(0)}\in\mathcal{X}_r} |\mathbf{x}'\mathbf{T}_{ij}|.$$

From the results in Section 3.3.1, S can further be expressed as

$$S = \sup_{(i,j)\in\Lambda} \sup_{\mathbf{x}_{(0)}\in V} |\mathbf{x}'\mathbf{T}_{ij}| \quad (6.27)$$

where V is defined in (3.15). So one can use the following simulation algorithm similar as before to determine the critical value c, whilst Liu *et al.* (2007b) also provided a different method for finding the critical constant c by using the numerical integration method of Genz and Bretz (2002).

Step 0. Determine $(\mathbf{X}_i'\mathbf{X}_i)^{-1}$, $i = 1, \cdots, k$.

Step 1. Simulate independent $\mathbf{Z}_i \sim \mathcal{N}_{p+1}(\mathbf{0}, (\mathbf{X}_i'\mathbf{X}_i)^{-1})$ $(i = 1, \cdots, k)$ and $\hat{\sigma}/\sigma \sim \sqrt{\chi_\nu^2/\nu}$.

Step 2. Compute S from (6.27)

Repeat Steps 1-2 R times to simulate R independent replicates of T, and set the $(1-\alpha)$ sample quantile $\hat{c}$ as c.

Example 6.7 For the drug stability example considered in Example 6.1, the 95% constant-width simultaneous confidence bands for all pairwise comparisons of the six regression lines are given by

$$\beta_{i,0} + \beta_{i,1}x_1 - \beta_{j,0} - \beta_{j,1}x_1 \in \hat{\beta}_{i,0} + \hat{\beta}_{i,1}x_1 - \hat{\beta}_{j,0} - \hat{\beta}_{j,1}x_1 \pm c\hat{\sigma}$$
$$\text{for } x_1 \in [0.0, 2.0] \text{ and for all } 1 \le i < j \le 6. \quad (6.28)$$

Based on 200,000 simulations, the critical value c is calculated to be $c = 6.362$.

In order to assess the maximum difference between the ith and jth regression lines based on the constant width bands, we calculate

$$\bar{D}_{ij} = \max_{x\in[0.0,2.0]} \left| \hat{\beta}_{i,0} + \hat{\beta}_{i,1}x_1 - \hat{\beta}_{j,0} - \hat{\beta}_{j,1}x_1 \pm c\hat{\sigma} \right|.$$

If and only if $\bar{D}_{ij} < \delta$, where δ is a pre-specified threshold, the ith and jth regression lines are declared equivalent and the two batches are pooled. For this particular data set, $\bar{D}_{ij}$ is larger than the D_{ij} calculated using the hyperbolic bands in Example 6.3 for the following six (i,j)'s: (2,1), (3,1), (3,2), (4,1), (4,2) and (4,3). But $\bar{D}_{ij} < D_{ij}$ for the other nine (i,j)'s. Over all, we have

$$\max_{1\le i<j\le 6} \bar{D}_{ij} = 2.612 < 2.886 = \max_{1\le i<j\le 6} D_{ij}.$$

So the constant width bands tend to have a better chance to establish equivalence

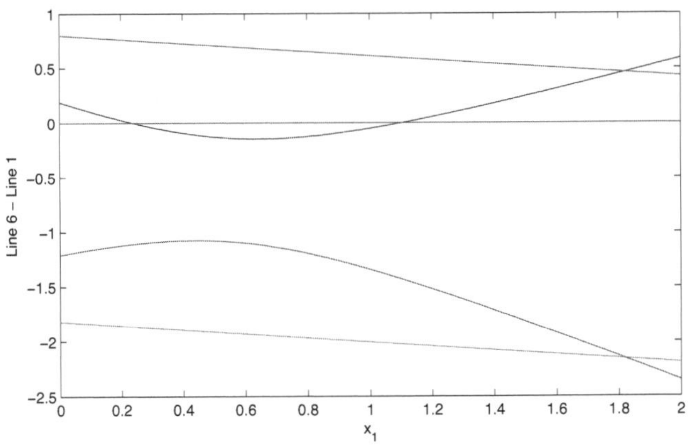

Figure 6.7: *The confidence bands for* $\mathbf{x}'\boldsymbol{\beta}_6 - \mathbf{x}'\boldsymbol{\beta}_1$

than the hyperbolic bands. Figure 6.7 plots the hyperbolic and constant width bands for $\mathbf{x}'\boldsymbol{\beta}_6 - \mathbf{x}'\boldsymbol{\beta}_1$ to illustrate the relative positions of the two bands often observed, from which one observes $\bar{D}_{1,6} = 2.191$ and $D_{1,6} = 2.351$. From this figure, one can see the reason why D_{ij} often tends to be be larger than $\bar{D}_{ij}$ is that a hyperbolic band tends to be wider than a constant width band near the ends of $x_1 \in [0.0, 2.0]$.

One can also compute the p-values p_{ij} for the constant width bands in (6.25) in order to assess whether the constant width band for $\mathbf{x}'\boldsymbol{\beta}_i - \mathbf{x}'\boldsymbol{\beta}_j$ contains the zero regression plane for $(i, j) \in \Lambda$. First, one calculates the observed value

$$s_{ij} = \sup_{\mathbf{x}_{(0)} \in \mathcal{X}_r} \frac{|\mathbf{x}'(\hat{\boldsymbol{\beta}}_i - \hat{\boldsymbol{\beta}}_j)|}{\hat{\sigma}} = \sup_{\mathbf{x}_{(0)} \in V} |\mathbf{x}'\mathbf{t}_{ij}|$$

where $\mathbf{t}_{ij} = (\hat{\boldsymbol{\beta}}_i - \hat{\boldsymbol{\beta}}_j)/\hat{\sigma}$ with $\hat{\boldsymbol{\beta}}_l$ and $\hat{\sigma}$ being the estimates of $\boldsymbol{\beta}_l$ and σ, respectively, based on the observed data set. Next, one computes

$$p_{ij} = \mathrm{P}\{S > s_{ij}\} \ \forall\, (i, j) \in \Lambda \tag{6.29}$$

by using simulation as before, where the random variable S is given in (6.27). Now one can use p_{ij} to judge whether the confidence band for $\mathbf{x}'\boldsymbol{\beta}_i - \mathbf{x}'\boldsymbol{\beta}_j$ in (6.25) contains the zero regression plane over $\mathbf{x}_{(0)} \in \mathcal{X}_r$ and so whether the two regression models are significantly different:

$$\mathbf{x}'\hat{\boldsymbol{\beta}}_i - \mathbf{x}'\hat{\boldsymbol{\beta}}_j \pm c\hat{\sigma} \text{ contains the zero plane } \iff p_{ij} > \alpha$$

for each $(i, j) \in \Lambda$.

A size α test of the hypotheses in (6.2) based on the simultaneous confidence bands in (6.25) is to

$$\text{reject } H_0 \Leftrightarrow \sup_{(i,j)\in\Lambda} \sup_{\mathbf{x}_{(0)}\in\mathcal{X}_r} \frac{|\mathbf{x}'(\hat{\beta}_i - \hat{\beta}_j)|}{\hat{\sigma}} > c.$$

This test can also be represented by using p-value. One first computes the observed value of the test statistic

$$s = \sup_{(i,j)\in\Lambda} \sup_{\mathbf{x}_{(0)}\in\mathcal{X}_r} \frac{|\mathbf{x}'(\hat{\beta}_i - \hat{\beta}_j)|}{\hat{\sigma}}$$

from the data available. One next calculates the p-value

$$p = \mathrm{P}\{S > s\}$$

where the random variable S is specified in (6.27). Finally, one rejects H_0 in (6.2) if and only if $p < \alpha$. Again, we have $p = \min_{(i,j)\in\Lambda} p_{ij}$. Hence $p < \alpha$ if and only if $p_{ij} < \alpha$ for at least one $(i,j) \in \Lambda$.

Example 6.8 For the sales value example considered in Example 6.6, if the constant width confidence bands in (6.25) are used for the comparisons of the second and third regression models with the first regression model then it is computed that $p_{2,1} = 0.3865$ and $p_{3,1} = 0.0117$ based on 200,000 simulations. One can therefore conclude, without plotting the confidence bands, that at $\alpha = 0.05$, the second (i.e., west region) and first (i.e., south region) regression models are not significantly different since $p_{2,1} > \alpha$, while the third (i.e., midwest region) and first (i.e., south region) regression models are significantly different since $p_{3,1} < \alpha$. Furthermore $p = \min\{p_{2,1}, p_{3,1}\} = 0.0117$ and so H_0 in (6.2), the coincidence of the three regression models, is rejected at any $\alpha > 0.0117$ by the test induced by the constant width bands in (6.25).

6.4 Bands for finite contrasts over ellipsoidal region

6.4.1 Hyperbolic bands

For the two-sided comparisons of the ith and jth regression models for all $(i,j) \in \Lambda$ one can use the simultaneous confidence bands:

$$\mathbf{x}'\beta_i - \mathbf{x}'\beta_j \in \mathbf{x}'\hat{\beta}_i - \mathbf{x}'\hat{\beta}_j \pm c\hat{\sigma}\sqrt{\mathbf{x}'\Delta_{ij}\mathbf{x}} \;\; \forall\, \mathbf{x}_{(0)} \in \mathcal{X}_{e(ij)} \text{ for all } (i,j) \in \Lambda, \quad (6.30)$$

where c is a critical constant chosen so that the simultaneous confidence level of the bands is equal to $1-\alpha$. Here

$$\mathcal{X}_{e(ij)} = \left\{ \mathbf{x}_{(0)} : \left(\mathbf{x}_{(0)} - \zeta_{ij}\right)' \left(\frac{\mathbf{V}_i^{-1}}{n_i} + \frac{\mathbf{V}_j^{-1}}{n_j} \right) \left(\mathbf{x}_{(0)} - \zeta_{ij}\right) \le a^2 \kappa_{ij} \right\} \quad (6.31)$$

where $a \geq 0$ is a given constant,

$$\mathbf{V}_i = \frac{1}{n_i}\left(\mathbf{X}'_{i(0)}\mathbf{X}_{i(0)} - n_i\bar{\mathbf{x}}_{i(0)}\bar{\mathbf{x}}'_{i(0)}\right),$$

$$\zeta_{ij} = \left(\frac{\mathbf{V}_i^{-1}}{n_i} + \frac{\mathbf{V}_j^{-1}}{n_j}\right)^{-1}\left(\frac{\mathbf{V}_i^{-1}}{n_i}\bar{\mathbf{x}}_{i(0)} + \frac{\mathbf{V}_j^{-1}}{n_j}\bar{\mathbf{x}}_{j(0)}\right) \text{ and}$$

$$\kappa_{ij} = \left(\frac{1}{n_i} + \frac{1}{n_j}\right) + \bar{\mathbf{x}}'_{i(0)}\frac{\mathbf{V}_i^{-1}}{n_i}\bar{\mathbf{x}}_{i(0)} + \bar{\mathbf{x}}'_{j(0)}\frac{\mathbf{V}_j^{-1}}{n_j}\bar{\mathbf{x}}_{j(0)} - \left(\bar{\mathbf{x}}'_{i(0)}\frac{\mathbf{V}_i^{-1}}{n_i} + \bar{\mathbf{x}}'_{j(0)}\frac{\mathbf{V}_j^{-1}}{n_j}\right)$$
$$\times\left(\frac{\mathbf{V}_i^{-1}}{n_i} + \frac{\mathbf{V}_j^{-1}}{n_j}\right)^{-1}\left(\frac{\mathbf{V}_i^{-1}}{n_i}\bar{\mathbf{x}}_{i(0)} + \frac{\mathbf{V}_j^{-1}}{n_j}\bar{\mathbf{x}}_{j(0)}\right).$$

The simultaneous confidence level of this band is given by $\mathrm{P}\{S < c\}$ where

$$S = \sup_{(i,j)\in\Lambda} S_{ij} \text{ where} \tag{6.32}$$
$$S_{ij} = \sup_{\mathbf{x}_{(0)}\in\mathcal{X}_{e(ij)}} \frac{|\mathbf{x}'(\hat{\beta}_i - \beta_i - \hat{\beta}_j + \beta_j)|}{\hat{\sigma}\sqrt{\mathbf{x}'\Delta_{ij}\mathbf{x}}}.$$

From expression (5.39) of Section 5.4,

$$S_{ij} = \sup_{\mathbf{v}\in\mathcal{V}_e} \frac{|\mathbf{v}'\mathbf{T}_{ij}|}{\|\mathbf{v}\|}$$
$$= \begin{cases} \|\mathbf{T}_{ij}\| & \text{if } |T_{ij0}| \geq \frac{1}{a}\|\mathbf{T}_{ij(0)}\| \\ \frac{|T_{ij0}| + a\|\mathbf{T}_{ij(0)}\|}{\sqrt{1+a^2}} & \text{otherwise} \end{cases} \tag{6.33}$$

where

$$\mathcal{V}_e = \left\{\mathbf{v} : \|\mathbf{v}\| \leq v_0\sqrt{1+a^2}\right\} \subset \Re^{p+1} \text{ and}$$
$$\mathbf{T}_{ij} = (\mathbf{z}_{ij}, \mathbf{Z}_{ij})^{-1}\Delta_{ij}{}^{-1}(\hat{\beta}_i - \beta_i - \hat{\beta}_j + \beta_j)/\hat{\sigma} \sim \mathcal{T}_{p+1,\nu}(\mathbf{0}, \mathbf{I}) \tag{6.34}$$

with $\mathbf{z}_{ij}$ and $\mathbf{Z}_{ij}$ being defined in the following way. Let $\mathbf{z}_{ij}{}^*$ be the first column of $\Delta_{ij}{}^{-1}$, and let the $(1,1)$-element of $\Delta_{ij}{}^{-1}$ be z^*_{ij0}. Then $\mathbf{z}_{ij} = \mathbf{z}_{ij}{}^*/\sqrt{z^*_{ij0}}$. Let $\mathbf{Q}_{ij} = \sqrt{\Delta_{ij}}$ and $\{\mathbf{V}_{ij1}, \cdots, \mathbf{V}_{ijp}\}$ be an orthonormal basis of the null space of $\mathbf{Q}_{ij}\mathbf{z}_{ij} \in \Re^{p+1}$. Then $\mathbf{Z}_{ij} = \mathbf{Q}_{ij}^{-1}(\mathbf{V}_{ij1}, \cdots, \mathbf{V}_{ijp})$.

So the random variable S can be computed straightforwardly by using (6.32) and (6.33), and one can use simulation to approximate the exact critical constant c in a similar way as before.

It is clear from (6.30) that the comparison of the ith and jth models is based on the confidence band

$$\mathbf{x}'\beta_i - \mathbf{x}'\beta_j \in \mathbf{x}'\hat{\beta}_i - \mathbf{x}'\hat{\beta}_j \pm c\hat{\sigma}\sqrt{\mathbf{x}'\Delta_{ij}\mathbf{x}} \;\; \forall\, \mathbf{x}_{(0)} \in \mathcal{X}_{e(ij)}. \tag{6.35}$$

Note that $\mathcal{X}_{e(ij)}$ is different for different $(i,j) \in \Lambda$. Furthermore, the pivot S_{ij} in (6.33) on which the (i,j)-comparison hinges has the same distribution for all $(i,j) \in \Lambda$; this implies that each (i,j)-comparison, $(i,j) \in \Lambda$, is allowed the same rate of type I error. These two properties of the confidence bands (6.30) are not shared by the confidence bands (6.19), which may have a good power to detect the differences between some $\mathbf{x}'\beta_i$'s but an excessively low power to detect the differences between some other $\mathbf{x}'\beta_i$'s. Possible remedies for this are either using different rectangular covariate regions or replacing the common critical constant c by $l_{ij}c$ for different $(i,j) \in \Lambda$; how to choose the different rectangular regions or the weights l_{ij} warrants further research.

One can compute the p-values associated with the confidence bands (6.30) in the following way. First, one calculates the observed values

$$s_{ij} = \sup_{\mathbf{x}_{(0)} \in \mathcal{X}_{e(ij)}} \frac{|\mathbf{x}'(\hat{\beta}_i - \hat{\beta}_j)|}{\hat{\sigma}\sqrt{\mathbf{x}'\Delta_{ij}\mathbf{x}}}$$

by using the expressions in (6.33) but with $\mathbf{T}_{ij} = (\mathbf{z}_{ij}, \mathbf{Z}_{ij})^{-1}\Delta_{ij}{}^{-1}(\hat{\beta}_i - \hat{\beta}_j)/\hat{\sigma}$ where $\hat{\beta}_l$ and $\hat{\sigma}$ are the estimates of β_l and σ calculated from the observed data. One then computes by simulation

$$p_{ij} = \mathrm{P}\{S > s_{ij}\} \ \forall (i,j) \in \Lambda$$

where the random variable S is defined in (6.32). If $p_{ij} < \alpha$ then the confidence band in (6.35) does not contain the zero regression hyper-plane completely over $\mathbf{x}_{(0)} \in \mathcal{X}_{e(ij)}$. If $p = \min_{(i,j)\in\Lambda} p_{ij} < \alpha$ then at least one of the confidence bands in (6.30) does not contain the zero regression hyper-plane completely over its covariate region $\mathbf{x}_{(0)} \in \mathcal{X}_{e(ij)}$, and so the null hypothesis of coincidence H_0 in (6.2) is rejected.

Example 6.9 Let us consider again the sales value data set given in Example 6.1, and the interest is still the comparisons of midwest and west with south. One can use the confidence bands in (6.30) with $\Lambda = \{(i,j) : i = 2,3 \text{ and } j = 1\}$. Suppose $\alpha = 0.05$ and $a = 0.94$. Then the two covariate regions $\mathcal{X}_{e(12)}$ and $\mathcal{X}_{e(13)}$ are plotted in Figure 6.8: $\mathcal{X}_{e(12)}$ given by the smaller ellipsoidal region and $\mathcal{X}_{e(13)}$ the larger ellipsoidal region. The rectangle in the figure is the observed covariate region. It is clear that the two ellipsoidal regions have different centers and sizes.

Based on 200,000 simulations, the critical constant c is calculated to be 3.203; this critical constant is slightly smaller than the critical constant 3.359 for the bands in (6.22) over the observed covariate rectangle. The associated p-values are calculated to be $p_{2,1} = 0.23666$ and $p_{3,1} = 0.00049$. Hence, at $\alpha = 0.05$, the confidence band for $(1,2)$-comparison contains the zero regression plane and the two regression models are not significantly different. But the confidence band for $(1,3)$-comparison does not contain the zero regression plane completely and the two regression models are significantly different. These conclusions are similar to those in Example 6.6 based on the bands over the observed covariate rectangle.

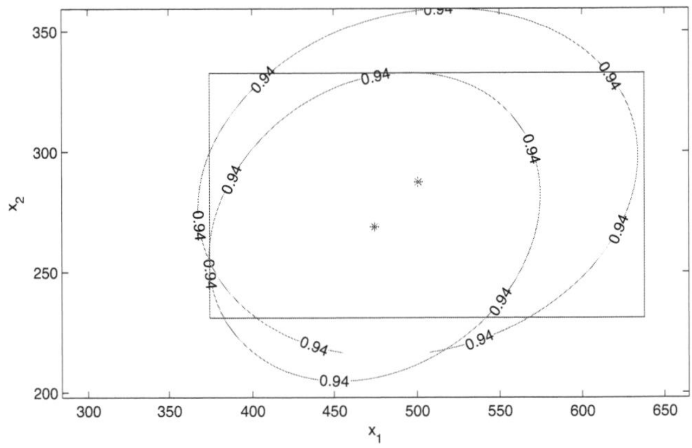

Figure 6.8: *The covariate regions* $\mathcal{X}_{e(12)}$ *and* $\mathcal{X}_{e(13)}$

6.4.2 Constant width bands

One can also use the following constant width confidence bands for two-sided comparison of the ith and jth regression models for all $(i,j) \in \Lambda$:

$$\mathbf{x}'\beta_i - \mathbf{x}'\beta_j \in \mathbf{x}'\hat{\beta}_i - \mathbf{x}'\hat{\beta}_j \pm c\hat{\sigma}\sqrt{\kappa_{ij}(1+a^2)} \ \ \forall\, \mathbf{x}_{(0)} \in \mathcal{X}_{e(ij)} \ \forall\, (i,j) \in \Lambda, \quad (6.36)$$

where c is a critical constant chosen so that the simultaneous confidence level of the bands is equal to $1-\alpha$. Here $\mathcal{X}_{e(ij)}$ and κ_{ij} are the same as in the hyperbolic bands (6.30).

The simultaneous confidence level of this set of bands is given by $P\{S < c\}$ where

$$S = \sup_{(i,j)\in\Lambda} S_{ij} \text{ where} \quad (6.37)$$

$$S_{ij} = \sup_{\mathbf{x}_{(0)}\in\mathcal{X}_{e(ij)}} \frac{|\mathbf{x}'(\hat{\beta}_i - \beta_i - \hat{\beta}_j + \beta_j)|}{\hat{\sigma}\sqrt{\kappa_{ij}(1+a^2)}}.$$

From expression (5.52) of Section 5.4.3,

$$S_{ij} = \frac{|T_{ij0}| + a\|\mathbf{T}_{ij(0)}\|}{\sqrt{1+a^2}} \quad (6.38)$$

where $\mathbf{T}_{ij}$ is defined in (6.34). So the random variable S can easily be computed by using (6.37) and (6.38), and the exact critical constant c can be approximated by using simulation in a similar way as before.

From (6.36), the comparison of the ith and jth models is based on the confidence band

$$\mathbf{x}'\beta_i - \mathbf{x}'\beta_j \in \mathbf{x}'\hat{\beta}_i - \mathbf{x}'\hat{\beta}_j \pm c\hat{\sigma}\sqrt{\kappa_{ij}(1+a^2)} \;\; \forall\, \mathbf{x}_{(0)} \in \mathcal{X}_{e(ij)}. \tag{6.39}$$

Again, the pivot S_{ij} in (6.38) on which the (i,j)-comparison depends has the same distribution for all $(i,j) \in \Lambda$, which implies that each (i,j)-comparison, $(i,j) \in \Lambda$, is allowed the same rate of type I error. Also note that the width of the confidence band for (i,j)-comparison varies with $(i,j) \in \Lambda$, while all the confidence bands in (6.25) have the same width. The confidence bands in (6.25) can be modified to devote different (or same) type I error rates for different (i,j)-comparisons by using different rectangular covariate regions or differently weighted critical constants $l_{ij}c$ for different (i,j)-comparisons. No research is done on how to choose different rectangular regions or the weights l_{ij} however.

One can compute the p-values associated with the confidence bands (6.36) as follows. First, one calculates the observed values

$$s_{ij} = \sup_{\mathbf{x}_{(0)} \in \mathcal{X}_{e(ij)}} \frac{|\mathbf{x}'(\hat{\beta}_i - \hat{\beta}_j)|}{\hat{\sigma}\sqrt{\kappa_{ij}(1+a^2)}}$$

by using the expression in (6.38) but with $\mathbf{T}_{ij} = (\mathbf{z}_{ij}, \mathbf{Z}_{ij})^{-1}\Delta_{ij}{}^{-1}(\hat{\beta}_i - \hat{\beta}_j)/\hat{\sigma}$ where $\hat{\beta}_l$ and $\hat{\sigma}$ are the estimates calculated from the observed data. One then computes by simulation

$$p_{ij} = \mathrm{P}\{S > s_{ij}\} \;\; \forall (i,j) \in \Lambda$$

where the random variable S is defined in (6.37). If $p_{ij} < \alpha$ then the confidence band in (6.39) does not contain the zero regression hyper-plane completely over $\mathbf{x}_{(0)} \in \mathcal{X}_{e(ij)}$. If $p = \min_{(i,j)\in\Lambda} p_{ij} < \alpha$ then at least one of the confidence bands in (6.36) does not contain the zero regression hyper-plane completely over its covariate region $\mathbf{x}_{(0)} \in \mathcal{X}_{e(ij)}$, and so the null hypothesis of coincidence H_0 in (6.2) is rejected.

Example 6.10 Continue from Example 6.9. One can also use the confidence bands in (6.36) with $\Lambda = \{(i,j) : i = 2,3 \text{ and } j = 1\}$ for the comparisons of midwest and west with south. For $\alpha = 0.05$ and $a = 0.94$, the critical constant c is calculated to be 3.153 based on 200,000 simulations. The associated p-values are calculated to be $p_{2,1} = 0.22945$ and $p_{3,1} = 0.00227$. Hence, at $\alpha = 0.05$, the confidence band for $(1,2)$-comparison contains the zero regression plane. But the confidence band for $(1,3)$-comparison does not contain the zero regression plane completely. These conclusions agree with those in Example 6.8.

One can construct one-sided hyperbolic or constant width simultaneous confidence bands over ellipsoidal covariate regions in a similar way. But no details are provided here.

6.5 Equivalence of more than two models

It is clear that, in the drug stability example discussed above, the goal is to assess the equivalence of the regression models rather than to detect the differences among the models. The simultaneous confidence bands considered in this chapter are versatile and can be applied to assess the equivalence of the models. But they are not particularly suitable for demonstrating equivalence, for similar reasons as pointed out in Section 5.5. In this section, we provide an intersection-union test for assessing the equivalence of all the k regression models over a given region of the covariates. This test has a greater chance to establish equivalence than the simultaneous confidence bands in this chapter.

Since the goal is to assess whether all the k regression models are equivalent in the sense that the maximum difference, over the given covariate region $\mathbf{x}_{(0)} \in \mathcal{X}$, between any two of the k models is strictly less than a given threshold δ:

$$\max_{1\le i\ne j\le k} \sup_{\mathbf{x}_{(0)}\in\mathcal{X}} |\mathbf{x}'\boldsymbol{\beta}_i - \mathbf{x}'\boldsymbol{\beta}_j| < \delta,$$

the hypotheses are set up as

$$\begin{aligned} H_0^E &: \max_{1\le i\ne j\le k} \sup_{\mathbf{x}_{(0)}\in\mathcal{X}} |\mathbf{x}'\boldsymbol{\beta}_i - \mathbf{x}'\boldsymbol{\beta}_j| \ge \delta \\ \text{against} \quad H_a^E &: \max_{1\le i\ne j\le k} \sup_{\mathbf{x}_{(0)}\in\mathcal{X}} |\mathbf{x}'\boldsymbol{\beta}_i - \mathbf{x}'\boldsymbol{\beta}_j| < \delta. \end{aligned} \tag{6.40}$$

If and only if the null hypothesis H_0^E is rejected, the equivalence of the k models can be claimed.

A size α test can be constructed by using the intersection-union principle of Berger (1982); see Appendix D. Let the equivalence hypotheses for the ith and jth models ($1 \le i \ne j \le k$) be

$$H_{0,ij}^E : \sup_{\mathbf{x}_{(0)}\in\mathcal{X}} |\mathbf{x}'\boldsymbol{\beta}_i - \mathbf{x}'\boldsymbol{\beta}_j| \ge \delta \ \text{ against } \ H_{a,ij}^E : \sup_{\mathbf{x}_{(0)}\in\mathcal{X}} |\mathbf{x}'\boldsymbol{\beta}_i - \mathbf{x}'\boldsymbol{\beta}_j| < \delta.$$

From Section 5.5.3, a size α test for these hypotheses rejects $H_{0,ij}^E$ if and only if

$$\begin{aligned} &\min_{\mathbf{x}_{(0)}\in\mathcal{X}} (\mathbf{x}'\hat{\boldsymbol{\beta}}_i - \mathbf{x}'\hat{\boldsymbol{\beta}}_j - t_\nu^\alpha \hat{\sigma}\sqrt{\mathbf{x}'\Delta_{ij}\mathbf{x}}\,) > -\delta \\ \text{and} \quad &\max_{\mathbf{x}_{(0)}\in\mathcal{X}} (\mathbf{x}'\hat{\boldsymbol{\beta}}_i - \mathbf{x}'\hat{\boldsymbol{\beta}}_j + t_\nu^\alpha \hat{\sigma}\sqrt{\mathbf{x}'\Delta_{ij}\mathbf{x}}\,) < \delta \end{aligned}$$

Notice that the hypotheses in (6.40) can be written as

$$H_0^E = \cup_{1\le i\ne j\le k} H_{0,ij}^E \ \text{ and } \ H_a^E = \cap_{1\le i\ne j\le k} H_{a,ij}^E.$$

Hence a size α intersection-union test of H_0^E against H_a^E rejects H_0^E if and only if

$$\begin{aligned} &\min_{1\le i\ne j\le k} \min_{\mathbf{x}_{(0)}\in\mathcal{X}} (\mathbf{x}'\hat{\boldsymbol{\beta}}_i - \mathbf{x}'\hat{\boldsymbol{\beta}}_j - t_\nu^\alpha \hat{\sigma}\sqrt{\mathbf{x}'\Delta_{ij}\mathbf{x}}) > -\delta \\ \text{and} \quad &\max_{1\le i\ne j\le k} \max_{\mathbf{x}_{(0)}\in\mathcal{X}} (\mathbf{x}'\hat{\boldsymbol{\beta}}_i - \mathbf{x}'\hat{\boldsymbol{\beta}}_j + t_\nu^\alpha \hat{\sigma}\sqrt{\mathbf{x}'\Delta_{ij}\mathbf{x}}) < \delta. \end{aligned} \tag{6.41}$$

The size α of the test is the probability of falsely claiming equivalence when the k models are not all equivalent. The power of the test is the probability of rejecting H_0 and so claiming the equivalence when the k models are equivalent, which depends on $\mathcal{X}$ and the true values of β_i and σ.

Example 6.11 Let's continue the drug stability example considered in this chapter. We can use the IU test in (6.41) to assess whether all six batches are equivalent and hence can be pooled to calculate one expiry date. For $\alpha = 0.05$ and the observations given in Table 6.1, we have

$$\min_{1\le i\ne j\le k} \min_{\mathbf{x}_{(0)}\in\mathcal{X}} \left(\mathbf{x}'\hat{\beta}_i - \mathbf{x}'\hat{\beta}_j - t_\nu^\alpha \hat{\sigma}\sqrt{\mathbf{x}'\Delta_{ij}\mathbf{x}}\,\right)$$

$$= \max_{1\le i\ne j\le k} \max_{\mathbf{x}_{(0)}\in\mathcal{X}} \left(\mathbf{x}'\hat{\beta}_i - \mathbf{x}'\hat{\beta}_j + t_\nu^\alpha \hat{\sigma}\sqrt{\mathbf{x}'\Delta_{ij}\mathbf{x}}\,\right) = 2.024.$$

Therefore if the threshold δ is larger than 2.024, all six batches can be claimed to be equivalent. From the numerical results given in Example 6.7, it is clear that the test (6.41) has a better chance of demonstrating the equivalence of all six batches than the hyperbolic or constant width simultaneous confidence bands.

The conclusions one may draw from test (6.41) are either all the k models are equivalent or not. A more useful procedure is to divide the k models into several groups: the models in the same group are equivalent and the models in different groups are not equivalent. Although the simultaneous confidence bands developed in this chapter can serve for this inferential need, they are not particularly pertinent for this purpose. For this particular purpose, one needs a multiple test to test the family of null hypotheses

$$\left\{H_{0,ij}^E:\ 1\le i\ne j\le k\right\}.$$

To devise such a multiple test that controls strongly the type I familywise error rate (i.e., the probability that any two non-equivalent models are falsely allocated into an equivalent group) at α is not an easy task, and there is no published work on this to date. The work of Bofinger and Bofinger (1995) is relevant, in which the simpler problem of dividing several normal means into the two equivalent and non-equivalent groups relative to a control normal mean is studied. It is also noteworthy that Hewett and Lababidi (1982) gave an intersection-union test for assessing the ordering of three regression straight lines on an interval.

Another interesting problem is to construct an upper confidence bound on

$$\max_{1\le i\ne j\le k} \max_{\mathbf{x}_{(0)}\in\mathcal{X}} |\mathbf{x}'\beta_i - \mathbf{x}'\beta_j|/\sigma,$$

which generalizes the result of Liu, Hayter and Wynn (2007) for $k = 2$. Note that Bofinger *et al.* (1993) constructed an upper confidence bound on

$$\max_{1\le i\ne j\le k} |\mu_i - \mu_j|/\sigma$$

for k normal populations $N(\mu_1, \sigma^2), \cdots, N(\mu_k, \sigma^2)$.

Finally, there is no published work on simultaneous confidence bands when the error variances σ^2 of the k models are different.

7

Confidence Bands for Polynomial Regression

For making inferences about polynomial regression models, the commonly used partial F tests are applied in the same ways as for linear regression models where the covariates have no functional relationships among themselves; the only place that the functional relationships are used is in the design matrix. This lack of full utilization of the special form of a polynomial regression model has some undesirable consequences. In this chapter various simultaneous confidence bands pertinent for making inferences for polynomial regression models are provided. They are more intuitive and more informative than the partial F tests. The ideas presented in this chapter can clearly be applied to other linear regression models where the covariates have functional relationships, such as response surface (cf. Khuri, 2006 and Meyers *et al.*, 2009) and mixture models (cf. Cornell, 1990). Some applications of simultaneous confidence bands to response-surface methodology are given in Sa and Edwards (1993) and Merchant, McCann and Edwards (1998).

7.1 Confidence bands for one model

Consider a standard (univariate) polynomial regression model

$$\mathbf{Y} = \mathbf{X}\beta + \mathbf{e} \tag{7.1}$$

where $\mathbf{Y} = (y_1, \cdots, y_n)'$ is a vector of observations, $\mathbf{X}$ is a $n \times (p+1)$ full column-rank design matrix with the lth $(1 \le l \le n)$ row given by $(1, x_l, \cdots, x_l^p)$, $\beta = (\beta_0, \cdots, \beta_p)'$ is a vector of unknown coefficients, and $\mathbf{e} = (e_1, \cdots, e_n)'$ is a vector of independent random errors with each $e_i \sim N(0, \sigma^2)$, where σ^2 is an unknown parameter. From Chapter 1, the least squares estimators of β and σ are given, respectively, by $\hat{\beta} = (\mathbf{X}'\mathbf{X})^{-1}\mathbf{X}'\mathbf{Y} \sim \mathcal{N}_{p+1}(\beta, \sigma^2(\mathbf{X}'\mathbf{X})^{-1})$ and $\hat{\sigma}^2 = \|\mathbf{Y} - \mathbf{X}\hat{\beta}\|^2/\nu \sim \sigma^2\chi^2_\nu/\nu$ where $\nu = n - p - 1$.

Throughout this chapter, let $\tilde{\mathbf{x}} = (1, x, \cdots, x^p)' \in \Re^{p+1}$, and denote $\mathbf{x} = (1, x_1, \cdots, x_p)'$ as before. We also assume that the order of the polynomial model (7.1), p, is larger than one in order to exclude the straight line regression models considered in Chapter 2. In a usual linear regression model, the covariates $x_1, \cdots, x_p$ do not have functional relationships among themselves. On the other hand, the covariates of the polynomial regression model (7.1) are $x, \cdots, x^p$, which depend on just one variable x. This important feature of a polynomial regression model should be taken advantage of in developing any inferential procedures that are pertinent to polynomial regression. Note, however, after $\mathbf{X}$ is determined, this feature is no longer used in, for example, the construction of the confidence set (7.2) below for β.

One important statistical inference is to assess where lies the true polynomial model $\tilde{\mathbf{x}}'\beta$ for x in a given range of interest, say (a,b). Theorem 1.2. gives the following exact $1-\alpha$ level confidence region for β:

$$C(\mathbf{Y},\mathbf{X}) = \left\{\beta : \frac{(\hat{\beta}-\beta)'(\mathbf{X}'\mathbf{X})(\hat{\beta}-\beta)/(p+1)}{\hat{\sigma}^2} \leq f^{\alpha}_{p+1,\nu}\right\}. \tag{7.2}$$

From this confidence region, any $\beta \in C(\mathbf{Y},\mathbf{X})$ is deemed to be a plausible value and so the corresponding $\tilde{\mathbf{x}}'\beta$ a plausible polynomial model.

From Theorem 3.2, this confidence region $C(\mathbf{Y},\mathbf{X})$ for β is equivalent to the following exact $1-\alpha$ level simultaneous confidence band for the regression hyper-plane $\mathbf{x}'\beta$:

$$\mathbf{x}'\beta \in \mathbf{x}'\hat{\beta} \pm \sqrt{(p+1)f^{\alpha}_{p+1,\nu}}\,\hat{\sigma}\sqrt{\mathbf{x}'(X'X)^{-1}\mathbf{x}}, \ \forall\, \mathbf{x} \in \Re^{p+1}. \tag{7.3}$$

From this confidence band, in order to judge whether $\tilde{\mathbf{x}}'\beta$ is a plausible model, one has to check whether $\mathbf{x}'\beta$ is contained in the confidence band (7.3) over the whole space of $\mathbf{x} \in \Re^{p+1}$.

This is not sensible since we are interested only in the polynomial regression model $\tilde{\mathbf{x}}'\beta$ and hence the behavior of the hyper-plane $\mathbf{x}'\beta$ only along the loci $\mathbf{x} = \tilde{\mathbf{x}}$ rather than over all $\mathbf{x} \in \Re^{p+1}$. Using the confidence region (7.2) or the equivalent confidence band (7.3) to assess whether a given polynomial regression model $\tilde{\mathbf{x}}'\beta$ is plausible has the following undesirable phenomenon. Suppose $\tilde{\mathbf{x}}'\beta_1$ and $\tilde{\mathbf{x}}'\beta_2$ are two given polynomial models and, furthermore, the first model $\tilde{\mathbf{x}}'\beta_1$ is closer than the second model $\tilde{\mathbf{x}}'\beta_2$ to the estimated model $\tilde{\mathbf{x}}'\hat{\beta}$ from the observed data $\mathbf{Y}$ and $\mathbf{X}$ in the sense that

$$\sup_{-\infty<x<\infty} \frac{|\tilde{\mathbf{x}}'\beta_1 - \tilde{\mathbf{x}}'\hat{\beta}|}{\hat{\sigma}\sqrt{\tilde{\mathbf{x}}'(X'X)^{-1}\tilde{\mathbf{x}}}} < \sup_{-\infty<x<\infty} \frac{|\tilde{\mathbf{x}}'\beta_2 - \tilde{\mathbf{x}}'\hat{\beta}|}{\hat{\sigma}\sqrt{\tilde{\mathbf{x}}'(X'X)^{-1}\tilde{\mathbf{x}}}}.$$

Intuitively, one would expect that any sensible inferential procedure would always claim the first model $\tilde{\mathbf{x}}'\beta_1$ is a plausible model if the second model $\tilde{\mathbf{x}}'\beta_2$ is claimed to be a plausible model. This, however, is not the case when using the confidence region (7.2) or the equivalent confidence band (7.3), as shown in the next subsection. The reason of this undesirable phenomenon is that the confidence band (7.3) is for the regression model $\mathbf{x}'\beta$, which does not use the special form of the polynomial regression model and so is not pertinent for making inference about the polynomial model $\tilde{\mathbf{x}}'\beta$ of interest.

7.1.1 Hyperbolic bands

An appropriate hyperbolic confidence band for the polynomial model $\tilde{\mathbf{x}}'\beta$ over a pre-specified interval $x \in (a,b)$ is given by

$$\tilde{\mathbf{x}}'\beta \in \tilde{\mathbf{x}}'\hat{\beta} \pm c\hat{\sigma}\sqrt{\tilde{\mathbf{x}}'(\mathbf{X}'\mathbf{X})^{-1}\tilde{\mathbf{x}}} \quad \forall\, x \in (a,b) \tag{7.4}$$

where c is a suitably chosen critical constant so that the simultaneous confidence level of this band is $1-\alpha$. It is clear that the value of c is strictly smaller than $\sqrt{(p+1)f^{\alpha}_{p+1,v}}$ used in the confidence band (7.3) since the confidence band (7.4) is over $\{\tilde{\mathbf{x}} = (1,x,\cdots,x^p)' : x \in (a,b)\}$, a set strictly smaller than $\Re^{p+1}$ even when $a = -\infty$ and $b = \infty$.

For the special case of $p = 2$ and $(a,b) = (-\infty,\infty)$, that is, for a quadratic polynomial regression model over the entire line $x \in (-\infty,\infty)$, Wynn and Bloomfield (1971) expressed the simultaneous confidence level of the band (7.4) as a three-dimensional integral of a trivariate t density function (cf. Dunnett and Sobel, 1954b), from which the exact value of c can be computed by using numerical quadrature. For $p = 2$ and $\mathbf{X}$ satisfying certain conditions (Conditions (17) of Spurrier, 1993), Spurrier (1992, 1993) expressed the simultaneous confidence level as a three-dimensional integral of a trivariate t density. Piegorsch (1986) constructed simultaneous confidence bands for a quadratic regression function with zero intercept over an interval of the covariate.

Wynn (1984) constructed a confidence band of neither hyperbolic nor constant width; it is extended from a set of simultaneous confidence intervals over a finite set of points x_i's in $(-\infty,\infty)$ at which all the estimators $(1,x_i,\cdots,x_i^p)\hat{\beta}$ are statistically independent. Spurrier (1993) compared under the average width criterion Wynn's (1984) band with band (7.4) but only for p=2 and some special situations.

Knafl *et al.* (1985) provided a method to construct an approximate two-sided hyperbolic band; this method is most suitable for a regression model that may have several predictors but all the predictors are given functions of only one variable x, such as the polynomial regression model (7.1). The basic idea is firstly to choose a suitable finite grid $G = \{g_1,\cdots,g_J\}$ over the interval $x \in (a,b)$, that is, $a = g_1 < \cdots < g_J = b$. Then the critical constant c is determined so that the simultaneous coverage probability of the confidence band covering the regression model only at all the $x = g_i$'s is guaranteed to be no less than $1-\alpha$; a simple up-crossing probability inequality was used to determine a conservative c for this purpose. Finally, the 'discrete' band over the grid G is extended to the whole interval $x \in (a,b)$ by interpolation, utilizing the functional form of the regression model. Strictly speaking, the band constructed in this way is not of hyperbolic shape. But if the grid G is sufficiently dense on $[a,b]$ then this band looks pretty similar to a hyperbolic band. On the other hand, for the up-crossing probability inequality to produce a not too conservative c value, the grid G should not be overly dense. This method was further developed by Rychlik (1992) to construct approximate hyperbolic bands for some more general linear regression models.

The famous result of Naiman (1986), which relates to the volume of a tube about a one-dimensional manifold in the unit sphere in $\Re^{p+1}$, can be used to provide a lower bound on the simultaneous confidence level and so a conservative critical constant c. McCann and Edwards (1996) applied Naiman's (1986) result to provide a set of conservative simultaneous confidence intervals for the pairwise differences of several normal means. In particular, Naiman's (1986) result gives the following lower bound on the confidence level of band (7.4):

$$1-\int_0^{1/c} \min\{F_{p-1,2}\left(2\left((ct)^{-2}-1\right)/(p-1)\right)\Lambda(\gamma)/\pi$$
$$+F_{p,1}\left(\left((ct)^{-2}-1\right)/p\right),\ 1\}f_T(t)dt$$

where $f_T(\cdot)$ denotes the pdf of $T := \sqrt{\chi_\nu^2/\nu}/\sqrt{\chi_{p+1}^2}$ with the two chi-square random variables χ_ν^2 and χ_{p+1}^2 being independent, and $\Lambda(\gamma) = \int_a^b \|\gamma'(x)\|dx$ is the length of the path γ (in the unit sphere centered at the origin in $\Re^{p+1}$) defined by

$$\gamma(x) = \frac{(\mathbf{X}'\mathbf{X})^{-1/2}\tilde{\mathbf{x}}}{\|(\mathbf{X}'\mathbf{X})^{-1/2}\tilde{\mathbf{x}}\|} \quad \text{for } x \in (a,b).$$

Tube volume theory was started by Hotelling (1939) and Weyl (1939) and has been studied by Naiman (1986, 1990), Johnstone and Siegmund (1989), Knowles and Siegmund (1989), Johansen and Johnstone (1990), Sun (1993) and Efron (1997) among others. Sun and Loader (1994) used these results to provide an approximate formula for the simultaneous confidence level of a two-sided hyperbolic simultaneous confidence band for a general curvilinear regression model, from which an approximate critical constant c can be computed. In particular, this can be applied to give the following approximate confidence level for band (7.4):

$$1-\frac{\kappa_0}{\pi}\left(1+\frac{c^2}{\nu}\right)^{-\nu/2} - \mathrm{P}\{|t_\nu| > c\}$$

where $\kappa_0 = \int_a^b \|T'(x)\|dx$ is the length of the path (in the unit sphere centered at the origin in $\Re^n$) defined by

$$T(x) = \frac{\mathbf{X}(\mathbf{X}'\mathbf{X})^{-1}\tilde{\mathbf{x}}}{\|\mathbf{X}(\mathbf{X}'\mathbf{X})^{-1}\tilde{\mathbf{x}}\|} \quad \text{for } x \in (a,b).$$

Note that $\kappa_0 = \Lambda(\gamma)$ in fact. Faraway and Sun (1995) considered the construction of approximate two-sided hyperbolic bands when the variances of the errors e_i may be different at different Y_i's. Piegorsch (1987b) considered the robustness of simultaneous confidence bands for a regression straight line considered in Chapter 2 while the true regression model is a quadratic polynomial function. Loader and

Sun (1997) investigated by simulation the robustness of tube formula-based confidence bands to normal-error assumption. The problem considered in Pirgorsch *et al.* (2005) is also related to simultaneous confidence bands for a quadratic polynomial regression model, whilst Hoel (1954), Halperin *et al.* (1967) and Gafarian (1978) constructed conservative confidence bands for a polynomial regression model.

For the general setting here, including the special case of $(a,b)=(-\infty,\infty)$, we use simulation in a similar way as in the previous chapters to approximate c, following Liu, Wynn and Hayter (2008). Kosorok and Qu (1999) used simulation to construct simultaneous confidence bands for polynomial regressions in a different context. The simultaneous confidence level of band (7.4) is given by $\mathrm{P}\{S \le c\}$ where

$$\begin{aligned} S &= \sup_{a<x<b} \frac{|\tilde{\mathbf{x}}'(\hat{\boldsymbol{\beta}}-\boldsymbol{\beta})/\hat{\sigma}|}{\sqrt{\tilde{\mathbf{x}}'(\mathbf{X}'\mathbf{X})^{-1}\tilde{\mathbf{x}}}} \\ &= \sup_{a<x<b} \frac{|\tilde{\mathbf{x}}'\mathbf{N}/(\hat{\sigma}/\sigma)|}{\sqrt{\tilde{\mathbf{x}}'(\mathbf{X}'\mathbf{X})^{-1}\tilde{\mathbf{x}}}} \\ &= K_{2h}(\mathbf{T},(\mathbf{X}'\mathbf{X})^{-1},(a,b)) \end{aligned} \tag{7.5}$$

where $\mathbf{N}=(\hat{\boldsymbol{\beta}}-\boldsymbol{\beta})/\sigma \sim \mathcal{N}_{p+1}(\mathbf{0},(\mathbf{X}'\mathbf{X})^{-1})$, $\mathbf{T}=\mathbf{N}/(\hat{\sigma}/\sigma) \sim \mathcal{T}_{p+1,v}(\mathbf{0},(\mathbf{X}'\mathbf{X})^{-1})$ and

$$K_{2h}(\mathbf{t},\Delta,(a,b)) = \sup_{a<x<b} \frac{|\tilde{\mathbf{x}}'\mathbf{t}|}{\sqrt{\tilde{\mathbf{x}}'\Delta\tilde{\mathbf{x}}}}. \tag{7.6}$$

An efficient algorithm for computing $K_{2h}(\mathbf{z},\Delta,(a,b))$ is provided in Appendix E.

To simulate one S, one simulates independent $\mathbf{N}=(\hat{\boldsymbol{\beta}}-\boldsymbol{\beta})/\sigma \sim \mathcal{N}_{p+1}(\mathbf{0},(\mathbf{X}'\mathbf{X})^{-1})$ and $\hat{\sigma}/\sigma \sim \sqrt{\chi^2_v/v}$, and calculate S from (7.5) by using the algorithm in Appendix E to compute $K_{2h}(\mathbf{T},(\mathbf{X}'\mathbf{X})^{-1},(a,b))$. We simulate R independent replicates $S_1,\cdots,S_R$ of S, and use the $\langle(1-\alpha)R\rangle$th largest S_i's, denoted as $\hat{c}$, as an approximation to c. The accuracy of $\hat{c}$ can be assessed in a similar way as before; see Appendix A.

The confidence band (7.4) induces a $1-\alpha$ level confidence region for $\boldsymbol{\beta}$:

$$C^*(\mathbf{Y},\mathbf{X},(a,b)) = \left\{\boldsymbol{\beta}:\ \sup_{a<x<b} \frac{|\tilde{\mathbf{x}}'(\hat{\boldsymbol{\beta}}-\boldsymbol{\beta})|}{\hat{\sigma}\sqrt{\tilde{\mathbf{x}}'(\mathbf{X}'\mathbf{X})^{-1}\tilde{\mathbf{x}}}} \le c\right\}. \tag{7.7}$$

Partition the union set of the confidence regions $C(\mathbf{Y},\mathbf{X})$ in (7.2) and $C^*(\mathbf{Y},\mathbf{X},(a,b))$ in (7.7) into three subsets:

$$\begin{aligned} S_1 &= C(\mathbf{Y},\mathbf{X}) - C^*(\mathbf{Y},\mathbf{X},(a,b)), \\ S_2 &= C(\mathbf{Y},\mathbf{X}) \cap C^*(\mathbf{Y},\mathbf{X},(a,b)), \\ S_3 &= C^*(\mathbf{Y},\mathbf{X},(a,b)) - C(\mathbf{Y},\mathbf{X}). \end{aligned}$$

It can be shown that all three subsets are non-empty when $p > 1$. If $\beta_1 \in S_3$ and $\beta_2 \in S_1$ then we have

$$\sup_{a<x<b} \frac{|\tilde{\mathbf{x}}'(\hat{\beta} - \beta_1)|}{\hat{\sigma}\sqrt{\tilde{\mathbf{x}}'(\mathbf{X}'\mathbf{X})^{-1}\tilde{\mathbf{x}}}} \leq c < \sup_{a<x<b} \frac{|\tilde{\mathbf{x}}'(\hat{\beta} - \beta_2)|}{\hat{\sigma}\sqrt{\tilde{\mathbf{x}}'(\mathbf{X}'\mathbf{X})^{-1}\tilde{\mathbf{x}}}}$$

since $\beta_1 \in C^*(\mathbf{Y}, \mathbf{X}, (a,b))$ but $\beta_2 \notin C^*(\mathbf{Y}, \mathbf{X}, (a,b))$. However, the model $\tilde{\mathbf{x}}'\beta_1$ is deemed not to be plausible but $\tilde{\mathbf{x}}'\beta_2$ is deemed to be plausible by the confidence region $C(\mathbf{Y}, \mathbf{X})$ since $\beta_1 \notin C(\mathbf{Y}, \mathbf{X})$ but $\beta_2 \notin C(\mathbf{Y}, \mathbf{X})$. This is the undesirable property of the confidence region $C(\mathbf{Y}, \mathbf{X})$ pointed out in the last subsection. The confidence region $C^*(\mathbf{Y}, \mathbf{X}, (a,b))$ does not have this problem due to its construction, which uses the functional relationships among the covariates of a polynomial regression model.

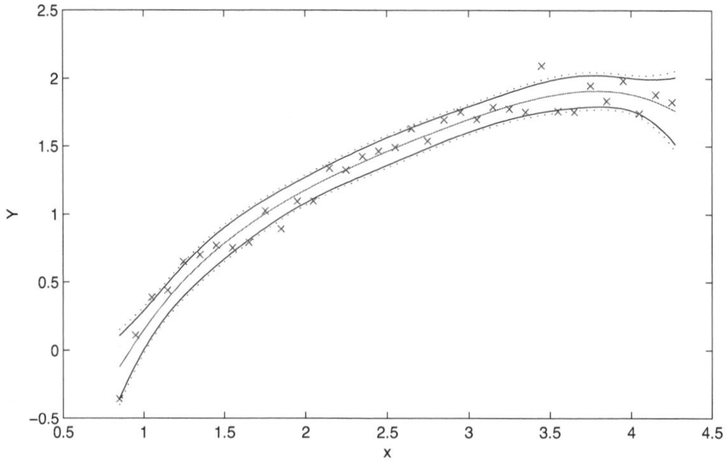

Figure 7.1: *The observations, fitted model and confidence bands for the perinatal data*

Example 7.1 Selvin (1998, p224) provided a data set on perinatal mortality (fetal deaths plus deaths within the first month of life) rate (PMR) and birth weight (BW) collected in California in 1998. The interest is on modelling how PMR changes with BW; Selvin (1998) considered fitting a 4th order polynomial regression model between $Y = \log(-\log(PMR))$ and $x = BW$. We will focus on the black infants only. Note that, in Selvin's data set, the first (corresponding to $BW < 0.85kg$) and the last (corresponding to $BW > 4.25kg$) observations are clearly different from the other 35 observations (which have BW in an interval of length 0.1kg and the middle point of the interval is used as the value of BW), and so discarded from our analysis below. Indeed the R^2 values for fitting a 4th order

Table 7.1: *Perinatal mortality data for black infants from Selvin (1998)*

i	x_i	Y_i	i	x_i	Y_i	i	x_i	Y_i
1	0.85	-0.3556	13	2.05	1.0972	25	3.25	1.7774
2	0.95	0.1089	14	2.15	1.3382	26	3.35	1.7538
3	1.05	0.3880	15	2.25	1.3254	27	3.45	2.0933
4	1.15	0.4399	16	2.35	1.4241	28	3.55	1.7594
5	1.25	0.6513	17	2.45	1.4632	29	3.65	1.7538
6	1.35	0.7022	18	2.55	1.4906	30	3.75	1.9478
7	1.45	0.7706	19	2.65	1.6324	31	3.85	1.8351
8	1.55	0.7523	20	2.75	1.5383	32	3.95	1.9830
9	1.65	0.7934	21	2.85	1.6955	33	4.05	1.7429
10	1.75	1.0233	22	2.95	1.7538	34	4.15	1.8827
11	1.85	0.8918	23	3.05	1.6998	35	4.25	1.8269
12	1.95	1.0959	24	3.15	1.7903			

polynomial regression model before and after deleting these two observations are markedly different, given by 90.9% and 97.1% respectively. The 35 observations (Y_i, x_i) are given in Table 7.1.

These observations and the fitted 4th order polynomial model, $Y = -2.861 + 4.809x - 2.316x^2 + 0.568x^3 - 0.054x^4$, are plotted in Figure 7.1. From the plot, it seems that a 4th order polynomial model fits the observation quite well at least over the observed range of $x \in (0.85, 4.25)$. To assess the plausible range of the 4th order polynomial (from which the observations are assumed to be generated from), we construct the 95% two-sided hyperbolic band (7.4) over $x \in (0.85, 4.25)$. The critical constant c is calculated to be 2.985 based on 100,000 simulations. This confidence band is given in Figure 7.1 by the two solid-line curves around the fitted model. From this confidence band, any polynomial model of order 4 or lower that falls completely inside this band over $x \in (0.85, 4.25)$ is deemed to be a plausible model. For example, both the least square quadratic and cubic fits to the observations fall completely inside the band, and so are plausible models. On the other hand, any polynomial model of order 4 or lower that falls outside the band for at least one $x \in (0.85, 0.425)$ is deemed not to be a plausible model. For example the least square straight line fit to the observations is not plausible.

The level 95% confidence band (7.4) over the whole range $x \in (-\infty, \infty)$ has the critical value $c = 3.032$. But a negative x value of birth weight is obviously meaningless here. The critical value of the F band (7.3) is given by $\sqrt{5 f_{5,30}^{0.05}} = 3.559$, which is over 19% larger than the critical value 2.985 of band (7.4). The F band restricted to $x_i = x^i$ over $x \in (0.85, 4.25)$ is given by the two dotted curves in Figure 7.1. Note that if $c = 3.559$ is used in the confidence band (7.4) with $(a, b) = (0.85, 4.25)$ then the simultaneous confidence level of the band is about 98.7%, considerably larger than the nominal level 95%.

In Example 7.1, one may be interested only in the upper limits of perinatal mortality rates over a given range of x. If one-sided inference about the polynomial model $\tilde{\mathbf{x}}'\beta$ is of interest, one can use one-sided simultaneous confidence bands. The lower hyperbolic confidence band has the form

$$\tilde{\mathbf{x}}'\beta > \tilde{\mathbf{x}}'\hat{\beta} - c\hat{\sigma}\sqrt{\tilde{\mathbf{x}}'(\mathbf{X}'\mathbf{X})^{-1}\tilde{\mathbf{x}}} \quad \forall\, x \in (a,b) \tag{7.8}$$

where c is the critical constant chosen so that the simultaneous confidence level of this band is $1-\alpha$. The upper hyperbolic confidence band can be constructed in a similar way and use the same critical constant c as the lower band (7.8).

For the general setting here, we again use simulation to approximate c. The simultaneous confidence level of the band (7.8) is given by $\mathrm{P}\{S \le c\}$ where

$$\begin{aligned} S &= \sup_{a<x<b} \frac{\tilde{\mathbf{x}}'(\hat{\beta}-\beta)/\hat{\sigma}}{\sqrt{\tilde{\mathbf{x}}'(\mathbf{X}'\mathbf{X})^{-1}\tilde{\mathbf{x}}}} \\ &= \sup_{a<x<b} \frac{\tilde{\mathbf{x}}'\mathbf{T}}{\sqrt{\tilde{\mathbf{x}}'(\mathbf{X}'\mathbf{X})^{-1}\tilde{\mathbf{x}}}} \\ &= K_{1h}(\mathbf{T}, (\mathbf{X}'\mathbf{X})^{-1}, (a,b)) \end{aligned} \tag{7.9}$$

where $\mathbf{T} = (\hat{\beta}-\beta)/\hat{\sigma}$ as in the two-sided case and

$$K_{1h}(\mathbf{t}, \Delta, (a,b)) = \sup_{a<x<b} \frac{\tilde{\mathbf{x}}'\mathbf{t}}{\sqrt{\tilde{\mathbf{x}}'\Delta\tilde{\mathbf{x}}}}.$$

An efficient algorithm for computing $K_{1h}(\mathbf{t}, \Delta, (a,b))$ is provided in Appendix E.

To simulate one S, one simulates $\mathbf{T} = (\hat{\beta}-\beta)/\hat{\sigma} \sim \mathcal{T}_{p+1,\nu}(\mathbf{0}, (\mathbf{X}'\mathbf{X})^{-1})$, and calculates S from (7.9). The critical constant c can therefore be computed by simulation as before.

Example 7.2 For the data set given in Example 7.1, suppose we are interested only in the upper bounds on the 4th order polynomial regression function over $x \in (0.85, 4.25)$. The critical constant of the 95% upper confidence band is computed to be 2.687 based on 100,000 simulation. This critical value is 10% smaller than the two-sided critical constant 2.987 given in Example 7.1. Figure 7.2 gives this upper confidence band (solid curve), with the two-sided confidence band given by the two dotted curves. As expected, the upper confidence band is below the upper part of the two-sided confidence band. The critical constant of the upper confidence band over $x \in (-\infty, \infty)$ is given by 2.7383.

7.1.2 *Constant width bands*

A constant width confidence band for the polynomial model $\tilde{\mathbf{x}}'\beta$ over a pre-specified interval $x \in (a,b)$ is given by

$$\tilde{\mathbf{x}}'\beta \in \tilde{\mathbf{x}}'\hat{\beta} \pm c\hat{\sigma}\,, \quad \forall\, x \in (a,b) \tag{7.10}$$

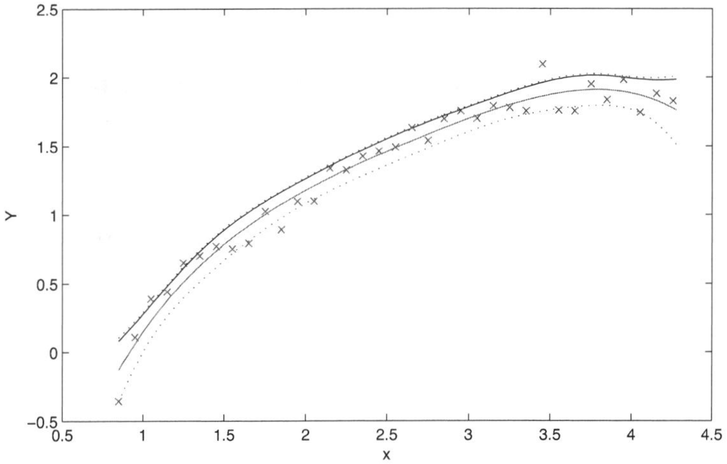

Figure 7.2: *The upper and two-sided confidence bands for the perinatal data*

where c is a suitably chosen critical constant so that the simultaneous confidence level of this band is $1-\alpha$. For the special case of $p=2$, Trout and Chow (1972, 1973) considered the constant width band and expressed the simultaneous confidence level as a three-dimensional integral of a trivariate t density.

For a general $p \geq 2$, simulation can again be used to approximate the critical constant c. The simultaneous confidence level of the band (7.10) is given by $\mathrm{P}\{S \leq c\}$ where

$$\begin{aligned} S &= \sup_{a<x<b} |\tilde{\mathbf{x}}'(\hat{\beta}-\beta)/\hat{\sigma}| \\ &= \sup_{a<x<b} |\tilde{\mathbf{x}}'\mathbf{T}| \\ &= K_{2c}(\mathbf{T},(a,b)) \end{aligned} \tag{7.11}$$

where $\mathbf{T}=(\hat{\beta}-\beta)/\hat{\sigma}$ as in the hyperbolic bands, and

$$K_{2c}(\mathbf{t},(a,b)) = \sup_{a<x<b} |\tilde{\mathbf{x}}'\mathbf{t}|.$$

An efficient algorithm for computing $K_{2c}(\mathbf{t},(a,b))$ is provided in Appendix E.

To simulate one S, one simulates $\mathbf{T} \sim \mathcal{T}_{p+1,\nu}(\mathbf{0},(\mathbf{X}'\mathbf{X})^{-1})$, and calculates S from (7.11) by using the algorithm in Appendix E. The critical constant c can therefore be computed by using simulation as before.

Example 7.3 For the problem considered in Example 7.1, one can also use a constant width band to bound the unknown 4th order polynomial regression function. The critical constant c of the 95% two-sided constant width band (7.10) over $x \in (0.85, 4.25)$ is computed to be 1.681 based on 100,000 simulations. Figure 7.3 gives the constant width band by the two solid curves, and the two-sided hyperbolic band by the two dotted curves. Similar to the observations made in the previous chapters, the hyperbolic band tends to be narrower than the constant width band in the middle part of the covariate interval $(0.85, 4.25)$, and vice versa near the two ends of the interval $(0.85, 4.25)$.

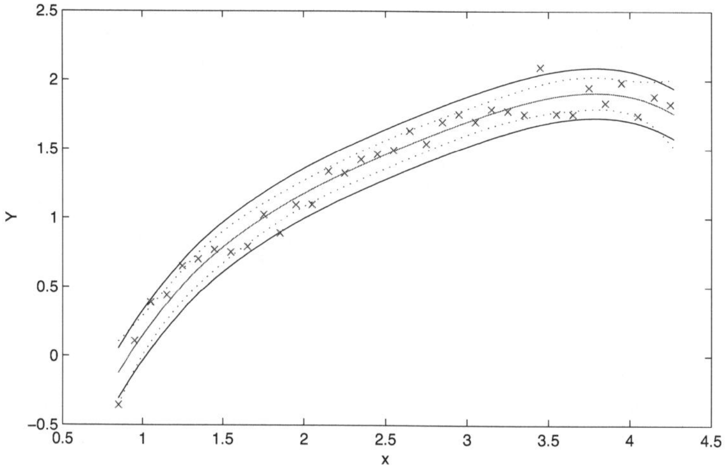

Figure 7.3 *The two-sided constant width and hyperbolic confidence bands for the perinatal data*

One can also construct a one-sided constant width band for making one-sided inferences. A lower constant width confidence band for the polynomial model $\tilde{\mathbf{x}}'\beta$ over a pre-specified interval $x \in (a, b)$ is given by

$$\tilde{\mathbf{x}}'\beta > \tilde{\mathbf{x}}'\hat{\beta} - c\hat{\sigma}, \ \forall x \in (a, b) \tag{7.12}$$

where c is a suitably chosen critical constant so that the simultaneous confidence level of this band is $1 - \alpha$. An upper constant width confidence band can be constructed in a similar way and uses the same critical constant c as the lower constant width confidence band (7.12).

The simultaneous confidence level of the band (7.12) is given by $P\{S \leq c\}$ where

$$S = \sup_{a<x<b} \tilde{\mathbf{x}}'(\hat{\beta} - \beta)/\hat{\sigma}$$

$$= \sup_{a<x<b} \tilde{\mathbf{x}}'\mathbf{T}$$
$$= K_{1c}(\mathbf{T},(a,b)) \qquad (7.13)$$

where $\mathbf{T} = (\hat{\boldsymbol{\beta}} - \boldsymbol{\beta})/\hat{\sigma}$ as before, and

$$K_{1c}(\mathbf{t},(a,b)) = \sup_{a<x<b} \tilde{\mathbf{x}}'\mathbf{t}.$$

An efficient algorithm for computing $K_{1c}(\mathbf{t},(a,b))$ is provided in Appendix E. So simulation can again be used to approximate the critical constant c.

Example 7.4 For the problem considered in Example 7.1, one can use a constant width upper band to bound the unknown 4th order polynomial regression function from above. The critical constant c of the 95% one-sided constant width band (7.12) over $x \in (0.85, 4.25)$ is computed to be 1.460 based on 100,000 simulations, which is over 13% smaller than the two-sided critical constant 1.681. Figure 7.4 gives the upper constant width band by the solid curve, and the two-sided constant width band by the two dotted curves.

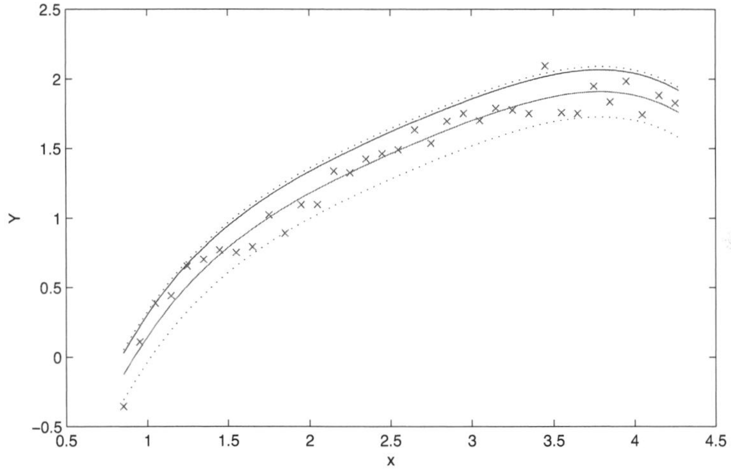

Figure 7.4: *The upper and two-sided constant width bands for the perinatal data*

7.2 Confidence bands for part of a polynomial model

Another important inferential problem is to assess whether the polynomial model (7.1) can be simplified. Specifically, if some of the coefficients β_i's are equal to zero then the corresponding terms x^i's have no effect on the response variable

Y, and so the model can be simplified. For a polynomial regression model, it is suggested (cf. Draper and Smith, 1998) that if the term x^i is included in the model then all the lower order terms $1, x, \cdots, x^{i-1}$ should be retained in the model too. So it is of particular interest for polynomial regression model (7.1) to assess whether $\boldsymbol{\beta}_2 = (\beta_{p-k+1}, \cdots, \beta_p)' = \mathbf{0}$ for some given $1 \leq k \leq p$. If $\boldsymbol{\beta}_2$ is zero then the terms $x^{p-k+1}, \cdots, x^p$ have no effect on the response variable Y and so model (7.1) reduces to the lower order polynomial model

$$\mathbf{Y} = \mathbf{X}_1 \boldsymbol{\beta}_1 + \mathbf{e} \tag{7.14}$$

where $\mathbf{X}_1$ is formed by the first $p-k+1$ columns of the matrix $\mathbf{X}$, and $\boldsymbol{\beta}_1 = (\beta_0, \cdots, \beta_{p-k})'$.

Throughout this section we assume $k \geq 2$. If $k = 1$ then all the tests become the t tests and the confidence regions become the t confidence intervals, and there is no improvement achieved by the new methods introduced in this section.

7.2.1 Partial F test approach

One may continue to apply the partial F test as in Chapter 4 to test the hypotheses

$$H_0: \boldsymbol{\beta}_2 = \mathbf{0} \quad \text{against} \quad H_a: \boldsymbol{\beta}_2 \neq \mathbf{0},$$

and H_0 is rejected if and only if

$$\begin{aligned} & \frac{(SS_R \text{ of model } (7.1) - SS_R \text{ of model } (7.14))/k}{\text{MS residual of model } (7.1)} \\ = & \frac{\hat{\boldsymbol{\beta}}_2' \mathbf{V}^{-1} \hat{\boldsymbol{\beta}}_2 / k}{\hat{\sigma}^2} > f_{k,\nu}^{\alpha}, \end{aligned} \tag{7.15}$$

where $\hat{\boldsymbol{\beta}}_2$ denotes the estimator of $\boldsymbol{\beta}_2 = (b_{p-k+1}, \cdots, b_p)'$, which is formed by the last k components of $\hat{\boldsymbol{\beta}}$ and has the distribution $\mathcal{N}(\boldsymbol{\beta}_2, \sigma^2 \mathbf{V})$ where $\mathbf{V}$ is a $k \times k$ partition matrix formed from the last k rows and the last k columns of $(\mathbf{X}'\mathbf{X})^{-1}$. Since $\mathbf{V}$ is non-singular, let $\mathbf{W}$ be the square root matrix of $\mathbf{V}$ satisfying $\mathbf{V} = \mathbf{W}^2$.

The partial F test can be derived from the exact $1-\alpha$ level confidence region for $\boldsymbol{\beta}_2$

$$\left\{ \boldsymbol{\beta}_2 : \frac{(\hat{\boldsymbol{\beta}}_2 - \boldsymbol{\beta}_2)' \mathbf{V}^{-1} (\hat{\boldsymbol{\beta}}_2 - \boldsymbol{\beta}_2)/k}{\hat{\sigma}^2} \leq f_{k,\nu}^{\alpha} \right\}; \tag{7.16}$$

H_0 is rejected if and only if the origin is outside this confidence region.

Denote $\tilde{\mathbf{x}}_2 = (x^{p-k+1}, \cdots, x^p)'$, the covariate terms corresponding to $\boldsymbol{\beta}_2$, and denote $\mathbf{x}_2 = (x_{p-k+1}, \cdots, x_p)$. Following from Theorems 4.1 and 4.2, the confidence region for $\boldsymbol{\beta}_2$ in (7.16) is equivalent to the following exact $1-\alpha$ level simultaneous confidence band for the hyper-plane $\mathbf{x}_2' \boldsymbol{\beta}_2$:

$$\mathbf{x}_2' \boldsymbol{\beta}_2 \in \mathbf{x}_2' \hat{\boldsymbol{\beta}}_2 \pm \sqrt{k f_{k,\nu}^{\alpha}}\, \hat{\sigma} \sqrt{\mathbf{x}_2' \mathbf{V} \mathbf{x}_2} \quad \forall\, \mathbf{x}_2 \in \Re^k. \tag{7.17}$$

The partial F test (7.15) rejects H_0 if and only if the zero hyper-plane $\mathbf{x}_2'\mathbf{0}$ is outside this confidence band for at least one $\mathbf{x}_2 \in \Re^k$.

As argued in Section 7.1, the confidence band (7.17) does not take fully into account the functional relationships among the covariates $x_i = x^i$ for $i = p-k+1, \cdots, p$ in the polynomial model (7.1). Consequently, the partial F test (7.15) or equivalently the confidence band (7.17) have some undesirable property. For example, even though one observation $\mathbf{Y}_1$, with the corresponding estimates $\hat{\beta}_{21}$ and $\hat{\sigma}_1$, indicates more strongly that $\tilde{\mathbf{x}}_2'\beta_2$ is closer to the zero polynomial function than another observation $\mathbf{Y}_2$, with the corresponding estimates $\hat{\beta}_{22}$ and $\hat{\sigma}_2$, in the sense that

$$\sup_{-\infty<x<\infty} \frac{|\tilde{\mathbf{x}}_2'\hat{\beta}_{21}|}{\hat{\sigma}_1\sqrt{\tilde{\mathbf{x}}_2'\mathbf{V}\tilde{\mathbf{x}}_2}} < \sup_{-\infty<x<\infty} \frac{|\tilde{\mathbf{x}}_2'\hat{\beta}_{22}|}{\hat{\sigma}_2\sqrt{\tilde{\mathbf{x}}_2'\mathbf{V}\tilde{\mathbf{x}}_2}},$$

the partial F test may reject H_0 when $\mathbf{Y}_1$ is observed but not reject H_0 when $\mathbf{Y}_2$ is observed. Such observations $\mathbf{Y}_1$ and $\mathbf{Y}_2$ are identified in the next subsection. The underlying reason is that the functional relationships among the covariate terms of $\tilde{\mathbf{x}}_2$ are not used fully in the partial F test (7.15) or the equivalent confidence band (7.17).

7.2.2 *Hyperbolic bands*

Since we are interested in assessing the magnitude of $\tilde{\mathbf{x}}_2'\beta_2$, often over a given range of x only, (a,b) say, it is natural to bound $\tilde{\mathbf{x}}_2'\beta_2$ by the hyperbolic confidence band

$$\tilde{\mathbf{x}}_2'\beta_2 \in \tilde{\mathbf{x}}_2'\hat{\beta}_2 \pm c\,\hat{\sigma}\sqrt{\tilde{\mathbf{x}}_2'\mathbf{V}\tilde{\mathbf{x}}_2} \quad \forall\, x \in (a,b) \tag{7.18}$$

where c is a suitably chosen critical constant so that the simultaneous confidence level of this band is $1-\alpha$.

To determine the critical constant c, note that the confidence level is given by $\mathrm{P}\{S \le c\}$ where

$$\begin{aligned} S &= \sup_{a<x<b} \frac{|\tilde{\mathbf{x}}_2'(\hat{\beta}_2-\beta_2)|}{\hat{\sigma}\sqrt{\tilde{\mathbf{x}}_2'\mathbf{V}\tilde{\mathbf{x}}}} \\ &= \sup_{a<x<b} \frac{|(1,\cdots,x^{k-1})(\hat{\beta}_2-\beta_2)|}{\hat{\sigma}\sqrt{(1,\cdots,x^{k-1})\mathbf{V}\begin{pmatrix}1\\ \vdots \\ x^{k-1}\end{pmatrix}}} \end{aligned} \tag{7.19}$$

by noting that $\tilde{\mathbf{x}}_2' = x^{p-k+1}(1,x,\cdots,x^{k-1})$. So the critical constant c is given by the critical constant c in the hyperbolic band (7.4) but with $p = k-1$ and $(\mathbf{X}'\mathbf{X})^{-1}$ replaced by $\mathbf{V}$. The critical constant c can therefore be computed by using the simulation method for the confidence band (7.4).

Note that the confidence band (7.18) for $\tilde{\mathbf{x}}_2'\beta_2$ is equivalent to the following exact $1-\alpha$ confidence region for β_2

$$\left\{\beta_2 : \sup_{a<x<b} \frac{|\tilde{\mathbf{x}}_2'(\hat{\beta}_2 - \beta_2)|}{\hat{\sigma}\sqrt{\tilde{\mathbf{x}}_2'\mathbf{V}\tilde{\mathbf{x}}}} \le c\right\}. \tag{7.20}$$

Let

$$C^*(\mathbf{X},k,(a,b)) = \left\{\mathbf{Y} = (Y_1,\cdots,Y_n)' : \sup_{a<x<b} \frac{|\tilde{\mathbf{x}}_2'\hat{\beta}_2|}{\hat{\sigma}\sqrt{\tilde{\mathbf{x}}_2'\mathbf{V}\tilde{\mathbf{x}}}} \le c\right\}$$

which is the set of observations $\mathbf{Y}$ for which the null hypothesis $H_0 : \beta_2 = \mathbf{0}$ is not rejected by the confidence region (7.20). Similarly, let

$$C(\mathbf{X},k) = \left\{\mathbf{Y} = (Y_1,\cdots,Y_n)' : \frac{\hat{\beta}_2'\mathbf{V}^{-1}\hat{\beta}_2/k}{\hat{\sigma}^2} \le f_{k,\nu}^{\alpha}\right\} \tag{7.21}$$

which is the set of observations $\mathbf{Y}$ for which the null hypothesis $H_0 : \beta_2 = \mathbf{0}$ is not rejected by the confidence region (7.16).

The union of the sets $C(\mathbf{X},k)$ and $C^*(\mathbf{X},k,(a,b))$ can be partitioned into three subsets:

$$S_1 = C(\mathbf{X},k) - C^*(\mathbf{X},k,(a,b)),$$
$$S_2 = C(\mathbf{X},k) \cap C^*(\mathbf{X},k,(a,b)),$$
$$S_3 = C^*(\mathbf{X},k,(a,b)) - C(\mathbf{X},k).$$

It can be shown that all three sets are non-empty when $k > 1$.

If $\mathbf{Y}_1 \in S_3$ with the corresponding estimates $\hat{\beta}_{21}$ and $\hat{\sigma}_1$, and $\mathbf{Y}_2 \in S_1$ with the corresponding estimates $\hat{\beta}_{22}$ and $\hat{\sigma}_2$, then we have that

$$\sup_{a<x<b} \frac{|\tilde{\mathbf{x}}_2'\hat{\beta}_{21}|}{\hat{\sigma}_1\sqrt{\tilde{\mathbf{x}}_2'\mathbf{V}\tilde{\mathbf{x}}_2}} \le c < \sup_{a<x<b} \frac{|\tilde{\mathbf{x}}_2'\hat{\beta}_{22}|}{\hat{\sigma}_2\sqrt{\tilde{\mathbf{x}}_2'\mathbf{V}\tilde{\mathbf{x}}_2}}$$

since $\mathbf{Y}_1 \in C^*(\mathbf{X},k,(a,b))$ but $\mathbf{Y}_2 \notin C^*(\mathbf{X},k,(a,b))$. So, intuitively, the estimated model $\tilde{\mathbf{x}}_2'\hat{\beta}_{21}$ when observing $\mathbf{Y}_1$ is closer to the zero polynomial function than the estimated model $\tilde{\mathbf{x}}_2'\hat{\beta}_{22}$ when observing $\mathbf{Y}_2$. However, the confidence region (7.16) or equivalently the partial F test (7.15) rejects $H_0 : \beta_2 = \mathbf{0}$ when observing $\mathbf{Y}_1$ but does not reject H_0 when observing $\mathbf{Y}_2$ since $\mathbf{Y}_1 \notin C(\mathbf{X},k)$ but $\mathbf{Y}_2 \in C(\mathbf{X},k)$. This is the undesirable property of the confidence region (7.16) or the partial F test (7.15) pointed out in the last subsection.

The size α test induced by the confidence region (7.20) or equivalently the confidence band (7.18) rejects H_0 if and only if

$$\sup_{a<x<b} \frac{|\tilde{\mathbf{x}}_2'\hat{\beta}_2|}{\hat{\sigma}\sqrt{\tilde{\mathbf{x}}_2'\mathbf{V}\tilde{\mathbf{x}}_2}} > c. \tag{7.22}$$

It does not have this undesirable property of the partial F test (7.15) since its underlying confidence band (7.18) utilizes the functional relationships among the covariate terms of the model $\tilde{\mathbf{x}}_2'\beta_2$. Since the confidence band (7.18) can easily be plotted for any p and k values, this test is more useful than the partial F test in (7.15).

Finally, it is noteworthy that confidence band (7.18) has the equivalent form

$$x^{-p+k-1}\tilde{\mathbf{x}}_2'\beta_2 \in x^{-p+k-1}\tilde{\mathbf{x}}_2'\hat{\beta}_2 \pm c\,\hat{\sigma}|x|^{-p+k-1}\sqrt{\tilde{\mathbf{x}}_2'\mathbf{V}\tilde{\mathbf{x}}_2}\,,\ \ \forall\, x \in (a,b) \quad (7.23)$$

where $x^{-p+k-1}\tilde{\mathbf{x}}_2 = (1,\cdots,x^{k-1})'$. This form is easier to plot than that in (7.18) which can be either very small for a small x or very large for a large x.

Example 7.5 We continue to consider the perinatal mortality (PM) data set given in Example 7.1. From the data points plotted in Figure 7.1, one may wonder whether a quadratic polynomial model is sufficient. Indeed the adjusted R^2 value of the least square 4th order polynomial model is 96.8%, while the corresponding value of the quadratic polynomial model is 96.5%. For this, one may test $H_0 : \beta_2 = (\beta_3, \beta_4)' = \mathbf{0}$ against $H_a : \beta_2 \neq \mathbf{0}$. Note that testing the null hypotheses $H_0 :\ \beta_3 = 0$ and $H_0 :\ \beta_4 = 0$ at 5% individually in the usual backward manner does not control the overall type I familywise error rate at 5%. The p-value of the partial F test (7.15) is 9.9% and so $H_0 : \beta_2 = \mathbf{0}$ cannot be rejected at 5% level.

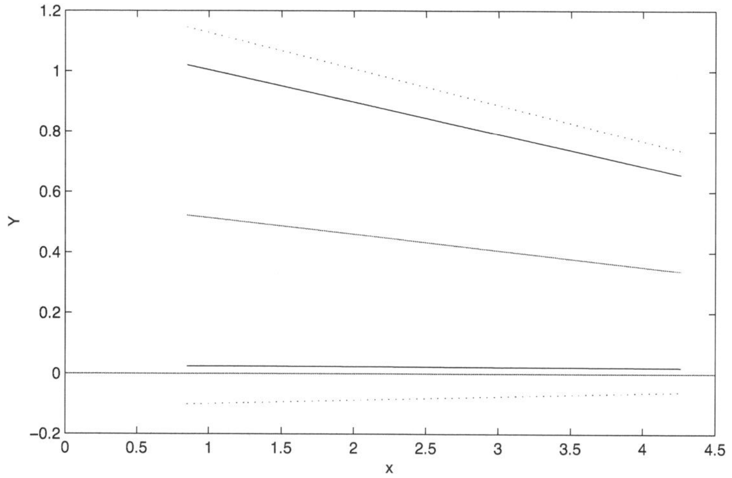

Figure 7.5: *The hyperbolic confidence band (7.23) for the perinatal data*

On the other hand, the test (7.22) with $(a,b) = (0.85, 4.25)$ has a p-value 3.8% and hence H_0 is rejected at $\alpha = 5\%$. In terms of confidence bands, the critical value c of band (7.18) for $\alpha = 5\%$ and $(a,b) = (0.85, 4.25)$ is given by 2.058 based on 100,000 simulation. Its equivalent form (7.23) is given in Figure 7.5 by the two solid lines. Since this band is completely above the zero line, the contribution of $\beta_3 x^3 + \beta_4 x^4$ is non-zero and, in fact, positive. So H_0 is clearly rejected from this confidence band. This agrees with the observed p-value 3.8% of course. On the other hand, the critical value of the F band (7.17) is given by $\sqrt{2 f_{2,30}^{0.05}} = 2.575$, which is about 25% larger than the critical value 2.058 of band (7.18). If this conservative critical value is used in the band (7.23), the band is plotted in Figure 7.5 by the two dotted lines. It contains the zero line completely as expected from the F band (7.17) which has a p-value 9.9%.

7.2.3   Constant width bands

One may also use the confidence band

$$\tilde{\mathbf{x}}_2' \boldsymbol{\beta}_2 \in \tilde{\mathbf{x}}_2' \hat{\boldsymbol{\beta}}_2 \pm c\,\hat{\sigma} |x|^{p-k+1} \quad \forall\, x \in (a,b) \tag{7.24}$$

to bound $\tilde{\mathbf{x}}_2' \boldsymbol{\beta}_2$ over $x \in (a,b)$, where c is a suitably chosen critical constant so that the simultaneous confidence level of this band is $1 - \alpha$. This confidence band is equivalent to the constant width band

$$x^{-p+k-1}\tilde{\mathbf{x}}_2' \boldsymbol{\beta}_2 \in x^{-p+k-1}\tilde{\mathbf{x}}_2' \hat{\boldsymbol{\beta}}_2 \pm c\,\hat{\sigma} \quad \forall\, x \in (a,b), \tag{7.25}$$

which is easier to plot than the band (7.24).

As before we can use simulation to approximate c. The simultaneous confidence level of the bands (7.24) or (7.25) is given by $P\{S \leq c\}$ where

$$\begin{aligned}
S &= \sup_{a<x<b} |(1,\cdots,x^{k-1})(\hat{\boldsymbol{\beta}}_2 - \boldsymbol{\beta}_2)/\hat{\sigma}| \\
&= \sup_{a<x<b} |(1,\cdots,x^{k-1})\mathbf{T}| \\
&= K_{2c}(\mathbf{T}, (a,b)) \tag{7.26}
\end{aligned}$$

where $\mathbf{T} = (\hat{\boldsymbol{\beta}}_2 - \boldsymbol{\beta}_2)/\hat{\sigma} \sim T_{k,\nu}(\mathbf{0}, \mathbf{V})$. To simulate one S, one simulates $\mathbf{T}$ from $T_{k,\nu}(\mathbf{0}, \mathbf{V})$ and calculates $K_{2c}(\mathbf{T}, (a,b))$ by using the efficient algorithm in Appendix E.

The confidence bands (7.24) or (7.25) induce a size α test for the hypotheses $H_0 : \boldsymbol{\beta}_2 = \mathbf{0}$ against $H_a : \boldsymbol{\beta}_2 \neq \mathbf{0}$; it rejects H_0 if and only if

$$\sup_{a<x<b} |(1,\cdots,x^{k-1})\hat{\boldsymbol{\beta}}_2|/\hat{\sigma} > c. \tag{7.27}$$

Example 7.6 Let us apply the constant width band (7.25) to the perinatal mortality (PM) data set given in Example 7.1 to assess whether a quadratic polynomial model is sufficient. For $\alpha = 5\%$ and $(a,b) = (0.85, 4.25)$, the critical constant is

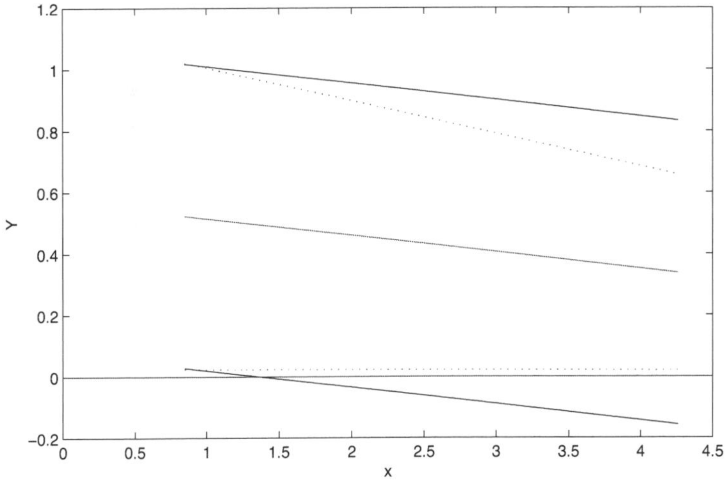

Figure 7.6: *The constant width band (7.25) for the perinatal data*

computed to be $c = 4.573$ based on 100,000 simulations, and the band is given by the two solid straight lines in Figure 7.6. Since this band does not contain the zero line over $x \in (0.85, 4.25)$, one can infer that $\beta_3 + \beta_4 x$ is unlikely to be zero and so $H_0 : \boldsymbol{\beta}_2 = \mathbf{0}$ is rejected. Indeed the p-value of the test (7.27) is calculated to be 0.038, which is smaller than $\alpha = 5\%$. Also given in Figure 7.6 is the hyperbolic band (7.23), given by the two dotted lines. It is clear from Figure 7.6 that, in this example, the constant width band (7.25) is slightly narrower than the hyperbolic band (7.23) only near $x = 0.85$, and so the hyperbolic band is more useful than the constant width band, even though the two corresponding tests have comparable p-values.

7.2.4 *Assessing equivalence to the zero function*

There may be a good reason, from subject knowledge for instance, to believe that $\tilde{\mathbf{x}}_2' \boldsymbol{\beta}_2$ should be close to the zero function. If the purpose of a study is to use the observed data to validate this, then the problem is better treated as an equivalence problem following similar arguments as in Section 4.3. Note that the primary concern in setting up the hypotheses $H_0 : \boldsymbol{\beta}_2 = \mathbf{0}$ against $H_a : \boldsymbol{\beta}_2 \neq \mathbf{0}$ is to establish that the null hypothesis H_0 is statistically significant and so some terms in $\tilde{\mathbf{x}}_2' \boldsymbol{\beta}_2$ do affect the response variable Y; non-rejection of H_0 does not imply $\boldsymbol{\beta}_2$ is equal to $\mathbf{0}$ or even close to $\mathbf{0}$.

For establishing equivalence, we set up the equivalence hypotheses as

$$H_0^E: \max_{a<x<b} |\tilde{\mathbf{x}}_2'\beta_2| \geq \delta \quad \text{against} \quad H_a^E: \max_{a<x<b} |\tilde{\mathbf{x}}_2'\beta_2| < \delta \tag{7.28}$$

where $\delta > 0$ is a pre-specified threshold of equivalence. The equivalence of $\tilde{\mathbf{x}}_2'\beta_2$ to the zero function over $x \in (a,b)$ is claimed if and only if the null hypothesis H_0^E is rejected. Since a hypotheses test guarantees the size at α, the chance of falsely claiming equivalence is no more than α.

Note that the equivalence hypotheses can be expressed as

$$H_0^E = \cup_{a<x<b} \{H_{0x}^{E+} \cup H_{0x}^{E-}\} \quad \text{against} \quad H_a^E = \cap_{a<x<b} \{H_{ax}^{E+} \cap H_{ax}^{E-}\}$$

where the individual hypotheses are given by

$$H_{0x}^{E+}: \tilde{\mathbf{x}}_2'\beta_2 \geq \delta, \quad H_{0x}^{E-}: \tilde{\mathbf{x}}_2'\beta_2 \leq -\delta$$
$$H_{ax}^{E+}: \tilde{\mathbf{x}}_2'\beta_2 < \delta, \quad H_{ax}^{E-}: \mathbf{x}_2'\beta_2 > -\delta.$$

The individual hypotheses H_{0x}^{E+} against H_{ax}^{E+} can be tested with the rejection region

$$\tilde{\mathbf{x}}_2'\hat{\beta}_2 + t_\nu^\alpha \hat{\sigma}\sqrt{\tilde{\mathbf{x}}_2'\mathbf{V}\tilde{\mathbf{x}}_2} < \delta.$$

The size of this individual test can be shown to be α in a similar way as in Section 4.3. Similarly, the individual hypotheses H_{0x}^{E-} against H_{ax}^{E-} can be tested with the rejection region

$$\tilde{\mathbf{x}}_2'\hat{\beta}_2 - t_\nu^\alpha \hat{\sigma}\sqrt{\tilde{\mathbf{x}}_2'\mathbf{V}\tilde{\mathbf{x}}_2} > -\delta,$$

and the size of this individual test is α. Hence the IU test (see Appendix D) of H_0^E against H_a^E based on these individual tests rejects H_0^E if and only if

$$-\delta < \min_{a<x<b} \{\tilde{\mathbf{x}}_2'\hat{\beta}_2 - t_\nu^\alpha \hat{\sigma}\sqrt{\tilde{\mathbf{x}}_2'\mathbf{V}\tilde{\mathbf{x}}_2}\} \text{ and } \max_{a<x<b} \{\tilde{\mathbf{x}}_2'\hat{\beta}_2 + t_\nu^\alpha \hat{\sigma}\sqrt{\tilde{\mathbf{x}}_2'\mathbf{V}\tilde{\mathbf{x}}_2}\} < \delta. \tag{7.29}$$

If one is to infer whether or not $\max_{a<x<b} |\tilde{\mathbf{x}}_2'\beta_2|$ is less than δ from the level $1-\alpha$ simultaneous confidence band (7.18), then equivalence can be claimed if and only if

$$-\delta < \min_{a<x<b} \{\tilde{\mathbf{x}}_2'\hat{\beta}_2 - c\hat{\sigma}\sqrt{\tilde{\mathbf{x}}_2'\mathbf{V}\tilde{\mathbf{x}}_2}\} \text{ and } \max_{a<x<b} \{\tilde{\mathbf{x}}_2'\hat{\beta}_2 + c\hat{\sigma}\sqrt{\tilde{\mathbf{x}}_2'\mathbf{V}\tilde{\mathbf{x}}_2}\} < \delta.$$

It is clear that this procedure has a smaller chance of establishing equivalence than the IU test (7.29) since the critical constant c of the simultaneous band (7.18) is larger than t_ν^α in (7.29).

Example 7.7 For the perinatal mortality (PM) data set given in Example 7.1 and $\alpha = 5\%$, $\tilde{\mathbf{x}}_2'\hat{\beta}_2 - t_\nu^\alpha \hat{\sigma}\sqrt{\tilde{\mathbf{x}}_2'\mathbf{V}\tilde{\mathbf{x}}_2}$ and $\tilde{\mathbf{x}}_2'\hat{\beta}_2 + t_\nu^\alpha \hat{\sigma}\sqrt{\tilde{\mathbf{x}}_2'\mathbf{V}\tilde{\mathbf{x}}_2}$ are plotted in Figure 7.7 by the two solid curves. Also plotted is the simultaneous confidence band (7.18)

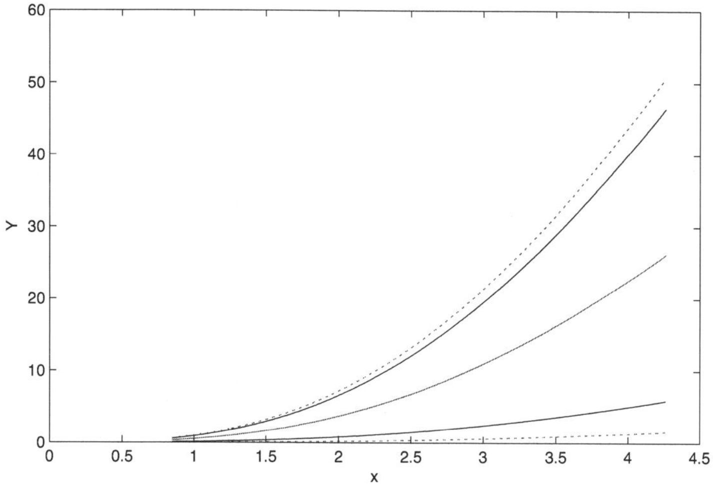

Figure 7.7: *The pointwise and simultaneous bands for the perinatal data*

over $x \in (0.85, 4.25)$, which is given by the two dotted curves. It is straightforward
to compute

$$\min_{a<x<b} \left\{ \tilde{\mathbf{x}}_2' \hat{\boldsymbol{\beta}}_2 - t_\nu^\alpha \hat{\sigma} \sqrt{\tilde{\mathbf{x}}_2' \mathbf{V} \tilde{\mathbf{x}}_2} \right\} = 0.069, \quad \max_{a<x<b} \left\{ \tilde{\mathbf{x}}_2' \hat{\boldsymbol{\beta}}_2 + t_\nu^\alpha \hat{\sigma} \sqrt{\tilde{\mathbf{x}}_2' \mathbf{V} \tilde{\mathbf{x}}_2} \right\} = 46.221.$$

So equivalence of $\tilde{\mathbf{x}}_2' \boldsymbol{\beta}_2$ to the zero function over $x \in (0.085, 4.25)$ can be claimed
if and only if the pre-specified threshold δ is larger than 46.221. Considering the
range of the observed Y_i given in Table 7.1, the threshold δ is most likely to be
much smaller than 46.221 and so equivalence cannot be claimed based on the
observed data.

7.3   Comparison of two polynomial models

Now we consider the problem of assessing whether two polynomial regression
models, which describe the relationships between a same response variable Y and
a same predict variable x, for two groups or treatments etc., are the same or not.
For example, Selvin (1998, p224) also provided a set of observations on perinatal
mortality (PM) and birth weight (BW) for a sample of white infants. The values
of $Y = \log(-\log(PMR))$ and $x = BW$ are given in Table 7.2.

These observations and the fitted 4th order polynomial model, $Y = -2.842 +
4.179x - 1.802x^2 + 0.421x^3 - 0.039x^4$, are plotted in Figure 7.8 by the crosses
and the solid curve respectively. From the plot, it seems that the 4th order polynomial model fits the observations well. Indeed the R^2 value of the fit is 0.994. The

Table 7.2: *Perinatal mortality data for white infants from Selvin (1998)*

i	x_i	Y_i	i	x_i	Y_i	i	x_i	Y_i
1	0.85	-0.4761	13	2.05	1.1204	25	3.25	1.7429
2	0.95	-0.1950	14	2.15	1.0919	26	3.35	1.8114
3	1.05	0.0849	15	2.25	1.2771	27	3.45	1.8269
4	1.15	0.2464	16	2.35	1.2771	28	3.55	1.8351
5	1.25	0.3791	17	2.45	1.3731	29	3.65	1.8351
6	1.35	0.4715	18	2.55	1.4241	30	3.75	1.8437
7	1.45	0.5364	19	2.65	1.4775	31	3.85	1.8527
8	1.55	0.6340	20	2.75	1.5165	32	3.95	1.8722
9	1.65	0.7391	21	2.85	1.6018	33	4.05	1.8437
10	1.75	0.7551	22	2.95	1.6751	34	4.15	1.8939
11	1.85	0.8042	23	3.05	1.6830	35	4.25	1.8527
12	1.95	0.9128	24	3.15	1.7429			

observations and fitted 4th order polynomial model for the black infants are also plotted in Figure 7.8 by the circles and the dotted curve, respectively. It is interesting to compare the two polynomial regression models to answer questions such as 'whether or not one group has significantly higher PM than the other group' or 'whether the two models are similar'.

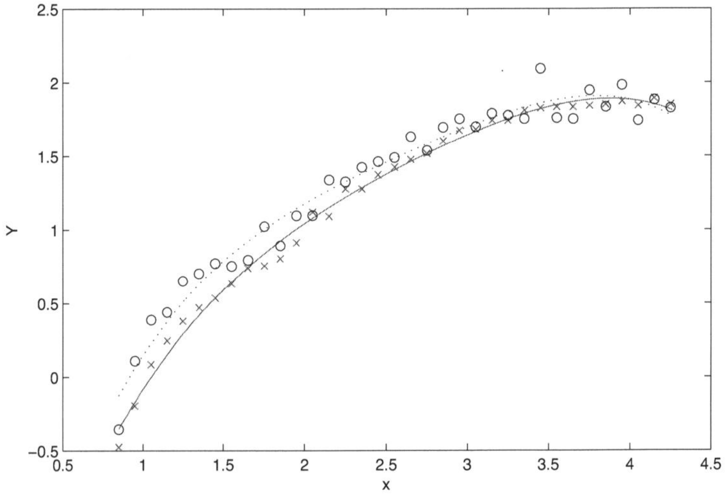

Figure 7.8: *The observations and fits of the perinatal data*

In general, suppose the two polynomial regression models of order $p > 1$ are given by

$$\mathbf{Y}_i = \mathbf{X}_i \beta_i + \mathbf{e}_i, \quad i = 1,2 \tag{7.30}$$

where $\mathbf{Y}_i = (Y_{i,1}, \cdots, Y_{i,n_i})'$ is a vector of observations, $\mathbf{X}_i$ is a $n_i \times (p+1)$ full column-rank design matrix with the lth $(1 \le l \le n_i)$ row given by $(1, x_{i,l}, \cdots, x_{i,l}^p)$, $\beta_i = (\beta_{i,0}, \cdots, \beta_{i,p})'$ is a vector of unknown regression coefficients, and $\mathbf{e}_i = (e_{i,1}, \cdots, e_{i,n_i})'$ is a vector of random errors with all the $\{e_{i,j} : j = 1, \cdots, n_i, i = 1,2\}$ being iid $N(0, \sigma^2)$ random variables, where σ^2 is an unknown parameter. Here $\mathbf{Y}_1$ and $\mathbf{Y}_2$ are two groups of observations that depend on the same covariate x via the polynomial regression models (7.30). The covariate values $(x_{i,1}, \cdots, x_{i,n_i})'$ for the two models $i = 1,2$ may be different.

The important question is whether the two models specified in (7.30) are similar or very different. If the two models are similar then only one polynomial model is necessary to describe the relationship between the response variable Y and the covariate x for both groups. On the other hand, if the two models are very different then one may want to further assess the differences between the two models and whether one model is higher than the other on average.

Applying the results in Section 5.1, one gets the least squares estimator of β_i, $\hat{\beta}_i = (\mathbf{X}_i'\mathbf{X}_i)^{-1}\mathbf{X}_i'\mathbf{Y}_i$ $(i = 1,2)$, and the pooled error mean square $\hat{\sigma}^2$, which is independent of the $\hat{\beta}_i$'s and has the distribution $\sigma^2\chi_\nu^2/\nu$ with $\nu = n_1 + n_2 - 2(p+1)$. Furthermore, for the purpose of comparing the two polynomial regression models, one can apply the partial F test in (5.6) or (5.7) to test the hypotheses

$$H_0 : \beta_1 = \beta_2 \quad \text{against} \quad H_a : \beta_1 \neq \beta_2; \tag{7.31}$$

this is applied in exactly the same way as for the comparison of two multiple linear regression models in which no functional relationships among the covariates exist. Clearly the polynomial form of the model functions is not fully utilized in this partial F test approach, and so it has similar undesirable property identified in the last two sections. Confidence band approach is a more intuitive, informative and appropriate alternative.

7.3.1 Hyperbolic bands

To assess the difference between the two polynomial regression models in both directions, one may use the following two-sided hyperbolic simultaneous confidence band over a pre-specified interval $x \in (a,b)$:

$$\tilde{\mathbf{x}}'\beta_1 - \tilde{\mathbf{x}}'\beta_2 \in \tilde{\mathbf{x}}'\hat{\beta}_1 - \tilde{\mathbf{x}}'\hat{\beta}_2 \pm c\hat{\sigma}\sqrt{\tilde{\mathbf{x}}'\Delta\tilde{\mathbf{x}}}\,, \quad \forall\, x \in (a,b) \tag{7.32}$$

where $\Delta = (\mathbf{X}_1'\mathbf{X}_1)^{-1} + (\mathbf{X}_2'\mathbf{X}_2)^{-1}$ and c is a critical constant chosen so that the simultaneous confidence level of the band is equal to $1-\alpha$.

For a general $p > 1$, we use simulation to approximate the value of c as before. Note that the simultaneous confidence level of the band is $P\{S < c\}$ where

$$S = \sup_{a<x<b} \frac{|\tilde{\mathbf{x}}'(\hat{\beta}_1 - \beta_1 - \hat{\beta}_2 + \beta_2)/\hat{\sigma}|}{\sqrt{\tilde{\mathbf{x}}'\Delta\tilde{\mathbf{x}}}}$$

$$= \sup_{a<x<b} \frac{|\tilde{\mathbf{x}}'\mathbf{T}|}{\sqrt{\tilde{\mathbf{x}}'\Delta\tilde{\mathbf{x}}}}$$

$$= K_{2h}(\mathbf{T},\Delta,(a,b)) \tag{7.33}$$

where $\mathbf{T} = (\hat{\beta}_1 - \beta_1 - \hat{\beta}_2 + \beta_2)/\hat{\sigma} \sim \mathcal{T}_{p+1,\nu}(\mathbf{0},\Delta)$, and $K_{2h}(\mathbf{T},\Delta,(a,b))$ is defined in (7.6). To simulate one S, one simulates $\mathbf{T} \sim \mathcal{T}_{p+1,\nu}(\mathbf{0},\Delta)$ and calculate S from (7.33) by using the algorithm in Appendix E.

The confidence band (7.32) can easily be plotted and visualized for any $p > 1$, from which information on the magnitude of difference between the two polynomial regression models is directly available. This confidence band induces the following test for the hypotheses in (7.31): reject H_0 if and only if the zero function $\tilde{\mathbf{x}}'\mathbf{0}$ is outside the band for at least one $x \in (a,b)$, that is,

$$\sup_{a<x<b} \frac{|\tilde{\mathbf{x}}'(\hat{\beta}_1 - \hat{\beta}_2)/\hat{\sigma}|}{\sqrt{\tilde{\mathbf{x}}'\Delta\tilde{\mathbf{x}}}} > c. \tag{7.34}$$

The size of this test is α, and its observed p-value can be approximated by simulation in a similar way as in Section 5.3.1.

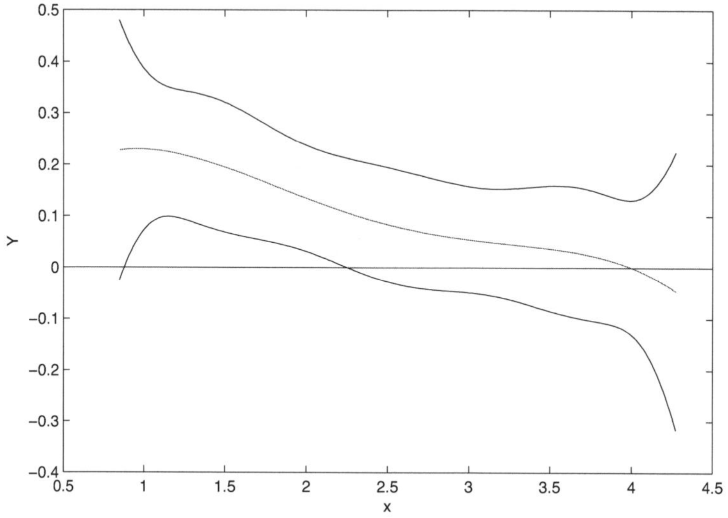

Figure 7.9: *The hyperbolic confidence band for* $\tilde{\mathbf{x}}'\beta_B - \tilde{\mathbf{x}}'\beta_W$

Example 7.8 For the perinatal mortality (PM) data set, it is interesting to compare the two polynomial regression functions for the black and white infants by using the confidence band (7.32). For $\alpha = 5\%$ and $(a,b) = (0.85,4.25)$, the critical constant is computed to be $c = 2.884$ based on 100,000 simulations, which is much smaller than the conservative critical value $\sqrt{(p+1)f^{\alpha}_{p+1,\nu}} = 3.441$. The confidence band is plotted in Figure 7.9 by the two solid curves that are symmetric about $\tilde{\mathbf{x}}'\hat{\beta}_B - \tilde{\mathbf{x}}'\hat{\beta}_W$, the solid curve in the middle. It is clear from the figure that the band excludes the zero function over $x \in (0.87, 2.25)$ and so $H_0 : \beta_W = \beta_B$ is rejected. This agrees with the observed p-value 8×10^{-5} of the test in (7.34). Furthermore, the confidence band indicates that, for $x \in (0.87, 2.25)$, the mean response of black infants $\tilde{\mathbf{x}}'\beta_B$ is significantly higher than the mean response of white infants $\tilde{\mathbf{x}}'\beta_W$; this information is not available from the test in (7.34) or the partial F test.

The confidence band (7.32) is suitable for two-sided inferences. If one is interested only in assessing whether one model has a higher mean response than the other model, such as whether black infants tend to have higher mean response than white infants, one-sided confidence bands are more pertinent. A lower hyperbolic confidence band over the interval $x \in (a,b)$ is given by

$$\tilde{\mathbf{x}}'\beta_1 - \tilde{\mathbf{x}}'\beta_2 > \tilde{\mathbf{x}}'\hat{\beta}_1 - \tilde{\mathbf{x}}'\hat{\beta}_2 - c\hat{\sigma}\sqrt{\tilde{\mathbf{x}}'\Delta\tilde{\mathbf{x}}} \ \ \forall\, x \in (a,b) \tag{7.35}$$

where the critical constant c is chosen so that the simultaneous confidence level of this band is equal to $1-\alpha$. An upper hyperbolic confidence band on $\tilde{\mathbf{x}}'\beta_1 - \tilde{\mathbf{x}}'\beta_2$ can be constructed in a similar way and, in particular, uses the same critical constant as the lower confidence band.

To determine the critical constant c by simulation, note that the simultaneous confidence level is given by $\mathrm{P}\{S < c\}$, where

$$\begin{aligned} S &= \sup_{a<x<b} \frac{\tilde{\mathbf{x}}'(\hat{\beta}_1 - \beta_1 - \hat{\beta}_2 + \beta_2)/\hat{\sigma}}{\sqrt{\tilde{\mathbf{x}}'\Delta\tilde{\mathbf{x}}}} \\ &= \sup_{a<x<b} \frac{\tilde{\mathbf{x}}'\mathbf{T}}{\sqrt{\tilde{\mathbf{x}}'\Delta\tilde{\mathbf{x}}}} \\ &= K_{1h}(\mathbf{T}, \Delta, (a,b)) \end{aligned} \tag{7.36}$$

where $\mathbf{T} = (\hat{\beta}_1 - \beta_1 - \hat{\beta}_2 + \beta_2)/\hat{\sigma}$ as in the two-sided case, and $K_{1h}(\mathbf{T}, \Delta, (a,b))$ is defined in (7.9). To simulate one S, one simulates one $\mathbf{T} \sim \mathcal{T}_{p+1,\nu}(\mathbf{0}, \Delta)$ and calculates S from (7.36) by using the algorithm in Appendix E.

From the confidence band (7.35), one can infer whether the mean response of the first model, $\tilde{\mathbf{x}}'\beta_1$, is greater than the mean response of the second model, $\tilde{\mathbf{x}}'\beta_2$, and by how much over the range of $x \in (a,b)$. Since the lower confidence band (7.35) is always above the lower part of the two-sided confidence band (7.32), the one-sided inferences from the band (7.35) are sharper than those from the band (7.32).

Example 7.9 For the perinatal mortality (PM) data set, we can use confidence band (7.35) to assess whether the mean response $\tilde{\mathbf{x}}'\beta_B$ of black infants tends to be higher than the mean response $\tilde{\mathbf{x}}'\beta_W$ of white infants. For $\alpha = 5\%$ and $(a,b) =$

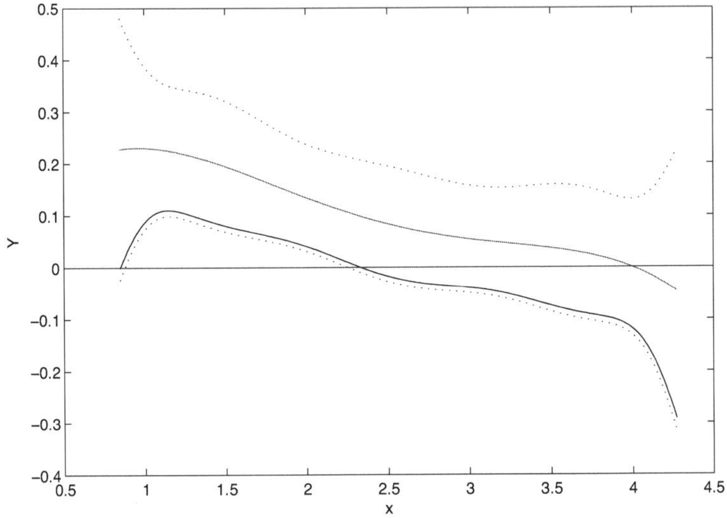

Figure 7.10: *The lower hyperbolic confidence band for* $\tilde{\mathbf{x}}'\beta_B - \tilde{\mathbf{x}}'\beta_W$

$(0.85, 4.25)$, the critical constant is computed to be $c = 2.618$ based on 100,000 simulations, which is smaller than the two-sided critical value $c = 2.884$. The lower confidence band is plotted in Figure 7.10 by the solid curve, while the two-sided confidence band is given by the two dotted curves. As expected, the lower confidence band is higher than the lower part of the two-sided band. From the lower confidence band, one can infer that, for $x \in (0.85, 2.3)$, the mean response of black infants $\tilde{\mathbf{x}}'\beta_B$ is significantly higher than the mean response of white infants $\tilde{\mathbf{x}}'\beta_W$.

7.3.2 Constant width bands

Constant width simultaneous confidence bands can also be used to bound the differences between $\tilde{\mathbf{x}}'\beta_1$ and $\tilde{\mathbf{x}}'\beta_2$ in either two or one directions. A two-sided constant width simultaneous confidence band over the interval $x \in (a, b)$ is given by

$$\tilde{\mathbf{x}}'\beta_1 - \tilde{\mathbf{x}}'\beta_2 \in \tilde{\mathbf{x}}'\hat{\beta}_1 - \tilde{\mathbf{x}}'\hat{\beta}_2 \pm c\hat{\sigma} \;\; \forall\, x \in (a, b) \tag{7.37}$$

where c is a critical constant chosen so that the simultaneous confidence level of the band is equal to $1 - \alpha$.

For a general $p > 1$, we use simulation to approximate the value of c. Note that the simultaneous confidence level of the band is given by $P\{S < c\}$ where

$$\begin{aligned} S &= \sup_{a<x<b} |\tilde{\mathbf{x}}'(\hat{\beta}_1 - \beta_1 - \hat{\beta}_2 + \beta_2)/\hat{\sigma}| \\ &= \sup_{a<x<b} |\tilde{\mathbf{x}}'\mathbf{T}| \\ &= K_{2c}(\mathbf{T}, (a,b))| \end{aligned} \tag{7.38}$$

where $\mathbf{T} = (\hat{\beta}_1 - \beta_1 - \hat{\beta}_2 + \beta_2)/\hat{\sigma} \sim \mathcal{T}_{p+1,\nu}(\mathbf{0}, \Delta)$ as before, and $K_{2c}(\mathbf{T}, (a,b))$ is defined in (7.11). So one S can be simulated by generating one $\mathbf{T}$ and calculating S from (7.38) by using the algorithm in Appendix E.

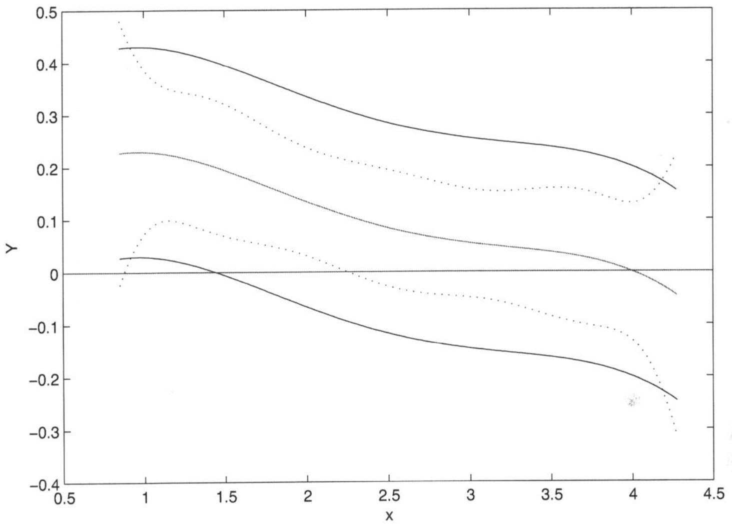

Figure 7.11: *The two-sided constant width confidence band for* $\tilde{\mathbf{x}}'\beta_B - \tilde{\mathbf{x}}'\beta_W$

Example 7.10 For the perinatal mortality (PM) data set, the critical constant c of the constant width band (7.37) is calculated to be $c = 2.331$ for $\alpha = 5\%$ and $(a,b) = (0.85, 4.25)$, based on 100,000 simulations. This confidence band for $\tilde{\mathbf{x}}'\beta_B - \tilde{\mathbf{x}}'\beta_W$ is plotted in Figure 7.11 by two solid curves, while the two-sided hyperbolic confidence band is given by the two dotted curves. Since the constant width band does not contain the zero function completely over $x \in (a,b)$, the two polynomial functions are significantly different; the p-value of the test induced by this confidence band is 0.021, which is smaller than $\alpha = 5\%$ as expected. Again, the constant width band is narrower than the hyperbolic band near the two ends of the interval $x \in (a,b)$ and vice versa in the middle of the interval.

A lower constant width simultaneous confidence band over the interval $x \in (a,b)$ is given by

$$\tilde{\mathbf{x}}'\beta_1 - \tilde{\mathbf{x}}'\beta_2 > \tilde{\mathbf{x}}'\hat{\beta}_1 - \tilde{\mathbf{x}}'\hat{\beta}_2 - c\hat{\sigma} \ \ \forall\, x \in (a,b) \tag{7.39}$$

where c is a critical constant chosen so that the simultaneous confidence level of the band is equal to $1-\alpha$.

The simultaneous confidence level of the band is $P\{S < c\}$ where

$$\begin{aligned} S &= \sup_{a<x<b} \tilde{\mathbf{x}}'(\hat{\beta}_1 - \beta_1 - \hat{\beta}_2 + \beta_2)/\hat{\sigma} \\ &= \sup_{a<x<b} \tilde{\mathbf{x}}'\mathbf{T} \\ &= K_{1c}(\mathbf{T},(a,b)) \end{aligned} \tag{7.40}$$

where $\mathbf{T} = (\hat{\beta}_1 - \beta_1 - \hat{\beta}_2 + \beta_2)/\hat{\sigma} \sim \mathcal{T}_{p+1,v}(\mathbf{0},\Delta)$. To simulate one S, one generates a $\mathbf{T} \sim \mathcal{T}_{p+1,v}(\mathbf{0},\Delta)$ and computes S from (7.40) by using the algorithm in Appendix E. Hence c can be approximated by simulation.

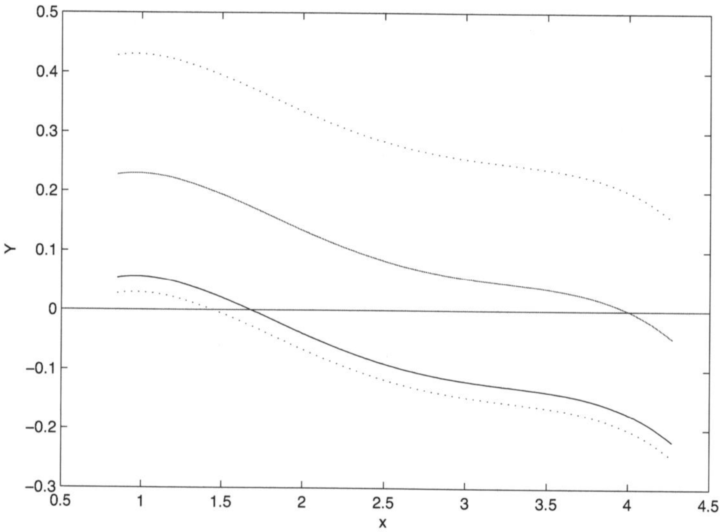

Figure 7.12: *The one-sided constant width confidence band for* $\tilde{\mathbf{x}}'\beta_B - \tilde{\mathbf{x}}'\beta_W$

Example 7.11 For the perinatal mortality (PM) data set, the critical constant c of the one-sided constant width band (7.39) is calculated to be $c = 2.021$ for $\alpha = 5\%$ and $(a,b) = (0.85, 4.25)$, based on 100,000 simulations. This confidence band for

$\tilde{\mathbf{x}}'\beta_B - \tilde{\mathbf{x}}'\beta_W$ is plotted in Figure 7.12 by the solid curve, while the two-sided constant width confidence band is given by the two dotted curves. It is clear that the lower band is higher than the lower part of the two-sided band as expected.

7.3.3 Non-superiority, non-inferiority and equivalence

For the perinatal mortality data, suppose the purpose of study is to demonstrate that $\tilde{\mathbf{x}}'\beta_B$ for black infants is not higher than $\tilde{\mathbf{x}}'\beta_W$ for white infants by a pre-specified threshold $\delta > 0$ and so $\tilde{\mathbf{x}}'\beta_B$ is non-superior to $\tilde{\mathbf{x}}'\beta_W$ over a given interval $x \in (a,b)$. The results of Section 5.5.1 can be applied to construct an upper confidence bound on $\max_{a<x<b}(\tilde{\mathbf{x}}'\beta_B - \tilde{\mathbf{x}}'\beta_W)$ for this purpose, which is more efficient than what can be deduced from an upper simultaneous confidence band on $\tilde{\mathbf{x}}'\beta_B - \tilde{\mathbf{x}}'\beta_W$ discussed in Sections 7.3.1 and 7.3.2.

In general, let $\delta > 0$ be a pre-specified threshold. The polynomial model function $\tilde{\mathbf{x}}'\beta_1$ is said to be non-superior to the polynomial model function $\tilde{\mathbf{x}}'\beta_2$ over the interval $x \in (a,b)$ if

$$\tilde{\mathbf{x}}'\beta_1 < \tilde{\mathbf{x}}'\beta_2 + \delta \ \forall\, x \in (a,b), \text{ i.e. } \max_{a<x<b}(\tilde{\mathbf{x}}'\beta_1 - \tilde{\mathbf{x}}'\beta_2) < \delta.$$

To derive an upper confidence bound on $\max_{a<x<b}(\tilde{\mathbf{x}}'\beta_1 - \tilde{\mathbf{x}}'\beta_2)$, let

$$U(\mathbf{Y}_1,\mathbf{Y}_2,x) = \tilde{\mathbf{x}}'\hat{\beta}_1 - \tilde{\mathbf{x}}'\hat{\beta}_2 + t_\nu^\alpha \sqrt{\tilde{\mathbf{x}}'\Delta\tilde{\mathbf{x}}}.$$

Note that $\mathrm{P}\{\tilde{\mathbf{x}}'\beta_1 - \tilde{\mathbf{x}}'\beta_2 < U(\mathbf{Y}_1,\mathbf{Y}_2,x)\} = 1-\alpha$ for each $-\infty < x < \infty$ and hence $U(\mathbf{Y}_1,\mathbf{Y}_2,x)$ is a pointwise upper confidence bound on $\tilde{\mathbf{x}}'\beta_1 - \tilde{\mathbf{x}}'\beta_2$. Now the following theorem follows directly from Theorem 5.1.

Theorem 7.1 Under the notations above, we have

$$\inf \mathrm{P}\left\{\tilde{\mathbf{x}}'\beta_1 - \tilde{\mathbf{x}}'\beta_2 \le \max_{a<x<b} U(\mathbf{Y}_1,\mathbf{Y}_2,x)\ \forall\, x \in (a,b)\right\} \ge 1-\alpha \tag{7.41}$$

where the inf is over all the possible values of the unknown parameters $\beta_1 \in \Re^{p+1}$, $\beta_2 \in \Re^{p+1}$ and $\sigma > 0$.

So $\max_{a<x<b} U(\mathbf{Y}_1,\mathbf{Y}_2,x)$ is a $1-\alpha$ upper simultaneous confidence band for $\tilde{\mathbf{x}}'\beta_1 - \tilde{\mathbf{x}}'\beta_2$ over $x \in (a,b)$, and it has the same value over $x \in (a,b)$. This upper confidence band can be used to bound $\max_{a<x<b}(\tilde{\mathbf{x}}'\beta_1 - \tilde{\mathbf{x}}'\beta_2)$ and, in particular, to claim the non-superiority of $\tilde{\mathbf{x}}'\beta_1$ to $\tilde{\mathbf{x}}'\beta_2$ over $x \in (a,b)$ if and only if

$$\max_{a<x<b} U(\mathbf{Y}_1,\mathbf{Y}_2,x) < \delta. \tag{7.42}$$

Example 7.12 For the perinatal mortality (PM) data set, the critical constant t_ν^α of the pointwise upper band $U(\mathbf{Y}_1,\mathbf{Y}_2,x)$ is 1.671 for $\alpha = 5\%$. This pointwise band

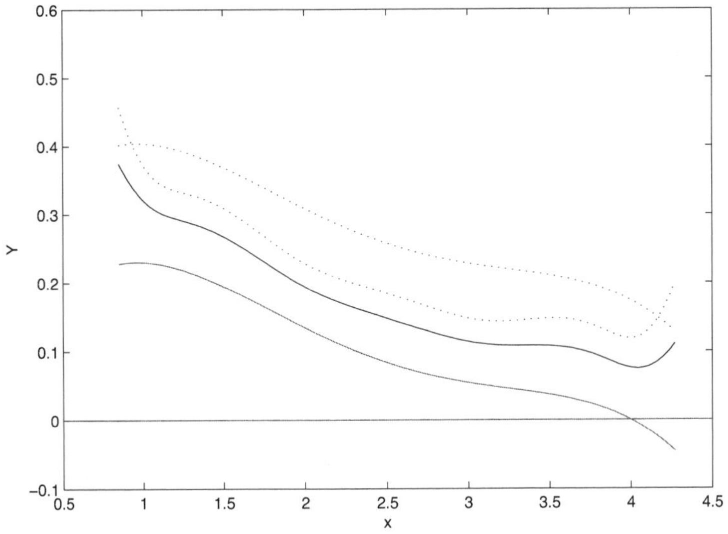

Figure 7.13: *The pointwise upper confidence band for* $\tilde{\mathbf{x}}'\beta_B - \tilde{\mathbf{x}}'\beta_W$

is plotted in Figure 7.13 by the solid curve, while the upper hyperbolic and constant width simultaneous confidence bands over $x \in (0.85, 4.25)$ are given by two dotted curves. We have $\max_{a<x<b} U(\mathbf{Y}_1, \mathbf{Y}_2, x) = 0.374$ and so non-superiority of $\tilde{\mathbf{x}}'\beta_B$ to $\tilde{\mathbf{x}}'\beta_W$ over $x \in (a,b)$ can be claimed if the pre-specified δ is larger than 0.374. To make the same non-superiority claim from the simultaneous hyperbolic or constant width bands, δ has to be larger than 0.457 or 0.402, respectively.

The polynomial model function $\tilde{\mathbf{x}}'\beta_1$ is said to be non-inferior to the polynomial model function $\tilde{\mathbf{x}}'\beta_2$ over the interval $x \in (a,b)$ if

$$\tilde{\mathbf{x}}'\beta_1 > \tilde{\mathbf{x}}'\beta_2 - \delta \ \forall\, x \in (a,b), \text{ i.e. } \min_{x\in(a,b)} (\tilde{\mathbf{x}}'\beta_1 - \tilde{\mathbf{x}}'\beta_2) > -\delta.$$

It is clear that the non-inferiority of $\tilde{\mathbf{x}}'\beta_1$ to $\tilde{\mathbf{x}}'\beta_2$ is just the non-superiority of $\tilde{\mathbf{x}}'\beta_2$ to $\tilde{\mathbf{x}}'\beta_1$, and so the non-inferiority of $\tilde{\mathbf{x}}'\beta_1$ to $\tilde{\mathbf{x}}'\beta_2$ can be established via the non-superiority of $\tilde{\mathbf{x}}'\beta_2$ to $\tilde{\mathbf{x}}'\beta_1$ by using Theorem 7.1.

The two model functions $\tilde{\mathbf{x}}'\beta_1$ and $\tilde{\mathbf{x}}'\beta_2$ are said to be equivalent if any one model is both non-superior and non-inferior to the other model, that is,

$$\max_{a<x<b} |\tilde{\mathbf{x}}'\beta_1 - \tilde{\mathbf{x}}'\beta_2| < \delta.$$

For the purpose of establishing equivalence of the two model functions, we set up the equivalence hypotheses

$$H_0^E: \max_{a<x<b} |\tilde{\mathbf{x}}'\beta_1 - \tilde{\mathbf{x}}'\beta_2| \geq \delta \iff H_a^E: \max_{a<x<b} |\tilde{\mathbf{x}}'\beta_1 - \tilde{\mathbf{x}}'\beta_2| < \delta,$$

and the equivalence can be claimed if and only if the null hypothesis H_0^E is rejected. An intersection-union test of Berger (1982; see Appendix D) can be constructed in a similar way as in Section 5.5.3. This test of size α rejects the null hypothesis H_0^E if and only if

$$-\delta < \min_{a<x<b} L_P(\mathbf{Y}_1, \mathbf{Y}_2, x) \text{ and } \max_{a<x<b} U_P(\mathbf{Y}_1, \mathbf{Y}_2, x) < \delta$$

where $L(\mathbf{Y}_1, \mathbf{Y}_2, x) = \tilde{\mathbf{x}}'\hat{\beta}_1 - \tilde{\mathbf{x}}'\hat{\beta}_2 - t_\nu^\alpha \sqrt{\tilde{\mathbf{x}}'\Delta\tilde{\mathbf{x}}}$.

Example 7.13 For the perinatal mortality (PM) data set, $\max_{a<x<b} U(\mathbf{Y}_1, \mathbf{Y}_2, x) = 0.374$ and $\min_{a<x<b} L(\mathbf{Y}_1, \mathbf{Y}_2, x) = -0.203$ for $\alpha = 5\%$ and $(a,b) = (0.85, 4.25)$. Hence the equivalence of $\tilde{\mathbf{x}}'\beta_B$ to $\tilde{\mathbf{x}}'\beta_W$ over $x \in (a,b)$ can be claimed if the pre-specified δ is larger than 0.374.

7.4 Comparison of more than two polynomial models

Consider the comparison of k ($k \geq 3$) polynomial regression models, which describe the relationship between the same response variable Y and the same covariate x for k different groups or treatments. Suppose the k polynomial regression models are specified by

$$\mathbf{Y}_i = \mathbf{X}_i\beta_i + \mathbf{e}_i, \quad i = 1, \cdots, k \tag{7.43}$$

where $\mathbf{Y}_i = (Y_{i1}, \cdots, Y_{in_i})'$ is a vector of random observations, $\mathbf{X}_i$ is a $n_i \times (p+1)$ full column-rank matrix with the lth $(1 \leq l \leq n_i)$ row given by $(1, x_{i,l}, \cdots, x_{i,l}^p)$, $\beta_i = (\beta_{i,0}, \cdots, \beta_{i,p})'$ is a vector of unknown regression coefficients, and $\mathbf{e}_i = (e_{i,1}, \cdots, e_{i,n_i})'$ is a vector of random errors with all the $\{e_{i,j}, j = 1, \cdots, n_i, i = 1, \cdots, k\}$ being iid $N(0, \sigma^2)$. Since $\mathbf{X}_i'\mathbf{X}_i$ is non-singular, the least squares estimator of β_i is given by $\hat{\beta}_i = (\mathbf{X}_i'\mathbf{X}_i)^{-1}\mathbf{X}_i'\mathbf{Y}_i, i = 1, \cdots, k$. Let $\hat{\sigma}^2$ denote the pooled error mean square with degrees of freedom $\nu = \sum_{i=1}^k (n_i - p - 1)$; $\hat{\sigma}^2$ has the distribution $\sigma^2\chi_\nu^2/\nu$ and is independent of the $\hat{\beta}_i$'s.

One common approach is to test the hypotheses

$$H_0: \beta_1 = \cdots = \beta_k \quad \text{against} \quad H_a: \text{not } H_0. \tag{7.44}$$

In particular, the partial F test in (1.7) or (1.11) is implemented in exactly the same way as given in Section 6.1 for k multiple linear regression models in which the p covariates x_i are assumed to have no functional relationships. Clearly, the relationships among the covariates of a polynomial model have not been taken fully

Table 7.3: *Perinatal mortality data for non-black and non-white infants*

i	x_i	Y_i	i	x_i	Y_i	i	x_i	Y_i
1	0.85	-0.4337	13	2.05	1.1329	25	3.25	1.5126
2	0.95	-0.2376	14	2.15	1.1161	26	3.35	1.5905
3	1.05	0.0508	15	2.25	1.1209	27	3.45	1.7130
4	1.15	0.4103	16	2.35	1.2330	28	3.55	1.6367
5	1.25	0.2727	17	2.45	1.1727	29	3.65	1.6358
6	1.35	0.5511	18	2.55	1.1759	30	3.75	1.8273
7	1.45	0.6673	19	2.65	1.4988	31	3.85	1.5862
8	1.55	0.6860	20	2.75	1.4763	32	3.95	1.6953
9	1.65	0.7556	21	2.85	1.3615	33	4.05	1.7289
10	1.75	0.8515	22	2.95	1.3539	34	4.15	1.7840
11	1.85	0.7294	23	3.05	1.5560	35	4.25	1.5445
12	1.95	1.0599	24	3.15	1.5707			

into account by using this partial F test, in addition to the problem that possible constraints on the covariates are not used by the partial F test as pointed out in Chapter 6. Furthermore, inferences that can be drawn from the partial F test are very limited. If H_0 is rejected then the k polynomial models are deemed to be significantly different, but no information is provided on which models are different, let alone by how much. If H_0 is not rejected then there is no sufficient statistical evidence to conclude that some regression models are different. Of course this means anything except that the k models are the same. Simultaneous confidence bands can provide much more intuitive and informative inferences.

Example 7.14 For the perinatal mortality (PM) data, assume that, in addition to the observations on black infants in Table 7.1 and on white infants in Table 7.2, a set of observations on 'other' (non-black and non-white) infants is also available. These observation are computer-generated from a 4th order polynomial regression model and given in Table 7.3. The interest is on the comparison of the three infant groups, black, white and other, to see whether the three polynomial regression functions are similar or very different. If one employs the partial F test for testing the hypotheses in (7.44), then the test statistic is calculated to be 10.273 with p-value 2.4×10^{-11}. So the F test is highly significant: it is highly unlikely that the null hypotheses H_0 is true. But no information is provided by the F test on which two models are significantly different and by how much.

7.4.1 *Hyperbolic bands for finite comparisons*

In many real problems, one is often interested in the direct comparisons between the k polynomial functions specified in (7.43) in the form of

$$\tilde{\mathbf{x}}'\beta_i - \tilde{\mathbf{x}}'\beta_j \;\; \forall x \in (a,b) \;\text{ and } \forall (i,j) \in \Lambda \tag{7.45}$$

where (a,b) is a given interval of the covariate x over which the comparisons of the models are of interest, and Λ is a given index set that determines the comparison of interest. The most common forms of Λ are: $\Lambda = \{(i,j) : 1 \le i \ne j \le k\}$ for the pairwise comparisons of the k models; $\Lambda = \{(i,j) : 2 \le i \le k, j = 1\}$ for the many-one comparisons of the second to kth models with the first model as the control; $\Lambda = \{(i,i+1) : 1 \le i \le k-1\})$ for the successive comparisons of the k regression models.

A set of hyperbolic simultaneous confidence bands appropriate for the comparisons in (7.45) is given by

$$\tilde{\mathbf{x}}'\beta_i - \tilde{\mathbf{x}}'\beta_j \in \tilde{\mathbf{x}}'\hat{\beta}_i - \tilde{\mathbf{x}}'\hat{\beta}_j \pm c\hat{\sigma}\sqrt{\tilde{\mathbf{x}}'\Delta_{ij}\tilde{\mathbf{x}}}, \ \forall\, x \in (a,b) \text{ and } \forall (i,j) \in \Lambda \quad (7.46)$$

where c is the critical constant required so that the confidence level of this set of simultaneous confidence bands is equal to $1-\alpha$, and $\Delta_{ij} = (\mathbf{X}_i'\mathbf{X}_i)^{-1} + (\mathbf{X}_j'\mathbf{X}_j)^{-1}$.

The critical constant c can be approximated by simulation as before. Note that the confidence level of the bands in (7.46) is given by $\mathrm{P}\{S < c\}$ where

$$\begin{aligned} S &= \sup_{(i,j)\in\Lambda} \sup_{a<x<b} \frac{|\tilde{\mathbf{x}}'((\hat{\beta}_i - \beta_i) - (\hat{\beta}_j - \beta_j))|}{\hat{\sigma}\sqrt{\tilde{\mathbf{x}}'\Delta_{ij}\tilde{\mathbf{x}}}} \\ &= \sup_{(i,j)\in\Lambda} \sup_{a<x<b} \frac{|\tilde{\mathbf{x}}'\mathbf{T}_{ij}|}{\sqrt{\tilde{\mathbf{x}}'\Delta_{ij}\tilde{\mathbf{x}}}} \\ &= \sup_{(i,j)\in\Lambda} K_{2h}(\mathbf{T}_{ij}, \Delta_{ij}, (a,b)) \qquad (7.47) \end{aligned}$$

where $K_{2h}(\mathbf{T}, \Delta, (a,b))$ is defined in (7.6), and $\mathbf{T}_{ij} = (\mathbf{N}_i - \mathbf{N}_j)/(\hat{\sigma}/\sigma)$ $(1 \le i \ne j \le k)$ with $\mathbf{N}_i = (\hat{\beta}_i - \beta_i)/\sigma$ $(i = 1, \cdots, k)$ being independent $\mathcal{N}_{p+1}(\mathbf{0}, (\mathbf{X}_i'\mathbf{X}_i)^{-1})$ random vectors independent of $\hat{\sigma}$.

Simulation of one random variable S can be implemented in the following way.

Step 0. determine $\Delta_{ij} = (\mathbf{X}_i'\mathbf{X}_i)^{-1} + (\mathbf{X}_j'\mathbf{X}_j)^{-1}$ for $(i,j) \in \Lambda$.

Step 1. simulate independent $\mathbf{N}_i \sim \mathcal{N}_{p+1}(\mathbf{0}, (X_i'X_i)^{-1}), i = 1, \cdots, k$ and $\hat{\sigma}/\sigma \sim \sqrt{\chi^2_\nu/\nu}$.

Step 2. calculate $\mathbf{T}_{ij} = (\mathbf{N}_i - \mathbf{N}_j)/(\hat{\sigma}/\sigma)$.

Step 3. compute $K_{2h}(\mathbf{T}_{ij}, \Delta_{ij}, (a,b))$ by using the algorithm in Appendix E.

Step 4. find S from (7.47).

Although the simultaneous confidence bands in (7.46) (and those developed below) can be used to test the hypotheses in (7.44) in a similar way as in Section 6.3, all the confidence bands for polynomial models can be easily plotted and so the tests and associated p-values are probably of less interest for polynomial regressions. Also, it is not necessary to use the same interval (a,b) or the same critical constant c for all $(i,j) \in \Lambda$ in the confidence bands in (7.46). Indeed, one may use different intervals (a,b) or differently weighted critical constants $l_{ij}c$

to deliberately allocate different (or same) type I error rates to different (i, j)-comparisons. Research on this is required.

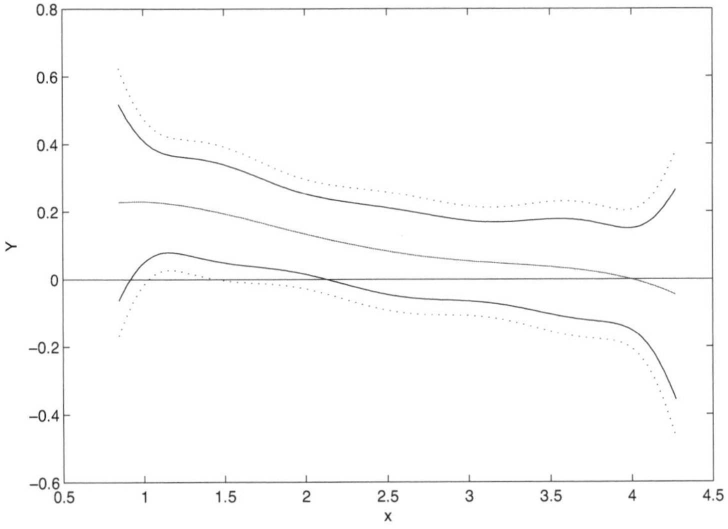

Figure 7.14: *The confidence bands for* $\tilde{\mathbf{x}}'\beta_B - \tilde{\mathbf{x}}'\beta_W$

Example 7.15 For the perinatal mortality (PM) data, simultaneous confidence bands for pairwise comparison of the three infant groups are constructed. In this case $\Lambda = \{(1,2),(2,3),(3,2)\}$, assuming $i = 1$ denotes the black infant group, $i = 2$ the white infant group, and $i = 3$ the other infant group. For $x \in (0.85, 4.25)$ and $\alpha = 0.05$, the critical constant c in (7.46) is calculated to be 3.228 based on 100,000 simulations. One can plot the three confidence bands for making inferences on pairwise comparisons among the three groups. For example, the confidence band for $\tilde{\mathbf{x}}'\beta_B - \tilde{\mathbf{x}}'\beta_W$ is plotted in Figure 7.14 by the two solid curves. From this confidence band, it is clear that the two model functions $\tilde{\mathbf{x}}'\beta_B$ and $\tilde{\mathbf{x}}'\beta_W$ are significantly different over $x \in (0.85, 4.25)$. In particular, $\tilde{\mathbf{x}}'\beta_B$ is higher than $\tilde{\mathbf{x}}'\beta_W$ for $x \in (0.9, 2.15)$ but $\tilde{\mathbf{x}}'\beta_B$ and $\tilde{\mathbf{x}}'\beta_W$ are not significantly different for x in the remaining part of the interval $(0.85, 4.25)$. Note that this confidence band is slightly wider than the confidence band given in Figure 7.9, since it is one of the set of three simultaneous confidence bands which have a simultaneous confidence level 95% and so the critical constant $c = 3.228$ is larger than the critical constant $c = 2.618$ of the band in Figure 7.9. The two dotted curves in Figure 7.14 present the band if the conservative critical constant in (6.8), $\sqrt{10 f_{10,90}^{0.05}} = 4.402$, is used in place of $c = 3.228$.

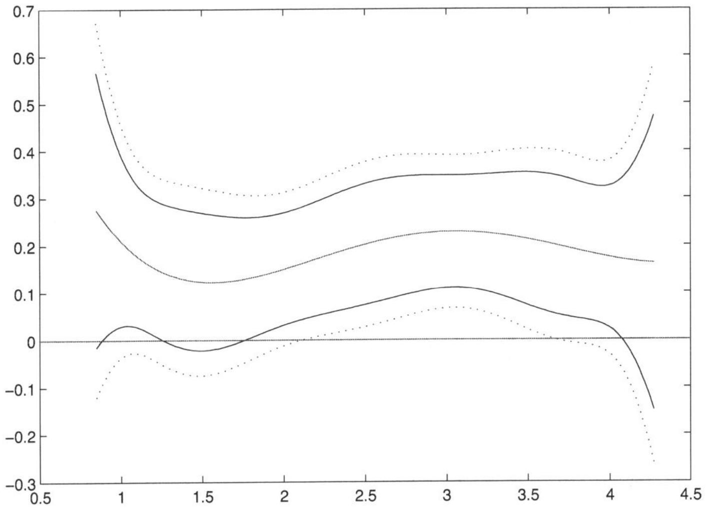

Figure 7.15: *The confidence bands for* $\tilde{\mathbf{x}}'\boldsymbol{\beta}_B - \tilde{\mathbf{x}}'\boldsymbol{\beta}_O$

The confidence bands for $\tilde{\mathbf{x}}'\boldsymbol{\beta}_B - \tilde{\mathbf{x}}'\boldsymbol{\beta}_O$ and $\tilde{\mathbf{x}}'\boldsymbol{\beta}_W - \tilde{\mathbf{x}}'\boldsymbol{\beta}_O$ are plotted in Figures 7.15 and 7.16 respectively. From these confidence bands, it is clear that all the three model functions $\tilde{\mathbf{x}}'\boldsymbol{\beta}_B$, $\tilde{\mathbf{x}}'\boldsymbol{\beta}_W$ and $\tilde{\mathbf{x}}'\boldsymbol{\beta}_O$ are significantly different over $x \in (0.85, 4.25)$.

If the interest is on the comparison between the black infant group and the other two infant groups, then $\Lambda = \{(1,2),(1,3)\}$ and the set of confidence bands in (7.46) consists of only the two confidence bands for $\tilde{\mathbf{x}}'\boldsymbol{\beta}_B - \tilde{\mathbf{x}}'\boldsymbol{\beta}_W$ and $\tilde{\mathbf{x}}'\boldsymbol{\beta}_B - \tilde{\mathbf{x}}'\boldsymbol{\beta}_O$. For $x \in (0.85, 4.25)$ and $\alpha = 0.05$, the critical constant c in (7.46) is calculated to be 3.097 based on 100,000 simulations. This critical value is smaller than the $c = 3.228$ for pairwise comparison. Again, one can plot the two confidence bands to make inferences on the comparisons $\tilde{\mathbf{x}}'\boldsymbol{\beta}_B - \tilde{\mathbf{x}}'\boldsymbol{\beta}_W$ and $\tilde{\mathbf{x}}'\boldsymbol{\beta}_B - \tilde{\mathbf{x}}'\boldsymbol{\beta}_0$.

All the confidence bands in (7.46) are two-sided. If one is interested in one-sided comparisons for some $(i, j) \in \Lambda$ and two-sided comparisons for the other $(i, j) \in \Lambda$, one can construct a set of simultaneous one-sided and two-sided confidence bands accordingly in order to be able to make sharper inferences. The critical constant required for this set of simultaneous confidence bands can be approximated by simulation in a similar way as before.

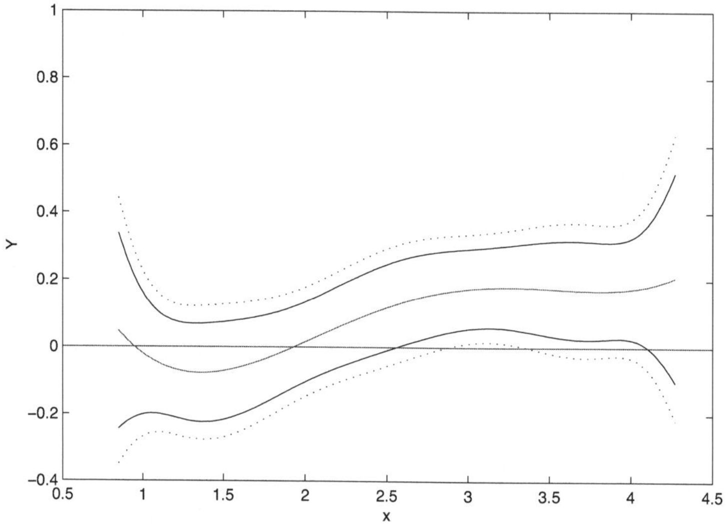

Figure 7.16: *The confidence bands for* $\tilde{\mathbf{x}}'\beta_W - \tilde{\mathbf{x}}'\beta_O$

7.4.2 *Constant width bands for finite comparisons*

A set of two-sided constant width simultaneous confidence bands for the comparisons in (7.45) is given by

$$\tilde{\mathbf{x}}'\beta_i - \tilde{\mathbf{x}}'\beta_j \in \tilde{\mathbf{x}}'\hat{\beta}_i - \tilde{\mathbf{x}}'\hat{\beta}_j \pm c\hat{\sigma}, \ \ \forall x \in (a,b) \ \text{ and } \forall (i,j) \in \Lambda \qquad (7.48)$$

where c is the critical constant chosen so that the simultaneous confidence level of this set of confidence bands is equal to $1-\alpha$.

The simultaneous confidence level of the bands in (7.48) is given by $P\{S < c\}$ where

$$\begin{aligned} S &= \sup_{(i,j)\in\Lambda} \sup_{a<x<b} \frac{|\tilde{\mathbf{x}}'((\hat{\beta}_i - \beta_i) - (\hat{\beta}_j - \beta_j))|}{\hat{\sigma}} \\ &= \sup_{(i,j)\in\Lambda} \sup_{a<x<b} |\tilde{\mathbf{x}}'\mathbf{T}_{ij}| \\ &= \sup_{(i,j)\in\Lambda} K_{2c}(\mathbf{T}_{ij}, (a,b)) \end{aligned} \qquad (7.49)$$

where $\mathbf{T}_{ij}$ is defined below (7.47) and $K_{2c}(\mathbf{T}_{ij}, (a,b))$ is defined in (7.11). Simulation of one random variable S can be implemented in the following way.

Step 1. simulate independent $\mathbf{N}_i \sim \mathcal{N}_{p+1}(\mathbf{0}, (X_i'X_i)^{-1}), i = 1, \cdots, k$ and $\hat{\sigma}/\sigma \sim \sqrt{\chi^2_\nu/\nu}$.

Step 2. calculate $\mathbf{T}_{ij} = (\mathbf{N}_i - \mathbf{N}_j)/(\hat{\sigma}/\sigma)$.

Step 3. compute $K_{2c}(\mathbf{T}_{ij}, (a,b))$ by using the algorithm in Appendix E.

Step 4. find S from (7.49).

The critical constant c can be approximated by simulation as before.

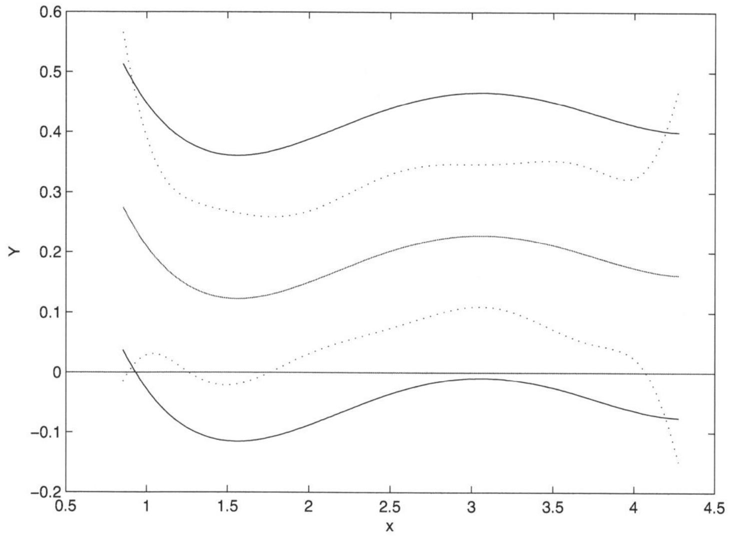

Figure 7.17: *The constant width band for* $\tilde{\mathbf{x}}'\beta_B - \tilde{\mathbf{x}}'\beta_O$

Example 7.16 Corresponding to the two-sided hyperbolic confidence bands constructed in Example 7.15, one can construct two-sided constant width confidence bands for pairwise comparison of the three infant groups. For $x \in (0.85, 4.25)$, $\Lambda = \{(1,2),(1,3),(2,3)\}$ and $\alpha = 0.05$, the critical constant c in (7.48) is 2.690 based on 100,000 simulations. The constant width band for $\tilde{\mathbf{x}}'\beta_B - \tilde{\mathbf{x}}'\beta_O$ is plotted in Figure 7.17 by the two solid curves, while the corresponding hyperbolic band is given by the two dotted curves. As this constant width band just excludes the zero function near $x = 0.85$, the null hypothesis of $\beta_B = \beta_O$ is only slightly significant; but the hyperbolic band indicates that $\tilde{\mathbf{x}}'\beta_B$ is higher than $\tilde{\mathbf{x}}'\beta_O$ over most of the interval $x \in (0.85, 4.25)$. The other two constant width bands for $\tilde{\mathbf{x}}'\beta_B - \tilde{\mathbf{x}}'\beta_W$ and $\tilde{\mathbf{x}}'\beta_W - \tilde{\mathbf{x}}'\beta_O$ include the zero function over $x \in (0.85, 4.25)$ and so the hyperbolic bands allow sharper inferences than the constant width bands, at least for this example.

Constant width simultaneous confidence bands for other types of comparison, such as treatment-control or successive comparisons, can readily be constructed.

If one-sided inferences are of interest then one-sided constant width simultaneous confidence bands can also be constructed in a similar way.

7.4.3 Equivalence of more than two polynomial models

If the main purpose of the comparisons of several polynomial models is to assess the equivalence of the models, then one can use the hypotheses test provided in this subsection. It has a greater chance to establish equivalence than the simultaneous confidence bands introduced in Sections 7.4.1 and 7.4.2 above; the simultaneous confidence bands are more suitable for detecting the differences among the polynomial models.

Suppose the goal is to assess whether all the k polynomial models are equivalent in the sense that the maximum difference over the given interval $x \in (a,b)$ between any two of the k models is strictly less than a given threshold δ, that is,

$$\max_{1\le i\neq j\le k}\max_{a<x<b}|\tilde{\mathbf{x}}'\beta_i - \tilde{\mathbf{x}}'\beta_j| < \delta.$$

Then the hypotheses are set up as

$$\begin{aligned} & H_0^E: \max_{1\le i\neq j\le k}\max_{a<x<b}|\tilde{\mathbf{x}}'\beta_i - \tilde{\mathbf{x}}'\beta_j| \ge \delta \\ \text{against}\quad & H_a^E: \max_{1\le i\neq j\le k}\max_{a<x<b}|\tilde{\mathbf{x}}'\beta_i - \tilde{\mathbf{x}}'\beta_j| < \delta. \end{aligned} \tag{7.50}$$

If and only if the null hypothesis H_0^E is rejected, the equivalence of the k models can be claimed. The probability of falsely claiming equivalence when the k models are actually not equivalent is controlled at the size of the test, α. A size α intersection-union test of Berger (1982; see Appendix D) can easily be devised in the following way.

Let the equivalence hypotheses for the ith and jth models ($1 \le i \neq j \le k$) be

$$H_{0,ij}^E: \max_{a<x<b}|\tilde{\mathbf{x}}'\beta_i - \tilde{\mathbf{x}}'\beta_j| \ge \delta \ \text{ against } \ H_{a,ij}^E: \max_{a<x<b}|\tilde{\mathbf{x}}'\beta_i - \tilde{\mathbf{x}}'\beta_j| < \delta.$$

From Section 7.3.3, a size α test for this pair of hypotheses rejects $H_{0,ij}^E$ if and only if

$$\begin{aligned} & \min_{a<x<b}\left(\tilde{\mathbf{x}}'\hat{\beta}_i - \tilde{\mathbf{x}}'\hat{\beta}_j - t_\nu^\alpha\sqrt{\tilde{\mathbf{x}}'\Delta_{ij}\tilde{\mathbf{x}}}\right) > -\delta \\ \text{and}\quad & \max_{a<x<b}\left(\tilde{\mathbf{x}}'\hat{\beta}_i - \tilde{\mathbf{x}}'\hat{\beta}_j + t_\nu^\alpha\sqrt{\tilde{\mathbf{x}}'\Delta_{ij}\tilde{\mathbf{x}}}\right) < \delta. \end{aligned} \tag{7.51}$$

Note that

$$H_0^E = \cup_{1\le i\neq j\le k} H_{0,ij}^E \ \text{ and } \ H_a^E = \cap_{1\le i\neq j\le k} H_{a,ij}^E.$$

So the size α intersection-union test rejects H_0^E if and only if $H_{0,ij}^E$ is rejected by the individual test (7.51) for all $1 \le i \neq j \le k$, that is,

$$\min_{1\le i\neq j\le k} \min_{a<x<b} \left(\tilde{\mathbf{x}}'\hat{\beta}_i - \tilde{\mathbf{x}}'\hat{\beta}_j - t_\nu^\alpha \sqrt{\tilde{\mathbf{x}}'\Delta_{ij}\tilde{\mathbf{x}}}\right) > -\delta$$

$$\text{and} \qquad \max_{1\le i\neq j\le k} \max_{a<x<b} \left(\tilde{\mathbf{x}}'\hat{\beta}_i - \tilde{\mathbf{x}}'\hat{\beta}_j + t_\nu^\alpha \sqrt{\tilde{\mathbf{x}}'\Delta_{ij}\tilde{\mathbf{x}}}\right) < \delta. \qquad (7.52)$$

Example 7.17 For the perinatal mortality data, if we want to assess whether the three polynomial regression functions corresponding to the three infant groups are sufficiently close to each other to be equivalent, we can use the test in (7.52). Note that $t_{90}^{0.05} = 1.662$ and so, for $x \in (0.85, 4.25)$, one calculates that

$$\min_{1\le i\neq j\le 3} \min_{a<x<b} \left(\tilde{\mathbf{x}}'\hat{\beta}_i - \tilde{\mathbf{x}}'\hat{\beta}_j - t_\nu^\alpha \sqrt{\tilde{\mathbf{x}}'\Delta_{ij}\tilde{\mathbf{x}}}\right) = -0.206$$

$$\text{and} \qquad \max_{1\le i\neq j\le 3} \max_{a<x<b} \left(\tilde{\mathbf{x}}'\hat{\beta}_i - \tilde{\mathbf{x}}'\hat{\beta}_j + t_\nu^\alpha \sqrt{\tilde{\mathbf{x}}'\Delta_{ij}\tilde{\mathbf{x}}}\right) = 0.425. \qquad (7.53)$$

Hence the equivalence of the three polynomial regression functions can be claimed if the pre-specified threshold δ is larger than 0.425.

Intersection-union tests for assessing equivalence in treatment-control or successive comparisons can easily be constructed in a similar way. The conclusions one may draw from the test (7.52) are that either all the k polynomial models are equivalent or not. If not all the k models are equivalent, the likely situation is that the k models can be divided into several groups: the models in the same group are equivalent and the models in different groups are not equivalent. One extreme situation is that there are k groups, with one model in each group, that is, no two models are equivalent. To devise a procedure that divides the k models into several equivalence groups according to the data observed and that controls the probability that any two non-equivalent models are falsely allocated into an equivalent group at a pre-specified α level is an important and challenging problem.

8

Confidence Bands for Logistic Regression

In this chapter, we consider simultaneous confidence bands for logistic regression analysis . The key idea is to combine the large sample asymptotic normality of the maximum likelihood estimators of the regression coefficients with the results for the normal-error linear regression models derived in the previous chapters. The methods given in this chapter are clearly applicable to other generalized linear models, linear mixed-effects models and generalized linear mixed-effects models for which asymptotic normality such as those given in (8.16), (8.21) and (8.28) below holds (cf. Pinheiro and Bates, 2000, and McCulloch and Searle, 2001). The logistic regression model is used mainly for the purpose of illustration.

8.1 Introduction to logistic regression

A logistic regression model describes the relationship between a binomial random variable Y and several independent variables $x_1, \cdots, x_p$. Let Y be the number of 'successes' out of m independent replications of a trial, with the success probability in each replication being p, and so Y is a binomial random variable with parameters m and p:

$$\mathrm{P}\{Y = y\} = \binom{m}{y} p^y (1-p)^{m-y}, \ y = 0, 1, \cdots, m. \tag{8.1}$$

Furthermore, the probability p depends on the setting of the trial specified by the covariates $x_1, \cdots, x_p$ in the form of

$$p = p(\mathbf{x}) = p(\mathbf{x}'\boldsymbol{\beta}) = \frac{\exp(\mathbf{x}'\boldsymbol{\beta})}{1+\exp(\mathbf{x}'\boldsymbol{\beta})} = \frac{1}{1+\exp(-\mathbf{x}'\boldsymbol{\beta})} \tag{8.2}$$

where $\mathbf{x} = (1, x_1, \cdots, x_p)'$, and $\boldsymbol{\beta} = (\beta_0, \beta_1, \cdots, \beta_p)'$ is a vector of unknown parameters and plays a similar role as the $\boldsymbol{\beta}$ in the normal-error regression model (1.1). The expression in (8.2) can be written in the alternative form

$$\ln(p/(1-p)) = \mathbf{x}'\boldsymbol{\beta}, \tag{8.3}$$

where $\ln(p/(1-p))$ is called the logit function.

Suppose that n independent Y random variables, $Y_1, \cdots, Y_n$, are observed, with the jth observation Y_j having covariate values (pattern) $x_{j1}, \cdots, x_{jp}$ and so

$$Y_j \sim \text{Binomial}(m_j, p(\mathbf{x}_{j:})) \tag{8.4}$$

where $\mathbf{x}_{j:} = (1, x_{j1}, \cdots, x_{jp})'$, $j = 1, \cdots, n$. Without loss of generality, assume that $\mathbf{x}_{j:}$ $(j = 1, \cdots, n)$ are all different; otherwise, use the sum of the Y_j's, instead of the individual Y_j's, with the same $\mathbf{x}_{j:}$, which still has a binomial distribution of the form (8.4). These n observations $(Y_j, m_j, \mathbf{x}_{j:})$ $(j = 1, \cdots, n)$ can be used to compute the maximum likelihood estimator (MLE) of β, the only unknown component of the logistic regression model.

To find the MLE of β, we write the likelihood function

$$L = \prod_{j=1}^{n} \left(\binom{m_j}{Y_j} p(\mathbf{x}_{j:})^{Y_j} (1 - p(\mathbf{x}_{j:})^{m_j - Y_j} \right)$$

and the log-likelihood function

$$\begin{aligned} l &= \sum_{j=1}^{n} \left(Y_j \ln \frac{p(\mathbf{x}_{j:})}{1 - p(\mathbf{x}_{j:})} + m_j \ln(1 - p(\mathbf{x}_{j:})) + \ln \binom{m_j}{Y_j} \right) \\ &= \sum_{j=1}^{n} \left(Y_j \mathbf{x}'_{j:} \beta - m_j \ln\left(1 + \exp\{\mathbf{x}'_{j:}\beta\}\right) + \ln \binom{m_j}{Y_j} \right) \\ &= \beta' \mathbf{X}' \mathbf{Y} - \sum_{j=1}^{n} m_j \ln\left(1 + \exp\{\mathbf{x}'_{j:}\beta\}\right) + \sum_{j=1}^{n} \ln \binom{m_j}{Y_j} \end{aligned} \tag{8.5}$$

where $\mathbf{Y} = (Y_1, \cdots, Y_n)'$ and $\mathbf{X} = (\mathbf{x}_{1:}, \cdots, \mathbf{x}_{n:})'$ as in the normal-error regression model (1.2). It follows immediately that the score statistic is given by

$$\mathbf{U} = \frac{\partial l}{\partial \beta} = \mathbf{X}'\mathbf{Y} - \sum_{j=1}^{n} m_j p(\mathbf{x}_{j:}) \mathbf{x}_{j:} \tag{8.6}$$

$$= \mathbf{X}'(\mathbf{Y} - E(\mathbf{Y})) \tag{8.7}$$

where $E(\mathbf{Y}) = (E(Y_1), \cdots, E(Y_n))' = (m_1 p(\mathbf{x}_{1:}), \cdots, m_n p(\mathbf{x}_{n:}))'$. The information matrix is then given by

$$\begin{aligned} \mathcal{I} = \text{Cov}(\mathbf{U}) &= \text{Cov}\left(\mathbf{X}'(\mathbf{Y} - E(\mathbf{Y}))\right) \\ &= \mathbf{X}'\text{Cov}(\mathbf{Y} - E(\mathbf{Y}))\mathbf{X} \\ &= \mathbf{X}'\mathbf{V}\mathbf{X} \end{aligned} \tag{8.8}$$

where $\mathbf{V}$ is the covariance matrix of $\mathbf{Y}$ and clearly given by

$$\mathbf{V} = \begin{pmatrix} m_1 p(\mathbf{x}_{1:})(1 - p(\mathbf{x}_{1:})) & & \\ & \ddots & \\ & & m_n p(\mathbf{x}_{n:})(1 - p(\mathbf{x}_{n:})) \end{pmatrix}$$

since $Y_1, \cdots, Y_n$ are independent random variables, each with a binomial distribution specified in (8.4). Alternatively, one can find the information matrix $\mathcal{I}$ by differentiating the score statistic $\mathbf{U}$ in (8.6) with respect to β to get the Hessian matrix

$$\mathcal{H} = \frac{\partial \mathbf{U}}{\partial \beta} = -\mathbf{X}'\mathbf{V}\mathbf{X}.$$

Since $\mathcal{H}$ is non-random, $\mathcal{I} = E(-\mathcal{H}) = -\mathcal{H} = \mathbf{X}'\mathbf{V}\mathbf{X}$.

The MLE of β can be obtained by solving the Fisher-scoring iterative equation

$$\beta^{(m)} = \beta^{(m-1)} + \left(\mathcal{I}^{(m-1)}\right)^{-1} \mathbf{U}^{(m-1)}; \tag{8.9}$$

this is also the Newton-Raphson iterative equation since $\mathcal{H} = -\mathcal{I}$ for logistic regression. One starts with some initial approximation $\beta^{(0)}$ of β to calculate $\mathcal{I}^{(0)}$ from (8.8) and $\mathbf{U}^{(0)}$ from (8.7), and then uses (8.9) to obtain $\beta^{(1)}$. One next uses $\beta^{(1)}$ to calculate $\mathcal{I}^{(1)}$ and $\mathbf{U}^{(1)}$ and then uses (8.9) to obtain $\beta^{(2)}$. Repeat this until $\beta^{(0)}, \beta^{(1)}, \beta^{(2)}, \cdots$ converge to some $\hat{\beta}$, which is the MLE of β. The MATLAB® function `glmfit`, which implements this iterative procedure, can be used to find $\hat{\beta}$.

The large sample approximate distribution of the MLE $\hat{\beta}$ is $\hat{\beta} \sim \mathcal{N}(\beta, \mathcal{I}^{-1})$. The information matrix $\mathcal{I}$ can be consistently estimated by $\hat{\mathcal{I}} = \mathcal{I}(\hat{\beta})$ by replacing the unknown β in $\mathcal{I}$ by the MLE $\hat{\beta}$, and hence

$$\hat{\beta} \sim \mathcal{N}(\beta, \hat{\mathcal{I}}^{-1}). \tag{8.10}$$

To assess how well the logistic regression model fits the observed data, one can use the deviance

$$\mathcal{D} = 2\sum_{j=1}^{n} \left(Y_j \ln\left(\frac{Y_j}{\hat{Y}_j}\right) + (m_j - Y_j)\ln\left(\frac{m_j - Y_j}{m_j - \hat{Y}_j}\right)\right) \tag{8.11}$$

where $\hat{Y}_j$ is the fitted value at $\mathbf{x}_{j:}$ and given by $\hat{Y}_j = m_j\hat{p}(\mathbf{x}_{j:}) = m_j\{1 + \exp\{-\mathbf{x}'_{j:}\hat{\beta}\}\}^{-1}$. If the Y_j's do follow the logistic model postulated then the large sample approximate distribution of the deviance is

$$\mathcal{D} \sim \chi^2_{n-p-1} \tag{8.12}$$

and the larger is the value of $\mathcal{D}$ the worse is the fit of the logistic regression model. Hence one can compare $\mathcal{D}$ with the critical constant $\chi^2_{n-p-1,\alpha}$ for some pre-specified small α, and $\mathcal{D} > \chi^2_{n-p-1,\alpha}$ indicates the model does not fit the data well.

For the same purpose, one can alternatively use the Pearson chi-squared statistic

$$X^2 = \sum_{j=1}^{n} \frac{(Y_j - m_j\hat{p}(\mathbf{x}_{j:}))^2}{m_j\hat{p}(\mathbf{x}_{j:})(1-\hat{p}(\mathbf{x}_{j:}))} \tag{8.13}$$

which can be shown (e.g., Dobson, 2002) to have the same large sample approximate distribution as $\mathcal{D}$. There is some evidence that χ^2_{n-p-1} approximates the distribution of X^2 better than that of $\mathcal{D}$ (Cressie and Read, 1989). But both the approximations are likely to be poor if the expected frequencies $m_j\hat{p}(\mathbf{x}_{j:})$ are too small (e.g., less than one). In particular, if $m_j = 1$ for $j = 1,\cdots,n$ then neither $\mathcal{D}$ nor X^2 provides a useful measure of fit (cf. Collett, 2003). In this case, one can use the Hosmer-Lemeshow statistic (Hosmer and Lemeshow, 1980) to assess the goodness of fit of the model, whose distribution is approximately chi-square but based on simulation. One can also use other diagnostics, such as residuals, to check the adequacy of the model; see e.g., Collett (2003) and McCullagh and Nelder (1989).

Next we consider the problem of testing the hypotheses

$$H_0 : \mathbf{A}\boldsymbol{\beta} = \mathbf{b} \text{ against } H_a : \mathbf{A}\boldsymbol{\beta} \neq \mathbf{b}$$

where $\mathbf{A}$ is a given $r \times (p+1)$ matrix having full row rank r, with $r \leq p+1$, and $\mathbf{b}$ is a given vector in $\Re^r$. For example, if one wants to assess whether several predictor variables, $x_{p-k+1},\cdots,x_p$ for example (where $1 \leq k \leq p$), are useful in explaining the variability in the Y value then one wants to know whether $\beta_i = 0$ for $i = p-k+1,\cdots,p$. In this case, $\mathbf{A}$ is set as the $k \times (p+1)$ matrix $(\mathbf{0},\mathbf{I}_k)$ and $\mathbf{b} = 0$.

The first test rejects H_0 if and only if

$$(\mathbf{A}\hat{\boldsymbol{\beta}} - \mathbf{b})'(\mathbf{A}\hat{\mathcal{I}}^{-1}\mathbf{A}')^{-1}(\mathbf{A}\hat{\boldsymbol{\beta}} - \mathbf{b}) > \chi^2_{r,\alpha}. \tag{8.14}$$

This test is of size α approximately since, under H_0, the test statistic has a large sample approximate χ^2_r distribution. This test is similar to the test (1.11) for the normal-error linear regression model.

The second test is based on deviance. Note that $\mathcal{D}$ is the deviance of the original logistic model. If H_0 is true then the original model will be reduced to a simpler model and let $\mathcal{D}_r$ denote the deviance of this reduced model. The null hypothesis H_0 is rejected if and only if

$$\mathcal{D}_r - \mathcal{D} > \chi^2_{r,\alpha}. \tag{8.15}$$

This test is again of size α approximately since, under H_0, the test statistic has a large sample approximate χ^2_r distribution. This test is similar to the test (1.7) for the normal-error linear regression model. So both tests have approximate size α, but it is suggested (cf. Dobson, 2002) that the second test often provides a better approximation than the first test.

Example 8.1 Myers *et al.* (2002, p114) provided an illustrative data set on a single quantal bioassay of a toxicity experiment. The interest is on the effect of different doses of nicotine on the common fruit fly by fitting a logistic regression model between the number of flies killed Y and $x_1 = \ln(\text{concentration } g/100cc)$. The seven observations (Y_j, m_j, x_j) are given in Table 8.1.

Table 8.1: *Toxicity data from Myers et al. (2002)*

j	Y_j number killed	m_j number of flies	x_j ln(concentration)
1	8	47	-2.3026
2	14	53	-1.8971
3	24	55	-1.6094
4	32	52	-1.2040
5	38	46	-0.6931
6	50	54	-0.3567
7	50	52	-0.0513

The MLE $\hat{\beta}$ is calculated to be $(\hat{\beta}_0, \hat{\beta}_1)' = (3.124, 2.128)'$ and hence the fitted model is

$$\hat{p}(\mathbf{x}) = \{1 + \exp(-\mathbf{x}'\hat{\beta})\} = \{1 + \exp(-3.124 - 2.128x_1)\}.$$

The observed Y_j/m_j and fitted $\hat{p}(\mathbf{x}_j)$ for $j = 1, \cdots 7$ are plotted in Figure 8.1. From the figure, it seems that the logistic regression model fits the observations very well. Indeed the deviance is $\mathcal{D} = 0.734$, which is very small in comparison with $\chi^2_{5,\alpha}$ for any conventional α value. Hence there is no evidence of lack of fit in the logistic model.

The approximate covariance matrix of $\hat{\beta}$ is

$$\hat{\mathcal{I}}^{-1} = \begin{pmatrix} 0.1122 & 0.0679 \\ 0.0679 & 0.0490 \end{pmatrix}$$

and so $\hat{\beta} \sim \mathcal{N}(\beta, \hat{\mathcal{I}}^{-1})$. From this, we have $\hat{\beta}_1 \sim N(\beta_1, 0.049)$ and so the test statistic in (8.14) for $H_0 : \beta_1 = 0$ against $H_1 : \beta_1 \neq 0$ is $2.128^2/0.049 = 92.403$. This value is highly significant in comparison with $\chi^2_{1,\alpha}$, which indicates that β_1 is very unlikely to be zero and hence the covariate x_1 is highly influential on Y.

One can use the test in (8.15) to test the same hypotheses. Under $H_0 : \beta_1 = 0$, one has $\hat{p}(\mathbf{x}_{1:}) = \cdots = \hat{p}(\mathbf{x}_{7:}) = \sum_{j=1}^7 Y_j / \sum_{j=1}^7 m_j = 0.602$ and so the deviance in (8.11) for this minimum model is $\mathcal{D}_r = 146.022$. The test statistic is therefore $\mathcal{D}_r - \mathcal{D} = 145.288$, which is again highly significant in comparison with $\chi^2_{1,\alpha}$.

8.2 Bands for one model

By using the large sample approximate distribution

$$\hat{\beta} \sim \mathcal{N}(\beta, \hat{\mathcal{I}}^{-1}), \tag{8.16}$$

the methods given in Chapters 2, 3 and 4 can be adapted to deal with a logistic regression model. In particular, one can construct a simultaneous confidence band for $\mathbf{x}'\beta$ over a given covariate region $\mathbf{x}_{(0)} \in \mathcal{X}$:

$$\mathbf{x}'\beta \in (L(\mathbf{x}, \mathbf{Y}),\ U(\mathbf{x}, \mathbf{Y})) \ \text{ for all } \mathbf{x}_{(0)} \in \mathcal{X}, \tag{8.17}$$

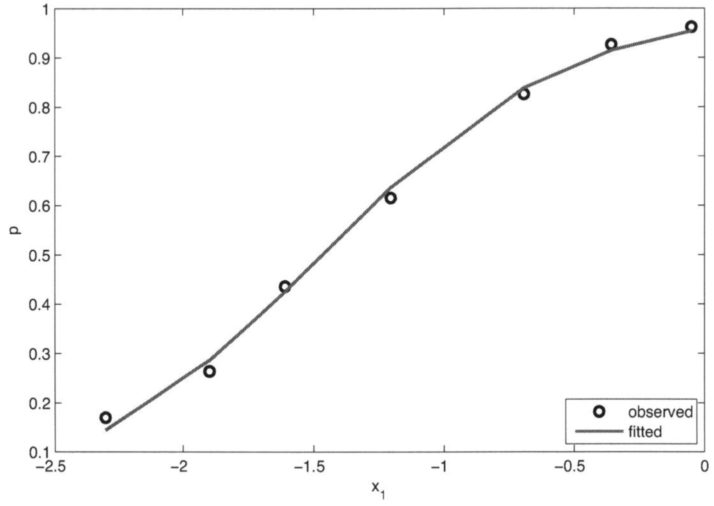

Figure 8.1: *The observed and fitted proportions for the toxicity data*

which has an approximate $1-\alpha$ confidence level. For example, by applying the two-sided hyperbolic confidence band (3.8), one has

$$\mathbf{x}'\beta \in \mathbf{x}'\hat{\beta} \pm c\sqrt{\mathbf{x}'\hat{\mathcal{I}}^{-1}\mathbf{x}} \;\text{ for all } \mathbf{x}_{(0)} \in \mathcal{X}_r \tag{8.18}$$

where the critical constant c can be computed by the simulation method provided in Section 3.2.1. Note, however, $\hat{\mathcal{I}}^{-1}$ and hence c depend on the observed random vector $\mathbf{Y}$, while the critical constant c in the band (3.8) does not depend on $\mathbf{Y}$. This is a major difference between simultaneous confidence bands for a normal-error linear regression model and a logistic regression model. When $\mathcal{X} = \Re^p$, it is clear that $c = \sqrt{\chi^2_{p+1,\alpha}}$ in (8.18). On the other hand, when $\mathcal{X}$ has just one point it is clear that $c = z^{\alpha/2}$ from a normal distribution.

The confidence band (8.17) is equivalent to the confidence band

$$p(\mathbf{x}) = (1+\exp(-\mathbf{x}'\beta))^{-1}$$
$$\in \left((1+\exp(-L(\mathbf{x},\mathbf{Y})))^{-1},\; (1+\exp(-U(\mathbf{x},\mathbf{Y})))^{-1}\right) \;\forall\, \mathbf{x}_{(0)} \in \mathcal{X} \tag{8.19}$$

since the logit link function $\ln(p/(1-p))$ is monotone increasing in $p \in (0,1)$. The confidence band (8.19) provides useful information on how the success probability $p(\mathbf{x})$ varies according to the covariates $\mathbf{x}_{(0)}$ in the region $\mathcal{X}$.

Example 8.2 For the data given in Example 8.1, we have computed the critical constant $c = 2.422$ ($s.e. = 0.0055$) of the 95% two-sided hyperbolic band

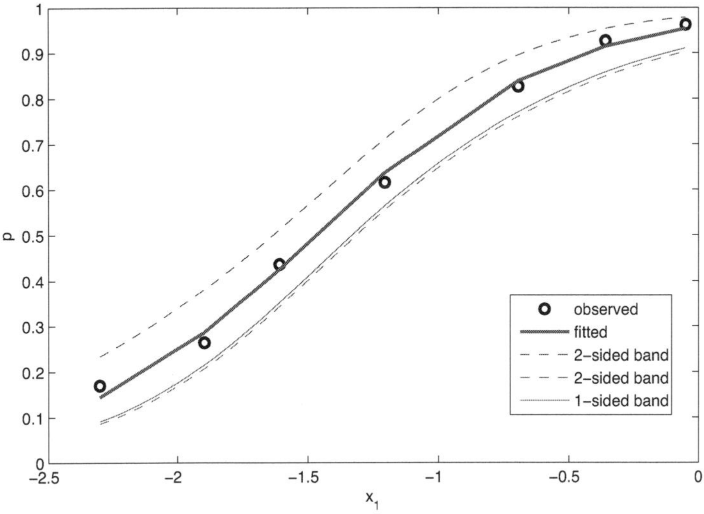

Figure 8.2: *The two-sided and one-sided hyperbolic bands for the toxicity data*

(8.18) over the observed range of the covariate $x_1 \in \mathcal{X}_r = (-2.303, -0.051)$, based on 100,000 simulations. The corresponding confidence band (8.19) is plotted in Figure 8.2. The 95% two-sided hyperbolic band over the whole range $x_1 \in (-\infty, \infty)$ has $c = \sqrt{\chi^2_{p+1,\alpha}} = 2.448$, which is larger than that of the band on $(-2.303, -0.051)$.

We have also computed the one-sided lower hyperbolic band

$$\mathbf{x}'\beta > \mathbf{x}'\hat{\beta} - c\sqrt{\mathbf{x}'\hat{\mathcal{I}}^{-1}\mathbf{x}} \ \text{ for all } \mathbf{x}_{(0)} \in \mathcal{X}_r$$

over the observed range of the covariate $x_1 \in (-2.303, -0.051)$. Based on 100,000 simulations, the critical constant c is 2.151, which is smaller than the critical value c of the two-sided hyperbolic band. The corresponding band (8.19) is also plotted in Figure 8.2.

We can also compute approximate $1-\alpha$ level two-sided or one-sided simultaneous constant width bands for $\mathbf{x}'\beta$, as given in Section 3.3. For the two-sided band

$$\mathbf{x}'\beta \in \mathbf{x}'\hat{\beta} \pm c \ \text{ for all } \mathbf{x}_{(0)} \in \mathcal{X}_r$$

where $\mathcal{X}_r = (-2.303, -0.051)$, the critical constant c for $\alpha = 5\%$ is 0.651 based on 100,000 simulations. The corresponding band (8.19) is plotted in Figure 8.3. For the one-sided lower band

$$\mathbf{x}'\beta > \mathbf{x}'\hat{\beta} - c \ \text{ for all } \mathbf{x}_{(0)} \in \mathcal{X}_r$$

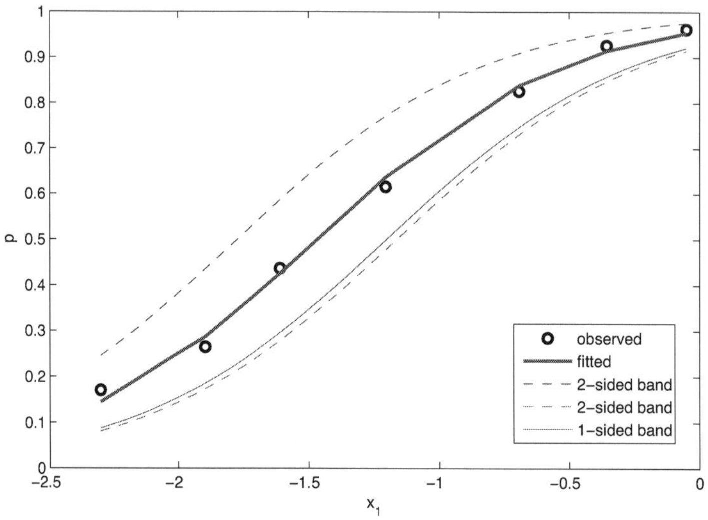

Figure 8.3: *The two-sided and one-sided constant width bands for the toxicity data*

where $\mathcal{X}_r = (-2.303, -0.051)$, the critical constant c for $\alpha = 5\%$ is 0.567 based on 100,000 simulations. The corresponding band (8.19) is plotted in Figure 8.3.

In order to compare the hyperbolic and constant width bands, the two-sided hyperbolic and constant width bands are plotted in Figure 8.4. It is clear from this figure that the constant width band is marginally narrower than the hyperbolic band near the upper limit of the covariate $x_1 = -0.051$, while the hyperbolic band is narrower (which can be very substantial) than the constant width band in most of the covariate region $x_1 \in (-2.303, -0.051)$. Hence the hyperbolic band seems to be more informative than the constant width band for at least this example.

As pointed out above, all these simultaneous confidence bands are of approximate confidence level $1 - \alpha$ based on the large sample approximate distribution (8.16). The true coverage probability of a band depends on the value of the regression coefficients $\boldsymbol{\beta}$, the design matrix $\mathbf{X}$ and the sample sizes $(m_1, \cdots, m_n)'$. While $\mathbf{X}$ and $(m_1, \cdots, m_n)'$ are given, $\boldsymbol{\beta}$ is unknown. It is often of interest to learn the true coverage probability of a chosen simultaneous confidence band when $\boldsymbol{\beta}$ takes values around $\hat{\boldsymbol{\beta}}$, assuming the postulated logistic regression model is correct. For a given $\boldsymbol{\beta}$, the true coverage probability of the band can be assessed by using simulation in the following way. First, simulate $(Y_1, \cdots, Y_n)'$ from the binomial distribution (8.4) for the given values of $(m_1, \cdots, m_n)'$, $\mathbf{X}$ and $\boldsymbol{\beta}$. Second, construct the confidence band using the observed $(Y_1, \cdots, Y_n)'$ and given $(m_1, \cdots, m_n)'$ and $\mathbf{X}$, which involves the computation of the critical value by using simulations. Third,

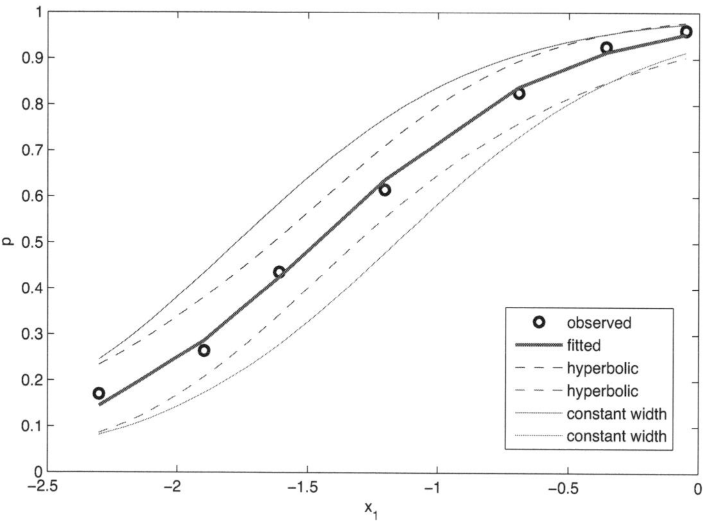

Figure 8.4: *The two-sided hyperbolic and constant width bands for the toxicity data*

check and record whether $\mathbf{x}'\boldsymbol{\beta}$ for the given $\boldsymbol{\beta}$ is contained in the confidence band constructed. Repeat these three steps many times, and the proportion of times that $\mathbf{x}'\boldsymbol{\beta}$ is contained in the confidence band is an approximation to the true coverage probability of the confidence band at the given $\boldsymbol{\beta}$. To make this approximation reasonably accurate, the number of simulations (of the three steps) should be reasonably large and, within each simulation, the number of simulations to determine the critical value should also be large. Hence the whole process can be computationally intensive, but there is no conceptual difficulty.

Hyperbolic simultaneous confidence bands for one logistic regression model have been considered in the statistical literature. Brand *et al.* (1973) gave the bands (8.18–8.19) for the special case of $p = 1$ (i.e., there is only one covariate) and $\mathcal{X}_r = \Re^1$, while Hauck (1983) gave the bands (8.18–8.19) for a general $p \geq 1$ and the special covariate region $\mathcal{X}_r = \Re^p$. Piegorsch and Casella (1988) provided a method to compute large sample conservative critical constant c in the hyperbolic band (8.18) for a finite rectangle covariate region $\mathcal{X}_r$ by using the hyperbolic bands over ellipsoidal regions for the normal-error linear regression model given in Casella and Strawderman (1980); see Section 3.4.1.2. Sun *et al.* (2000) generalized the method of Sun and Load (1994) to provide some two-sided hyperbolic confidence bands for a generalized linear regression model. The construction of these bands involves large sample normality and bias corrections, and these bands are expected to have a closer-to $1-\alpha$ confidence level than a band without bias corrections (such as the two-sided hyperbolic band given in Example 8.2). However, an extensive simulation study in Lin (2008), on true coverage probability (by using the simulation outlined in the last paragraph) of the type 4 confidence band recommended by Sun *et al.* (2000) for a logistic regression model with only one covariate constrained in a finite interval, shows no evidence that Sun *et al.*'s (2000) band has a better coverage probability than the two-sided hyperbolic band given in (8.18). Al-Saidy *et al.* (2003) and Nitcheva *et al.* (2005) constructed simultaneous confidence bands for some risk functions which also used large sample normality of the MLEs of the parameters involved in the risk functions.

8.3 Bands for comparing two models

As in Chapter 5, one may also consider the comparison of two logistic regression models, which describe the relationship between a same response variable Y and a same set of predict variables $x_1, \cdots, x_p$, for two groups or two treatments etc.

In general, suppose the two logistic regression models are specified by

$$Y_{ij} \sim \text{Binomial}\left(m_{ij}, \left(1+\exp(-\mathbf{x}_{ij:}'\beta_i)\right)\right), \quad j=1,\cdots,n_i; \quad i=1,2 \tag{8.20}$$

where Y_{ij}, m_{ij} and $\mathbf{x}_{ij:} = (1, x_{ij1}, \cdots, x_{ijp})'$ for $j = 1, \cdots, n_i$ are observations from the ith model, and β_i are the regression coefficients of the ith model, $i = 1, 2$. All the Y_{ij} are assumed to be independent random variables.

The n_i observations $(Y_{ij}, \mathbf{x}_{ij:}, m_{ij})$ $(j = 1, \cdots, n_i)$ can be used to compute the maximum likelihood estimator (MLE) of β_i, $\hat{\beta}_i$, $i = 1, 2$. By using the large sample approximate distribution

$$\hat{\beta}_i \sim \mathcal{N}(\beta_i, \hat{\mathcal{I}}_i^{-1}), \quad i=1,2 \tag{8.21}$$

and the methods given in chapter 5, one can construct various simultaneous confidence bands for $\mathbf{x}'\beta_2 - \mathbf{x}'\beta_1$ over a given covariate region $\mathbf{x}_{(0)} \in \mathcal{X}$ for the comparison of the two logistic regression models.

Example 8.3 A data set compiled from some social surveys is given in Collett (2003, Table 1.8) and the interest is in the relationship of education and sex to attitudes towards the role of women in society. Each respondent was asked if he or she agreed or disagreed with the statement 'Women should take care of running their homes and leave running the country up to men'. The responses of 1305 male respondents and 1566 females are reproduced in Table 8.2.

Suppose we fit one logistic regression model for male between Y and the level of education x_1, the only covariate, then, using only the observations on males, it is calculated that

$$\hat{\beta}_M = \begin{pmatrix} 2.098 \\ -0.234 \end{pmatrix}, \quad \hat{\mathcal{I}}_M^{-1} = \begin{pmatrix} 0.0555 & -0.0046 \\ -0.0046 & 0.0004 \end{pmatrix}, \quad \mathcal{D}_M = 18.945.$$

Table 8.2: *Social survey data from Collett (2003)*

Years of	Responses of males		Responses of females	
education (x_1)	Agree (Y)	Disagree	Agree (Y)	Disagree
0	4	2	4	2
1	2	0	1	0
2	4	0	0	0
3	6	3	6	1
4	5	5	10	0
5	13	7	14	7
6	25	9	17	5
7	27	15	26	16
8	75	49	91	36
9	29	29	30	35
10	32	45	55	67
11	36	59	50	62
12	115	245	190	403
13	31	70	17	92
14	28	79	18	81
15	9	23	7	34
16	15	110	13	115
17	3	29	3	28
18	1	28	0	21
19	2	13	1	2
20	3	20	2	4

Note that the deviance of this model is $\mathcal{D}_M = 18.945$ on 19 d.f., and so this logistic regression model fits the observations very well. The observed frequencies $Agree/(Agree + Disagree)$ and the fitted logistic curve $(1+\exp(-\mathbf{x}'\hat{\beta}_M))^{-1}$ are plotted in Figure 8.5. In a similar way, we can fit a logistic regression model for female based on the observations on females only. It is computed that

$$\hat{\beta}_F = \begin{pmatrix} 3.003 \\ -0.315 \end{pmatrix}, \quad \hat{\mathcal{I}}_F^{-1} = \begin{pmatrix} 0.0742 & -0.0063 \\ -0.0063 & 0.0006 \end{pmatrix}, \quad \mathcal{D}_F = 38.158.$$

Note that the deviance of this model is $\mathcal{D}_F = 38.158$ on 18 d.f., which indicates that this logistic regression model may not fit the observations very well. The observed frequencies $Agree/(Agree + Disagree)$ and the fitted logistic curve $(1+\exp(-\mathbf{x}'\hat{\beta}_F))^{-1}$ are also plotted in Figure 8.5. It is argued in Collett (2003) that the large deviance $\mathcal{D}_F$ for female is largely due to the first and last two observed frequencies, namely, $4/6$, $1/3$ and $2/6$. As these are based on relatively few individuals, $\mathcal{D}_F$ may not follow the large sample χ^2_{18} distribution accurately and so the deviance may not be a reliable measure of lack of fit of the model.

The plots given in Figure 8.5 show that the observed frequencies do not deviate systematically from the fitted model for either males or females.

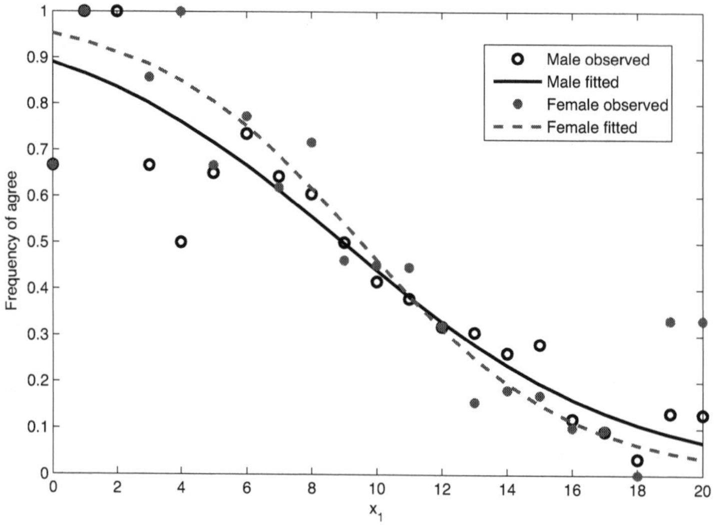

Figure 8.5: *The observed frequencies and fitted models for the social survey data*

Instead of fitting one model for each gender group separately, one can fit one bigger logistic regression model, which encompasses the two separate models for male and female, by introducing the dummy variable

$$z = \begin{cases} 1 & \text{if } Y \text{ is for male} \\ 0 & \text{if } Y \text{ is for female} \end{cases}$$

Then each Y has the binomial distribution with parameters m and success probability $p(\mathbf{x}) = p(\mathbf{x}'\beta)$ where

$$\mathbf{x}'\beta = \beta_0 + \beta_1 x_1 + \beta_2 z + \beta_3 x_1 * z. \tag{8.22}$$

This model implies $\mathbf{x}'\beta = (\beta_0 + \beta_2) + (\beta_1 + \beta_3)x_1$ for male and $\mathbf{x}'\beta = \beta_0 + \beta_1 x_1$ for female. Fitting this bigger model, we have

$$(\hat{\beta}_0, \hat{\beta}_1, \hat{\beta}_2, \hat{\beta}_3)' = (3.003, -0.315, -0.905, 0.081)', \quad \mathcal{D} = 57.103.$$

Hence $(\hat{\beta}_0 + \hat{\beta}_2, \hat{\beta}_1 + \hat{\beta}_3)' = (2.098, -0.234) = \hat{\beta}_M$, $(\hat{\beta}_0, \hat{\beta}_1)' = (3.003, -0.315) = \hat{\beta}_F$ and $\mathcal{D} = \mathcal{D}_M + \mathcal{D}_F$ as expected.

One can test the hypotheses $H_0 : \beta_M = \beta_F$ against $H_a : \beta_M \neq \beta_F$ for the two individual models or, equivalently, the hypotheses $H_0 : \beta_2 = \beta_3 = 0$ against *not* H_0

for the model (8.22), as a mean of comparing the two individual logistic regression models. If H_0 is rejected, one can infer that the two models are statistically significantly different. If H_0 is not rejected, all one can say is that the two models are not statistically significantly different, even though this is often mis-interpreted as that the two models are the same.

One can use the test in (8.14) to test the hypotheses. In this case, the test statistic on the left of the inequality in (8.14) is given by

$$(\hat{\beta}_M - \hat{\beta}_F)' \left(\hat{\mathcal{I}}_M^{-1} + \hat{\mathcal{I}}_F^{-1}\right)^{-1} (\hat{\beta}_M - \hat{\beta}_F) = 6.871,$$

which has a large sample approximate distribution χ^2_2. So the approximate p-value is 0.0322. Alternatively, one can use the test (8.15), and in this example, the test statistic on the left of the inequality in (8.15) is given by

$$\mathcal{D}_r - \mathcal{D} = 64.025 - 57.103 = 6.923$$

which also has a large sample approximate distribution χ^2_2. So the approximate p-value is 0.0314. Hence both tests have similar p-values that are smaller than $\alpha = 0.05$. So at significance level $\alpha = 0.05$, H_0 is rejected and the two models are significantly different.

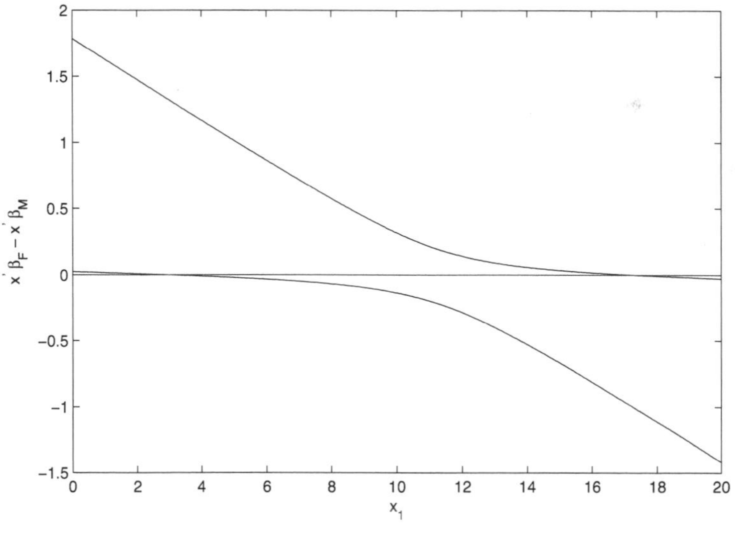

Figure 8.6: *The confidence band for* $\mathbf{x}'\beta_F - \mathbf{x}'\beta_M$ *for the social survey data*

A simultaneous confidence band on $\mathbf{x}'\beta_F - \mathbf{x}'\beta_M$ provides more insights on where the two regression models are significantly different over the covariate range of interest. Let us construct a 95% two-sided simultaneous hyperbolic confidence band for $\mathbf{x}'\beta_F - \mathbf{x}'\beta_M$ over the observed covariate range $x_1 \in [0,20]$:

$$\mathbf{x}'\beta_F - \mathbf{x}'\beta_M \in \mathbf{x}'\hat{\beta}_F - \mathbf{x}'\hat{\beta}_M \pm c\sqrt{\mathbf{x}'(\hat{\mathcal{I}}_F^{-1} + \hat{\mathcal{I}}_M^{-1})\mathbf{x}} \text{ for all } x_1 \in [0,20].$$

Using the simulation method given in Section 5.3, it is calculated that $c = 2.445$ based on 100,000 simulations. This simultaneous confidence band is plotted in Figure 8.6. It can be seen from the confidence band that females and males have significantly different responses only for $x_1 \in [0,3]$ and $x_1 \in [17.35,20]$. Furthermore, for $x_1 \in [0,3]$, i.e., among those who have relatively less education, females tend to agree with the statement significantly more strongly than males. But for $x_1 \in [17.35,20]$, i.e., among those who have relatively more education, males tend to agree with the statement significantly more strongly than females.

It is noteworthy that that the large sample critical constant c for the band over the whole line $x_1 \in (-\infty,\infty)$ is given by $\sqrt{\chi^2_{2,\alpha}} = 2.448$, which is only slightly larger than the critical constant $c = 2.445$ for the band over $x_1 \in [0,20]$. This is because the range $x_1 \in [0,20]$ is quite wide. If this range shrinks towards one point then the large sample critical constant c approaches the normal constant $z_{\alpha/2} = 1.960$, which is substantially smaller than $\sqrt{\chi^2_{2,\alpha}} = 2.448$. Finally, one may also use other simultaneous confidence bands, such as two-sided constant width bands or one-sided bands, to suit the specific need of inference.

Thus far we have constructed simultaneous confidence bands on $\mathbf{x}'\beta_2 - \mathbf{x}'\beta_1$ in order to compare the two logistic regression models. There are real problems in which it is more relevant to construct simultaneous confidence bands on $p(\mathbf{x}'\beta_2) - p(\mathbf{x}'\beta_1)$ in order to quantify the differences between the two response probabilities over a covariate region of interest. Let $q(x) = p'(x)$, the derivative of the function $p(x)$. By using the delta method (cf. Serfling, 1980), we have the following large sample approximations:

$$\begin{aligned} & (p(\mathbf{x}'\hat{\beta}_2) - p(\mathbf{x}'\hat{\beta}_1)) - (p(\mathbf{x}'\beta_2) - p(\mathbf{x}'\beta_1)) \\ \approx\ & q(\mathbf{x}'\beta_2)(\mathbf{x}'\hat{\beta}_2 - \mathbf{x}'\beta_2) - q(\mathbf{x}'\beta_1)(\mathbf{x}'\hat{\beta}_1 - \mathbf{x}'\beta_1) \\ \sim\ & N(0, \mathbf{x}'V_{\mathbf{x}}\mathbf{x}). \end{aligned} \tag{8.23}$$

where $V_{\mathbf{x}} = \left(q(\mathbf{x}'\hat{\beta}_2)\right)^2 \hat{\mathcal{I}}_2^{-1} + \left(q(\mathbf{x}'\hat{\beta}_1)\right)^2 \hat{\mathcal{I}}_1^{-1}$. One can construct the following two-sided hyperbolic band

$$p(\mathbf{x}'\beta_2) - p(\mathbf{x}'\beta_1) \in p(\mathbf{x}'\hat{\beta}_2) - p(\mathbf{x}'\hat{\beta}_1) \pm c\sqrt{\mathbf{x}'V_{\mathbf{x}}\mathbf{x}} \text{ for all } \mathbf{x}_{(0)} \in \mathcal{X}.$$

The confidence level of this band is given by $P\{T \le c\}$ where

$$T = \max_{\mathbf{x}_{(0)} \in \mathcal{X}} \frac{|(p(\mathbf{x}'\hat{\beta}_2) - p(\mathbf{x}'\hat{\beta}_1)) - (p(\mathbf{x}'\beta_2) - p(\mathbf{x}'\beta_1))|}{\sqrt{\mathbf{x}'V_{\mathbf{x}}\mathbf{x}}}$$

$$\approx \max_{\mathbf{x}_{(0)}\in\mathcal{X}} \frac{|\mathbf{x}'(q(\mathbf{x}'\hat{\beta}_2)(\hat{\beta}_2-\beta_2)-q(\mathbf{x}'\hat{\beta}_1)(\hat{\beta}_1-\beta_1))|}{\sqrt{\mathbf{x}'V_{\mathbf{x}}\mathbf{x}}}$$

$$\approx \max_{\mathbf{x}_{(0)}\in\mathcal{X}} \frac{|\mathbf{x}'(q(\mathbf{x}'\hat{\beta}_2)\hat{\mathcal{I}}_2^{-1/2}\mathbf{N}_2-q(\mathbf{x}'\hat{\beta}_1)\hat{\mathcal{I}}_1^{-1/2}\mathbf{N}_1)|}{\sqrt{\mathbf{x}'V_{\mathbf{x}}\mathbf{x}}} \qquad (8.24)$$

where $\mathbf{N}_1$ and $\mathbf{N}_2$ are two independent $\mathcal{N}_{p+1}(\mathbf{0},\mathbf{I})$ random vectors. So, in principle, the large sample approximate critical constant c can be found by simulating a large number of the random variable T from the expression (8.24). This idea can also be used to construct an approximate confidence band for a general parametric regression model or for the difference between two general parametric regression models; see Gsteiger, Bretz and Liu (2010). Note, however, the maximization involved in (8.24) is different from the maximization we have met before, due to the fact that the covariance matrix $\mathbf{V}_{\mathbf{x}}$ depends on $\mathbf{x}$. In particular, when $\mathcal{X}=\Re^p$, the expression in (8.24) does not equal the random variable $\sqrt{\chi^2_{p+1}}$ and so it is not clear whether $\sqrt{\chi^2_{p+1,\alpha}}$ is still a large sample conservative critical value. More research is required in this direction.

One can also use the following method to construct a large sample conservative simultaneous confidence band for $p(\mathbf{x}'\beta_2)-p(\mathbf{x}'\beta_1)$ over a given covariate region $\mathcal{X}$. Construct a $\sqrt{1-\alpha}$ level simultaneous confidence band

$$L_i(\mathbf{Y}_i,\mathbf{x}) < p(\mathbf{x}'\beta_i) < U_i(\mathbf{Y}_i,\mathbf{x}) \ \ \forall\, \mathbf{x}\in\mathcal{X}$$

for each $1\le i\le 2$ by using the methods in Section 8.2. Then it is clear that the confidence band

$$L_2(\mathbf{Y}_2,\mathbf{x})-U_1(\mathbf{Y}_1,\mathbf{x}) < p(\mathbf{x}'\beta_2)-p(\mathbf{x}'\beta_1) < U_2(\mathbf{Y}_2,\mathbf{x})-L_1(\mathbf{Y}_1,\mathbf{x}) \ \ \forall\, \mathbf{x}\in\mathcal{X}$$

has a simultaneous confidence level of at least $1-\alpha$. Again, detailed study is necessary to assess how conservative this band is.

As for the normal-error linear regression models, the methodologies considered above are primarily for detecting and quantifying the differences between the two models. In many problems, the purpose of research is to show that the two models do not differ, either in one direction or in both directions, by more than $\delta>0$, a pre-specified threshold, over a given covariate region $\mathcal{X}$, and therefore the two models can be claimed to be equivalently over $\mathbf{x}_{(0)}\in\mathcal{X}$.

To show one-sided or two-sided equivalence of $\mathbf{x}'\beta_1$ and $\mathbf{x}'\beta_2$, the methodologies developed in Chapter 5 can be applied directly by using the large sample asymptotic normal distributions of $\hat{\beta}_1$ and $\hat{\beta}_2$. Hence we concentrate on the equivalence of $p(\mathbf{x}'\beta_1)$ and $p(\mathbf{x}'\beta_2)$ over a given covariate region $\mathcal{X}$ below.

Model one $p(\mathbf{x}'\beta_1)$ is defined to be non-superior to model two $p(\mathbf{x}'\beta_2)$ over a given covariate region $\mathbf{x}_{(0)}\in\mathcal{X}$ if

$$p(\mathbf{x}'\beta_1) < p(\mathbf{x}'\beta_2)+\delta \ \ \forall \mathbf{x}_{(0)}\in\mathcal{X}, \text{ i.e. } \max_{\mathbf{x}_{(0)}\in\mathcal{X}}(p(\mathbf{x}'\beta_1)-p(\mathbf{x}'\beta_2)) < \delta$$

where $\delta > 0$ is pre-specified. From Theorem 5.1, an upper confidence bound on $\max_{\mathbf{x}_{(0)}\in\mathcal{X}}(p(\mathbf{x}'\boldsymbol{\beta}_1) - p(\mathbf{x}'\boldsymbol{\beta}_2))$ that allows the assessment of the non-superior of $p(\mathbf{x}'\boldsymbol{\beta}_1)$ to $p(\mathbf{x}'\boldsymbol{\beta}_2)$ over $\mathbf{x}_{(0)} \in \mathcal{X}$ requires only the construction of pointwise upper bound $U_p(\mathbf{Y}_1,\mathbf{Y}_2,\mathbf{x})$ for each $\mathbf{x}_{(0)} \in \mathcal{X}$. By using the large sample approximate distribution given in (8.23), we can construct the approximate $1-\alpha$ pointwise upper bound

$$U_p(\mathbf{Y}_1,\mathbf{Y}_2,\mathbf{x}) = p(\mathbf{x}'\hat{\boldsymbol{\beta}}_1) - p(\mathbf{x}'\hat{\boldsymbol{\beta}}_2) + z^{\alpha}\sqrt{\mathbf{x}'V_{\mathbf{x}}\mathbf{x}} \ \text{ for each } \mathbf{x}_{(0)} \in \mathcal{X} \qquad (8.25)$$

where z^{α} is the upper α point of a standard normal distribution, and $V_{\mathbf{x}}$ is given below expression (8.23). The non-superiority of $p(\mathbf{x}'\boldsymbol{\beta}_1)$ to $p(\mathbf{x}'\boldsymbol{\beta}_2)$ over $\mathbf{x}_{(0)} \in \mathcal{X}$ can be claimed if and only if

$$\max_{\mathbf{x}_{(0)}\in\mathcal{X}} U_p(\mathbf{Y}_1,\mathbf{Y}_2,\mathbf{x}) < \delta.$$

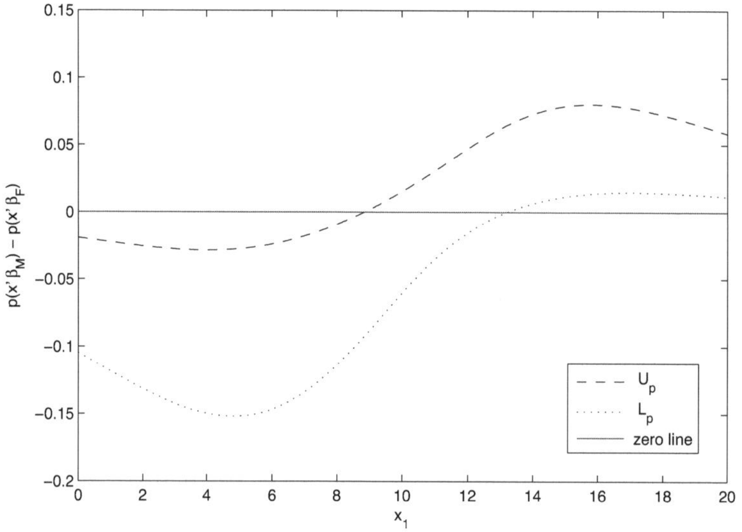

Figure 8.7: *The pointwise bands U_p and L_p for the social survey data*

Similarly, model one $p(\mathbf{x}'\boldsymbol{\beta}_1)$ is defined to be non-inferior to model two $p(\mathbf{x}'\boldsymbol{\beta}_2)$ over a given covariate region $\mathbf{x}_{(0)} \in \mathcal{X}$ if

$$p(\mathbf{x}'\boldsymbol{\beta}_1) > p(\mathbf{x}'\boldsymbol{\beta}_2) - \delta \ \ \forall \mathbf{x}_{(0)} \in \mathcal{X}, \text{ i.e. } \min_{\mathbf{x}_{(0)}\in\mathcal{X}} \{p(\mathbf{x}'\boldsymbol{\beta}_1) - p(\mathbf{x}'\boldsymbol{\beta}_2)\} > -\delta.$$

Let

$$L_p(\mathbf{Y}_1,\mathbf{Y}_2,\mathbf{x}) = p(\mathbf{x}'\hat{\boldsymbol{\beta}}_1) - p(\mathbf{x}'\hat{\boldsymbol{\beta}}_2) - z^{\alpha}\sqrt{\mathbf{x}'V_{\mathbf{x}}\mathbf{x}} \ \text{ for each } \mathbf{x}_{(0)} \in \mathcal{X}. \qquad (8.26)$$

Then from Theorem 5.2 non-inferiority of $p(\mathbf{x}'\beta_1)$ to $p(\mathbf{x}'\beta_2)$ over $\mathbf{x}_{(0)} \in \mathcal{X}$ can be claimed if and only if

$$\min_{\mathbf{x}_{(0)} \in \mathcal{X}} L_p(\mathbf{Y}_1, \mathbf{Y}_2, \mathbf{x}) > -\delta.$$

Finally, the models $p(\mathbf{x}'\beta_1)$ and $p(\mathbf{x}'\beta_2)$ are defined to be equivalent over a given covariate region $\mathbf{x}_{(0)} \in \mathcal{X}$ if

$$|p(\mathbf{x}'\beta_1) - p(\mathbf{x}'\beta_2)| < \delta \ \ \forall \mathbf{x}_{(0)} \in \mathcal{X}, \text{ i.e. } \max_{\mathbf{x}_{(0)} \in \mathcal{X}} |p(\mathbf{x}'\beta_1) - p(\mathbf{x}'\beta_2)| < \delta.$$

To assess equivalence, we set up the equivalence hypotheses

$$H_0^E : \max_{\mathbf{x}_{(0)} \in \mathcal{X}} |p(\mathbf{x}'\beta_1) - p(\mathbf{x}'\beta_2)| \geq \delta \iff H_a^E : \max_{\mathbf{x}_{(0)} \in \mathcal{X}} |p(\mathbf{x}'\beta_1) - p(\mathbf{x}'\beta_2)| < \delta.$$

So the equivalence of the two models can be claimed if and only if the null hypothesis H_0^E is rejected. An approximate size α intersection-union test (see Section 5.5.3) rejects the non-equivalence null hypothesis H_0^E if and only if

$$-\delta < \min_{\mathbf{x}_{(0)} \in \mathcal{X}} L_P(\mathbf{Y}_1, \mathbf{Y}_2, \mathbf{x}) \text{ and } \max_{\mathbf{x}_{(0)} \in \mathcal{X}} U_P(\mathbf{Y}_1, \mathbf{Y}_2, \mathbf{x}) < \delta.$$

Example 8.4 Continuing from Example 8.3, one can assess whether the two logistic regression models corresponding to the male and female groups are equivalent. For this purpose, we have computed $U_P(\mathbf{Y}_1, \mathbf{Y}_2, \mathbf{x})$ and $L_P(\mathbf{Y}_1, \mathbf{Y}_2, \mathbf{x})$ for $\alpha = 0.05$, which are plotted over $x_1 \in [0, 20]$ in Figure 8.7. In particular, we have

$$\max_{\mathbf{x}_{(0)} \in \mathcal{X}} U_P(\mathbf{Y}_1, \mathbf{Y}_2, \mathbf{x}) = 0.080 \text{ and } \min_{\mathbf{x}_{(0)} \in \mathcal{X}} L_P(\mathbf{Y}_1, \mathbf{Y}_2, \mathbf{x}) = -0.152.$$

So unless the pre-specified threshold δ is greater than 0.152, the null hypothesis H_0^E cannot be rejected and so the equivalence of the two models cannot be claimed.

8.4 Bands for comparing more than two models

As in Chapter 6, one can use simultaneous confidence bands for comparisons of more than two logistic regression models, which describe the relationship between a same response variable Y and a same set of predictor variables $x_1, \cdots, x_p$, for several groups or treatments, etc.

In general, suppose that the $k(\geq 3)$ logistic regression models are specified by

$$Y_{ij} \sim \text{Binomial}\left(m_{ij}, (1 + \exp(-\mathbf{x}'_{ij:}\beta_i))\right), \ \ j = 1, \cdots, n_i; \ \ i = 1, \cdots, k \quad (8.27)$$

where Y_{ij}, m_{ij} and $\mathbf{x}_{ij:} = (1, x_{ij1}, \cdots, x_{ijp})'$ for $j = 1, \cdots, n_i$ are observations from the ith model, and β_i are the regression coefficients of the ith model, $i = 1, \cdots, k$. All the Y_{ij} are assumed to be independent random variables.

Table 8.3: *Numbers of mice responding to four analgesic compounds*

Compound	$x_1 = \log_e(\text{Dose})$	Y =Number responded	m =Number tested
Morphine	0.38	22	110
Morphine	1.20	57	115
Morphine	1.75	84	123
Morphine	2.01	106	135
Amidone	0.39	15	67
Amidone	1.02	51	98
Amidone	1.43	70	108
Amidone	1.82	89	115
Pethidine	1.39	11	68
Pethidine	1.82	24	72
Pethidine	1.98	25	87
Pethidine	2.65	57	92
Pethidine	2.74	57	90
Phenadoxone	-0.22	37	98
Phenadoxone	0.43	59	87
Phenadoxone	1.20	82	90

The n_i observations $(Y_{ij}, \mathbf{x}_{ij:}, m_{ij})$ $(j = 1, \cdots, n_i)$ can be used to compute the maximum likelihood estimator (MLE) of $\boldsymbol{\beta}_i$ of the ith model, $\hat{\boldsymbol{\beta}}_i$, $i = 1, \cdots, k$. By using the large sample approximate distribution

$$\hat{\boldsymbol{\beta}}_i \sim \mathcal{N}(\boldsymbol{\beta}_i, \hat{\mathcal{I}}_i^{-1}), \quad i = 1, \cdots, k \tag{8.28}$$

and the methods given in Chapter 6, one can construct various simultaneous confidence bands for a pre-specified set of contrasts of the $\mathbf{x}'\boldsymbol{\beta}_i$'s, e.g., all contrasts, all pairwise comparisons and many-one comparisons among the logistic regression models, over a given covariate region $\mathbf{x}_{(0)} \in \mathcal{X}$.

Example 8.5 Table 8.3 provides a data set on an experiment that compared the analgesic potencies of amidone, pethidine and phenadoxone with that of morphine. The numbers of mice tested and responded at several doses of each drug recorded.

Fit one logistic regression model for each of the four compounds, by using only the observations on that compound, to get

$$\hat{\boldsymbol{\beta}}_M = \begin{pmatrix} -1.987 \\ 1.612 \end{pmatrix}, \quad \hat{\mathcal{I}}_M^{-1} = \begin{pmatrix} 0.0723 & -0.0443 \\ -0.0443 & 0.0318 \end{pmatrix}, \quad \mathcal{D}_M = 0.198$$

$$\hat{\boldsymbol{\beta}}_A = \begin{pmatrix} -1.755 \\ 1.670 \end{pmatrix}, \quad \hat{\mathcal{I}}_A^{-1} = \begin{pmatrix} 0.0991 & -0.0700 \\ -0.0700 & 0.0565 \end{pmatrix}, \quad \mathcal{D}_A = 0.737$$

$$\hat{\beta}_{PE} = \begin{pmatrix} -3.907 \\ 1.635 \end{pmatrix}, \ \hat{\mathcal{I}}_M^{-1} = \begin{pmatrix} 0.2700 & -0.1152 \\ -0.1152 & 0.0513 \end{pmatrix}, \ \mathcal{D}_{PE} = 2.028$$

$$\hat{\beta}_{PH} = \begin{pmatrix} -0.077 \\ 1.977 \end{pmatrix}, \ \hat{\mathcal{I}}_M^{-1} = \begin{pmatrix} 0.0251 & -0.0196 \\ -0.0196 & 0.0804 \end{pmatrix}, \ \mathcal{D}_{PH} = 0.026.$$

It is clear from the four deviances that each logistic model fits the observations very well. The observed frequencies and fitted logistic regression model for each of the four compounds are plotted in Figure 8.8.

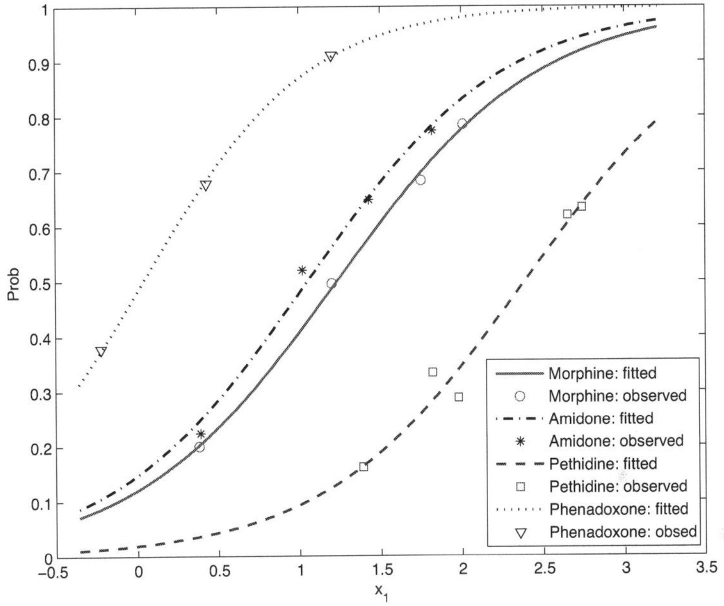

Figure 8.8: *The observed frequencies and fitted models for the four compounds*

Instead of fitting one model for each compound separately, one can fit one bigger logistic regression model, which encompasses the four individual models for the four compounds, by introducing three dummy variables

$$z_M = \begin{cases} 1 & \text{if } Y \text{ is from Morphine} \\ 0 & \text{otherwise} \end{cases}$$
$$z_A = \begin{cases} 1 & \text{if } Y \text{ is from Amidone} \\ 0 & \text{otherwise} \end{cases}$$
$$z_{PE} = \begin{cases} 1 & \text{if } Y \text{ is from Pethidine} \\ 0 & \text{otherwise .} \end{cases}$$

Then each Y, Number responding, has a binomial distribution with parameters m

and success probability $p(\mathbf{x}) = p(\mathbf{x}'\beta)$ where

$$\mathbf{x}'\beta = \beta_0 + \beta_1 x_1 + \beta_2 z_M + \beta_3 z_A + \beta_4 z_{PE} + \beta_5 x_1 * z_M + \beta_6 x_1 * z_A + \beta_7 x_1 * z_{PE}. \quad (8.29)$$

This model implies the four individual models

$$\begin{aligned}
(1,x_1)\beta_M &= (\beta_0+\beta_2)+(\beta_1+\beta_5)x_1 \text{ for Morphine} \\
(1,x_1)\beta_A &= (\beta_0+\beta_3)+(\beta_1+\beta_6)x_1 \text{ for Amidone} \\
(1,x_1)\beta_{PE} &= (\beta_0+\beta_4)+(\beta_1+\beta_7)x_1 \text{ for Pethidine} \\
(1,x_1)\beta_{PH} &= \beta_0+\beta_1 x_1 \text{ for Phenadoxone.}
\end{aligned}$$

Fitting this bigger model, we have

$$\begin{aligned}
(\hat\beta_0,\cdots,\hat\beta_7)' &= (-0.077, 1.977, -1.909, -1.678, -3.830, -0.365, -0.307, -0.342)' \\
\text{and } \mathcal{D} &= 2.989.
\end{aligned}$$

As expected, $\mathcal{D} = \mathcal{D}_M + \mathcal{D}_A + \mathcal{D}_{PE} + \mathcal{D}_{PH}$ and $(\hat\beta_0 + \hat\beta_2, \hat\beta_1 + \hat\beta_5)' = (-1.987, 1.612) = \hat\beta_M$ etc. Hence the two methods of fitting models give the same results.

One possible way to compare the analgesic potencies of the four compounds is to test the hypotheses $H_0 : \beta_M = \beta_A = \beta_{PE} = \beta_{PH}$ against H_a : not H_0. If H_0 is rejected, one can infer that the four compounds have statistically significantly different potencies. If H_0 is not rejected, all one can say is that the four compounds do not show statistically significantly different potencies. In terms of the model (8.29), the hypotheses become

$$H_0 : \mathbf{A}\beta = \mathbf{0} \text{ against } H_a : \mathbf{A}\beta \neq \mathbf{0}$$

where $\mathbf{A}$ is a 6×8 matrix given by $(\mathbf{0}, \mathbf{I}_6)$. The test statistic on the left of the inequality in (8.14) is calculated to be 211.41, which has a large sample approximate distribution χ^2_6. So the test statistic value 211.41 is very highly significant. Alternatively, one can use the test (8.15). In this example, model (8.29) will be reduced under the null hypothesis H_0 to

$$\mathbf{x}'\beta = \beta_0 + \beta_1 x_1$$

for all the four compounds, the deviance for which is computed to be $\mathcal{D}_r = 271.036$. Hence the test statistic on the left of the inequality in (8.15) is given by

$$\mathcal{D}_r - \mathcal{D} = 271.036 - 2.989 = 268.047$$

which also has a large sample approximate distribution χ^2_6. This test statistic value is also very highly significant. So the null hypothesis H_0 is rejected by both tests (8.14) and (8.15) very convincingly, and one can therefore conclude that the four logistic regression models for the four compounds are highly unlikely to be the same.

To make more detailed comparisons of the four models for the four compounds, one can use simultaneous confidence bands. For instance, if one wants to make all pairwise comparisons among the four models then one can construct the following simultaneous confidence bands

$$\mathbf{x}'\beta_i - \mathbf{x}'\beta_j \in \mathbf{x}'\hat{\beta}_i - \mathbf{x}'\hat{\beta}_j \pm c\sqrt{\mathbf{x}'(\hat{\mathcal{I}}_i^{-1} + \hat{\mathcal{I}}_j^{-1})\mathbf{x}}$$
$$\text{for all } x_1 \in [a,b] \text{ and for all } i \neq j \text{ and } i,j \in \{M, A, PE, PH\} \quad (8.30)$$

where $[a,b]$ is chosen as the observed range $[-0.22, 2.74]$, and critical constant c is chosen so that the large sample approximate simultaneous confidence level is 95%. Based on 200,000 simulations, the critical constant is computed to be $c = 2.965$. This value is considerably smaller than the large sample conservative value $\sqrt{qf_{q,\nu}^{\alpha}} = 3.549$ given in (6.8), where $q = (k-1)(p+1) = 3 \times 2 = 6$, $\nu = \infty$ and $\alpha = 0.05$ in this example.

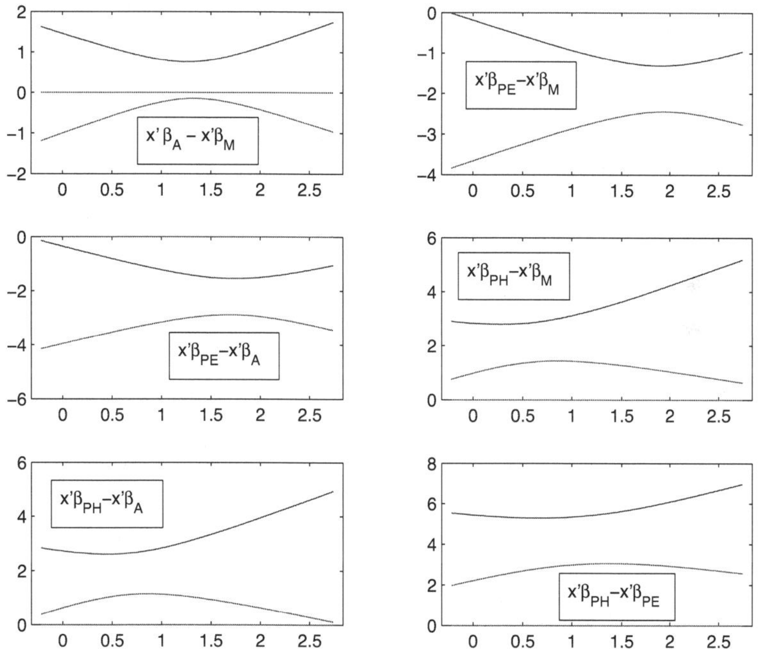

Figure 8.9: *The confidence bands for pairwise comparisons of the four compounds*

With $c = 2.965$, all six confidence bands in (8.30) are plotted in Figure 8.9, and comparison between any two models can be made from the corresponding confidence band. For example, the regression function of Phenadoxone, $p(\mathbf{x}'\beta_{PH})$, is significantly higher than the other three regression functions over the observed range of the covariate $x_1 = \log(\text{Dose}) \in [-0.22, 2.74]$. The regression function of Pethidine, $p(\mathbf{x}'\beta_{PE})$, is significantly lower than the regression functions $p(\mathbf{x}'\beta_A)$ and $p(\mathbf{x}'\beta_M)$ over the observed range of the covariate. But there is no statistically significant difference between the regression functions of Amidone and Mophine over $x_1 \in [-0.22, 2.74]$. These more detailed inferences are not available from the hypotheses tests and are consistent with the fitted regression functions plotted in Figure 8.8.

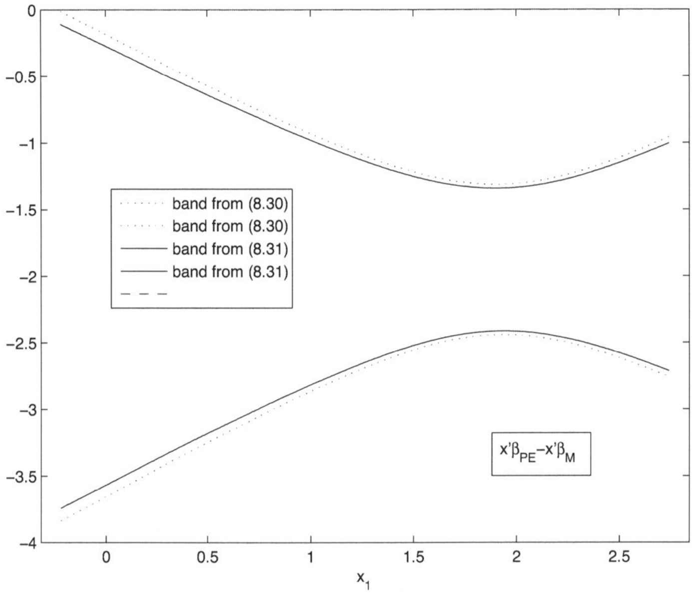

Figure 8.10: *The confidence bands for* $\mathbf{x}'\beta_{PE} - \mathbf{x}'\beta_M$

If one is interested only in the comparisons of Amidone, Pethidine and Phenadoxone with the control Mophine, then one can use the following set of simultaneous confidence bands

$$\mathbf{x}'\beta_i - \mathbf{x}'\beta_M \in \mathbf{x}'\hat{\beta}_i - \mathbf{x}'\hat{\beta}_M \pm c\sqrt{\mathbf{x}'(\hat{\mathcal{I}}_i^{-1} + \hat{\mathcal{I}}_M^{-1})\mathbf{x}}$$
$$\text{for all } x_1 \in [a,b] \text{ and for all } i \in \{A, PE, PH\} \tag{8.31}$$

where $[a,b] = [-0.22, 2.74]$ as before, and critical constant c is chosen so that the large sample approximate simultaneous confidence level is 95%. This set of confidence bands has less confidence bands than the set of confidence bands in (8.30), and so the critical constant c is smaller than that in (8.30). Sharper inferences concerning the three comparisons can therefore be made from the bands in (8.31) than those in (8.30). Based on 200,000 simulations, the critical constant is computed to be $c = 2.815$. This value is smaller than the $c = 2.965$ in (3.30) as expected. Figure 8.10 presents the confidence bands for $\mathbf{x}'\beta_{PE} - \mathbf{x}'\beta_M$ from (8.30) and (8.31). The band from (8.31) is narrower and so allows better inference than the band from (8.30).

Sometimes it is desirable to construct simultaneous confidence bands on $p(\mathbf{x}'\beta_i) - p(\mathbf{x}'\beta_j)$ for $\mathbf{x}_{(0)} \in \mathcal{X}$ and for all $(i, j) \in \Lambda$ in order to quantify directly the differences between the response probabilities $p(\mathbf{x}'\beta_i)$ and $p(\mathbf{x}'\beta_j)$ over $\mathbf{x}_{(0)} \in \mathcal{X}$ for all $(i, j) \in \Lambda$. The two ideas given in the last section on the comparison of two models are also applicable here, but an extra maximization over $(i, j) \in \Lambda$ is required. This warrants further research.

Sometimes one may hope to establish the equivalence of the response probabilities $p(\mathbf{x}'\beta_i)$ for $1 \le i \le k$ over a given covariate region $\mathbf{x}_{(0)} \in \mathcal{X}$ in the sense of

$$\max_{1\le i\ne j\le k}\max_{\mathbf{x}_{(0)}\in\mathcal{X}} |p(\mathbf{x}'\beta_i) - p(\mathbf{x}'\beta_j)| < \delta$$

where $\delta > 0$ is a pre-specified threshold. For this, one can set up the equivalence hypotheses

$$\begin{aligned} & H_0^E: \quad \max_{1\le i\ne j\le k}\max_{\mathbf{x}_{(0)}\in\mathcal{X}} |p(\mathbf{x}'\beta_i) - p(\mathbf{x}'\beta_j)| \ge \delta \\ \Longleftrightarrow \quad & H_a^E: \quad \max_{1\le i\ne j\le k}\max_{\mathbf{x}_{(0)}\in\mathcal{X}} |p(\mathbf{x}'\beta_i) - p(\mathbf{x}'\beta_j)| < \delta. \end{aligned}$$

The equivalence of all the k models can be claimed if and only if the null hypothesis H_0^E is rejected. Following the results of Sections 6.5 and 8.3, a large sample approximate size α intersection-union test rejects the non-equivalence null hypothesis H_0^E if and only if

$$\begin{aligned} & -\delta < \min_{1\le i\ne j\le k}\min_{\mathbf{x}_{(0)}\in\mathcal{X}} L_P(\mathbf{Y}_i, \mathbf{Y}_j, \mathbf{x}) \\ \text{and} \quad & \max_{1\le i\ne j\le k}\max_{\mathbf{x}_{(0)}\in\mathcal{X}} U_P(\mathbf{Y}_i, \mathbf{Y}_j, \mathbf{x}) < \delta \end{aligned}$$

where L_P and U_P are defined in (8.26) and (8.25) respectively.

Appendices

A

Approximation of the percentile of a random variable

As we have seen in the main part of this book, the critical constant c of a $1-\alpha$ simultaneous confidence band can be expressed as the $100(1-\alpha)$ percentile of a random variable S, and this population percentile can be approximated by the sample percentile by using simulation. We first simulate R independent replicates $S_1, \cdots, S_R$ of the random variable S, and then use the $\langle(1-\alpha)R\rangle$th largest S_i, $\hat{c}$, as an approximation to the exact critical constant c.

As we shall see from the discussions below, $\hat{c}$ can be as close to c as one wants if the number of replications R is sufficiently large. This simulation approach can therefore be regarded as exact, considering the computation power of modern computers and the available software. Edwards and Berry (1987) considered similar methods to approximate the critical constants in multiple tests. Westfall and Young (1993) considered simulation-based multiple tests. Beran (1988, 1990) proposed some simulation methods for the construction of balanced simultaneous confidence sets.

Our simulation approach is different from the usual bootstrap or Markov chain Monte Carlo (MCMC) methods. The bootstrap method usually samples from an estimated distribution of S; see e.g., Hall (1992), Efron and Tibshirani (1993) and Shao and Tu (1995). The random variates generated by a MCMC method are usually not independent; see e.g., Gamerman and Lopes (2006) and O'Hagan and Forster (2004). But in our approach the S_i's are independent and generated from the exact distribution of S.

A.1 Accuracy of the sample percentile

It is well known that the sample percentile $\hat{c}$ converges to the population percentile c almost surely as the simulation size R approaches infinity. The effect of R on the accuracy of $\hat{c}$ can be assessed via the asymptotic normality of $\hat{c}$. It is known (see e.g., Serfling, 1980) that, under quite weak conditions, $\hat{c}$ is asymptotically normal with mean c and standard error

$$\text{s.e.} = \sqrt{\frac{\alpha(1-\alpha)}{g^2(c)R}}, \qquad (A.1)$$

where $g(c)$ is the density function of S evaluated at c.

Although the value $g(c)$ is not available, it can be estimated by using a density estimation method and the simulated values of S: $S_1, \cdots, S_R$. We approximate $g(c)$ by the kernel density estimator

$$g(c) \approx \frac{1}{Rh\sqrt{2\pi}} \sum_{i=1}^{R} \exp\left(-\left(\frac{\hat{c} - S_i}{h} \right)^2 /2 \right) \tag{A.2}$$

where S_i is the ith simulated value, and h is the smoothing parameter. The value of h can be chosen in various ways, and we have used Silverman (1986, expression (3.28)) to determine the value of h. Several other values of h that we have tried in a few examples gave similar results for the standard error. The standard error of $\hat{c}$ can also be computed using bootstrap. The bootstrap estimates of the standard error and those obtained using (A.1) and (A,2) were very close in a number of examples that we tried, but the one based on (A.1) and (A.2) was easier to compute.

The standard error can be used to determine the simulation size R in order that $\hat{c} - c$ is within a pre-specified tolerance. For example, if one requires $|\hat{c} - c| \leq 0.009$, then the 3-sigma rule suggests that R should be such that the standard error is about 0.003. From the R replicates that have already been generated, calculate the standard error from (A.1) and (A.2). If it is smaller than 0.003 then no more replicates are required. Otherwise generate a batch of new replicates that are independent of the available replicates, and use the aggregated replicates to compute a new estimation of the standard error. Repeat this process until the desired accuracy is achieved.

From our experience, if $R = 50,000$ then $\hat{c}$ varies often only at the second decimal place, and if $R = 100,000$ then $\hat{c}$ is often accurate to the second decimal place.

A.2 Accuracy of the simultaneous confidence level

Alternatively, one can assess the accuracy of $\hat{c}$ by considering the simultaneous confidence level of the band that uses the approximate critical value $\hat{c}$. Of course the target simultaneous confidence level is $G(c) = P\{S \leq c\} = 1 - \alpha$. But the actual simultaneous confidence level from using $\hat{c}$ is given by $G(\hat{c}) = P\{S \leq \hat{c}|\hat{c}\}$; note that $S, S_1, \cdots, S_R$ are i.i.d. random variables with the same distribution function $G(\cdot)$, and $\hat{c} = S_{\langle (1-\alpha)R \rangle}$ depends only on $S_1, \cdots, S_R$ and so is independent of S. Since $\hat{c}$ is random, so is $G(\hat{c})$. Edwards and Berry (1987) gave the exact distribution, mean and variance of $G(\hat{c})$, which allows one to assess by how much $G(\hat{c})$ may deviate from the target value $1 - \alpha$.

Let $q = \langle (1 - \alpha)R \rangle$. Note that $G(S), G(S_1), \cdots, G(S_R)$ are i.i.d. random variables uniformly distributed on the interval $(0, 1)$, and

$$G(\hat{c}) = P\{S \leq \hat{c}|\hat{c}\} = P\{G(S) \leq G(S_q)|S_q\} = G(S_q)$$

where $G(S_q)$ is just the qth largest observation from a sample of R i.i.d. observations from the uniform distribution on the interval $(0, 1)$. From this, Edwards and Berry (1987, Lemma 2) showed that $G(\hat{c})$ has a Type I beta distribution with parameters $R-q-1$ and q. In particular, the mean and variance of $G(\hat{c})$ are given respectively by $1-\alpha$ and $\alpha(1-\alpha)/(R+2)$, and for a large R the distribution of $G(\hat{c})$ is essentially normal.

So if one wants $G(\hat{c})$ to be within $(1-\alpha)\pm\varepsilon$ with a probability of β, then the number of simulations R should be about $(z^{0.5(1-\beta)}/\varepsilon)^2\alpha(1-\alpha)$ where z^α is the upper α-point of the standard normal distribution. For example, if $\beta=0.99$, $\alpha=0.05$ and $\varepsilon=0.001$, then R should be about 320,000. Table 1 of Edwards and Berry (1987) provides values of R for given values of β and ε.

A.3 A computation saving method

In many situations, the random variable S is the magnitude of the projection of a random vector $\mathbf{T}$ onto a cone. While $\mathbf{T}$ is often very easy to simulate, it is often time-consuming to compute the projection of $\mathbf{T}$. It is also known that the required critical constant c is larger than some known constant t. For example, the c for the band in (3.8) is larger than the corresponding critical constant of the pointwise two-sided confidence interval $t_\nu^{\alpha/2}$, and the c for the band in (3.12) is larger than the corresponding critical constant of the pointwise one-sided confidence interval t_ν^α.

Then in a simulation replication that has the magnitude of $\mathbf{T}$, $\|\mathbf{T}\|$, less than or equal to t, the projection of $\mathbf{T}$ and so the exact value of S do not need to be computed. Instead, for this replication, the 'modified' value of S is simply recorded as t while the exact value of S is clearly no larger than t. It is clear that this modification to the exact value of S that is no larger than t does not change the sample quantile $\hat{c}$ that is sought.

For example, we know from (3.11) that

$$S=\max\{\|\pi(\mathbf{T},\mathbf{P},\mathcal{X}_r)\|,\|\pi(-\mathbf{T},\mathbf{P},\mathcal{X}_r)\|\}.$$

We also know that the required critical value c in this case is greater than $t_\nu^{\alpha/2}$. Thus, in a simulation replication, if $\|\mathbf{T}\|\le t_\nu^{\alpha/2}$ then $S\le\|\mathbf{T}\|\le t_\nu^{\alpha/2}$. For this simulation replication, it suffices for us to record $S=t_\nu^{\alpha/2}$ and not to compute the two projections $\pi(\mathbf{T},\mathbf{P},\mathcal{X}_r)$ and $\pi(-\mathbf{T},\mathbf{P},\mathcal{X}_r)$. Hence out of R simulation replications, there are on average about ξR replications that require the time-consuming computation of the two projections, where

$$\xi=P\{\|\mathbf{T}\|>t_\nu^{\alpha/2}\}=P\left\{F_{p+1,\nu}>\left(t_\nu^{\alpha/2}\right)^2/(p+1)\right\}.$$

For the infant data set in Example 3.2, $\xi=0.408$. This method can therefore save substantial amount of computation time.

B

Computation of projection $\pi(\mathbf{t}, \mathbf{P}, \mathcal{X}_r)$

We need to find the $\mathbf{v} \in \Re^{p+1}$ that minimizes $\|\mathbf{v} - \mathbf{t}\|^2$ subject to the constraints $\mathbf{v} \in C(\mathbf{P}, \mathcal{X}_r)$, where $C(\mathbf{P}, \mathcal{X}_r)$ is defined below expression (3.9). Let $\mathbf{e}_j \in \Re^{p+1}$ have the jth element equal to one and the remaining elements all equal to zero.

The objective function to minimize, $\|\mathbf{v} - \mathbf{t}\|^2$, can be expressed as $\mathbf{v}'\mathbf{v} - 2\mathbf{t}'\mathbf{v} + \mathbf{t}'\mathbf{t}$, which is equivalent to

$$\frac{1}{2}\mathbf{v}'\mathbf{v} - \mathbf{t}'\mathbf{v}. \tag{B.1}$$

From the definition of the cone $C(\mathbf{P}, \mathcal{X}_r)$, $\mathbf{v} \in C(\mathbf{P}, \mathcal{X}_r)$ implies that $\mathbf{v} = \lambda \mathbf{P}\mathbf{x}$ or, equivalently, $\mathbf{P}^{-1}\mathbf{v} = \lambda \mathbf{x} = (\lambda, \lambda x_1, \cdots, \lambda x_p)'$, for some $\mathbf{x}_{(0)} \in \mathcal{X}_r$ and $\lambda \geq 0$. Hence $\mathbf{e}_1'\mathbf{P}^{-1}\mathbf{v} = \lambda \geq 0$ and $a_j \leq \mathbf{e}_{j+1}'\mathbf{P}^{-1}\mathbf{v}/\mathbf{e}_1'\mathbf{P}^{-1}\mathbf{v} = x_j \leq b_j$ for $j = 1, \cdots, p$ or, equivalently,

$$\begin{aligned}
&-\mathbf{e}_1'\mathbf{P}^{-1}\mathbf{v} \leq 0 \\
&(\mathbf{e}_{j+1}' - b_j\mathbf{e}_1')\mathbf{P}^{-1}\mathbf{v} \leq 0 \ \text{ for } j = 1, \cdots, p \\
&(a_j\mathbf{e}_1' - \mathbf{e}_{j+1}')\mathbf{P}^{-1}\mathbf{v} \leq 0 \ \text{ for } j = 1, \cdots, p
\end{aligned}$$

These constraints can be expressed as

$$A\mathbf{v} \leq \mathbf{0} \tag{B.2}$$

where the $(2p+1) \times (p+1)$ matrix A is given by

$$\begin{pmatrix}
(\mathbf{e}_2' - b_1\mathbf{e}_1')\mathbf{P}^{-1} \\
(a_1\mathbf{e}_1' - \mathbf{e}_2')\mathbf{P}^{-1} \\
\cdots \\
(\mathbf{e}_{p+1}' - b_p\mathbf{e}_1')\mathbf{P}^{-1} \\
(a_p\mathbf{e}_1' - \mathbf{e}_{p+1}')\mathbf{P}^{-1} \\
-\mathbf{e}_1'\mathbf{P}^{-1}
\end{pmatrix}.$$

The problem of minimizing the objective function in (B.1) under the constraints in (B.2) is a standard quadratic programming problem and can be solved in a finite number of steps. There are many algorithms available for solving this problem numerically (see e.g., Fletcher, 1987, and Hu, 1998), and we have used the MATLAB® built-in function `quadprog(H,f,A,b)` for this purpose.

C

Computation of projection $\pi^*(\mathbf{t}, \mathbf{W}, \mathcal{X}_2)$

We need to find the $\mathbf{v} \in \Re^k$ that minimizes $\|\mathbf{v} - \mathbf{t}_2\|^2$ subject to the constraints $\mathbf{v} \in C(\mathbf{W}, \mathcal{X}_2)$, where $\mathbf{t}_2 \in \Re^k$ is given and $C(\mathbf{W}, \mathcal{X}_2)$ is defined in (4.9). Let $\mathbf{e}_j \in \Re^k$ have the jth element equal to one and the remaining elements all equal to zero.

From the definition of the cone $C(\mathbf{W}, \mathcal{X}_2)$, $\mathbf{v} \in C(\mathbf{W}, \mathcal{X}_2)$ implies that $\mathbf{v} = \lambda \mathbf{W}\mathbf{x}_2$ or, equivalently, $\mathbf{W}^{-1}\mathbf{v} = \lambda \mathbf{x}_2 = \lambda(x_{p-k+1}, \cdots, x_p)'$, for some $\mathbf{x}_2 \in \mathcal{X}_2$ and $\lambda \geq 0$. Hence, in terms of $\mathbf{u} = (\mathbf{v}', \lambda)'$, the constraints become

$$
\begin{aligned}
&(\mathbf{e}_j'\mathbf{W}^{-1}, -b_{p-k+j})\mathbf{u} \leq 0 \;\text{ for } j = 1, \cdots, k \\
&(-\mathbf{e}_j'\mathbf{W}^{-1}, a_{p-k+j})\mathbf{u} \leq 0 \;\text{ for } j = 1, \cdots, k \\
&(\mathbf{0}', -1)\mathbf{u} \leq 0.
\end{aligned}
$$

These constraints can be expressed as

$$
A\mathbf{u} \leq \mathbf{0} \tag{C.1}
$$

where the $(2k+1) \times (k+1)$ matrix A is given by

$$
\begin{pmatrix}
\mathbf{e}_1'\mathbf{W}^{-1}, & -b_{p-k+1} \\
-\mathbf{e}_1'\mathbf{W}^{-1}, & a_{p-k+1} \\
\cdots, & \cdots \\
\mathbf{e}_k'\mathbf{W}^{-1}, & -b_p \\
-\mathbf{e}_k'\mathbf{W}^{-1}, & a_p \\
\mathbf{0}', & -1
\end{pmatrix}.
$$

The objective function to minimize, $\|\mathbf{v} - \mathbf{t}_2\|^2$, can be expressed in terms of $\mathbf{u} = (\mathbf{v}', \lambda)'$ as

$$
\begin{aligned}
&\mathbf{v}'\mathbf{v} - 2\mathbf{t}_2'\mathbf{v} + \mathbf{t}_2'\mathbf{t}_2 \\
= &\begin{pmatrix} \mathbf{v} \\ \lambda \end{pmatrix}' \begin{pmatrix} \mathbf{I}_k & \mathbf{0} \\ \mathbf{0}' & 0 \end{pmatrix} \begin{pmatrix} \mathbf{v} \\ \lambda \end{pmatrix} - 2 \begin{pmatrix} \mathbf{t}_2 \\ 0 \end{pmatrix}' \begin{pmatrix} \mathbf{v} \\ \lambda \end{pmatrix} + \mathbf{t}_2'\mathbf{t}_2,
\end{aligned}
$$

which is equivalent to

$$
\frac{1}{2}\mathbf{u}' \begin{pmatrix} \mathbf{I}_k & \mathbf{0} \\ \mathbf{0}' & 0 \end{pmatrix} \mathbf{u} - \begin{pmatrix} \mathbf{t}_2 \\ 0 \end{pmatrix}' \mathbf{u}. \tag{C.2}
$$

The problem of minimizing the objective function of $\mathbf{u}$ in (C.2) under the constraints in (C.1) is a standard quadratic programming problem which can be solved in a finite number of steps. We have used the MATLAB® routine `quadprog(H,f,A,b)` for this purpose. Note, however, that the matrix H here is given by $\begin{pmatrix} \mathbf{I}_k & \mathbf{0} \\ \mathbf{0}' & 0 \end{pmatrix}$ and so singular; this is different from finding $\pi(\mathbf{t},\mathbf{P},\mathcal{X}_r)$ in Appendix B where the matrix H is given by the identity matrix and so non-singular. Jamshidian *et al.* (2007) gave a specific quadratic-programming based algorithm which is slightly more efficient than the MATLAB routine `quadprog(H,f,A,b)`.

After finding $\mathbf{u} = (\mathbf{v}',\lambda)'$ from the quadratic programming problem above, the required $\pi^*(\mathbf{t}_2,\mathbf{W},\mathcal{X}_2)$ is simply given by the first k components of $\mathbf{u}$.

D

Principle of intersection-union test

The idea of intersection-union (IU) tests appeared in Gleser (1973) and has been made popular by Berger (1982). Casella and Berger (1990, Chapter 8) provided a very readable introduction. These tests have been found to be very useful in bio-equivalence studies; see e.g., Hsu *et al.* (1994), Berger and Hsu (1996) and Wellek (2003).

Suppose that one wants to test some hypotheses about unknown parameter (vector) θ based on random observation $\mathbf{Y}$. Assume the null hypothesis can be expressed as the union of a set of individual null hypotheses:

$$H_0 = \cup_{\gamma \in \Gamma} H_{0\gamma}$$

where Γ is an arbitrary index set which may be finite or infinite, and $H_{0\gamma}$ is an individual null hypothesis. Correspondingly, the alternative hypothesis can be written as

$$H_a = \cap_{\gamma \in \Gamma} H_{a\gamma}$$

where $H_{a\gamma}$ is the alternative hypothesis corresponding to the null hypothesis $H_{0\gamma}$. Suppose that a size α test T_γ, with rejection region R_γ, is available for testing

$$H_{0\gamma} \text{ against } H_{a\gamma},$$

for each $\gamma \in \Gamma$. Then the IU test of H_0 against H_a has the rejection region $R = \cap_{\gamma \in \Gamma} R_\gamma$, i.e. H_0 is rejected if and only if $H_{0\gamma}$ is rejected by individual test T_γ for all $\gamma \in \Gamma$. Crucially, the size of this IU test is also α.

Theorem D.1 If each individual test T_γ is of size α for all $\gamma \in \Gamma$ then the IU test is also of size α.

Proof D.1 *When the unknown parameter θ takes a value such that H_0 is true, then $H_{0\gamma}$ is true for some $\gamma \in \Gamma$ since $H_0 = \cup_{\gamma \in \Gamma} H_{0\gamma}$. For such a value of θ and $\gamma \in \Gamma$, we then have*

$$P\{\mathbf{Y} \in R\} \leq P\{\mathbf{Y} \in R_\gamma\} \leq \alpha \tag{D.1}$$

since, for each $\gamma \in \Gamma$, $R \subset R_\gamma$ and the individual test T_γ is of size α. Expression (D.1) means that the size of the intersection-union test is controlled at α. ∎

E

Computation of the K-functions in Chapter 7

E.1 Computation of $K_{2h}(\mathbf{t},\Delta,(a,b))$ and $K_{1h}(\mathbf{t},\Delta,(a,b))$

The definition of $K_{2h}(\mathbf{t},\Delta,(a,b))$ is given in (7.6) by

$$K_{2h}(\mathbf{t},\Delta,(a,b)) = \sup_{a<x<b} \frac{|\tilde{\mathbf{x}}'\mathbf{t}|}{\sqrt{\tilde{\mathbf{x}}'\Delta\tilde{\mathbf{x}}}} \tag{E.1}$$

where $\mathbf{t}$ is a given $(p+1)$-vector, Δ a given non-singular matrix, (a,b) a given interval and $\tilde{\mathbf{x}} = (1,x,\cdots,x^p)'$. The definition of $K_{1h}(\mathbf{t},\Delta,(a,b))$ is given in (7.9) by

$$K_{1h}(\mathbf{t},\Delta,(a,b)) = \sup_{a<x<b} \frac{\tilde{\mathbf{x}}'\mathbf{t}}{\sqrt{\tilde{\mathbf{x}}'\Delta\tilde{\mathbf{x}}}}. \tag{E.2}$$

Note that both the supremums in (E.1) and (E.2) can be attained only possibly at a, b and the stationary points of the function

$$g_h(x) = \frac{\tilde{\mathbf{x}}'\mathbf{t}}{\sqrt{\tilde{\mathbf{x}}'\Delta\tilde{\mathbf{x}}}}.$$

Now

$$\frac{dg_h(x)}{dx} = \left(\left(\frac{d\tilde{\mathbf{x}}'}{dx}\right) \mathbf{t}(\tilde{\mathbf{x}}'\Delta\tilde{\mathbf{x}}) - (\tilde{\mathbf{x}}'\mathbf{t}) \left(\frac{d\tilde{\mathbf{x}}'}{dx}\right) \Delta\tilde{\mathbf{x}} \right) (\tilde{\mathbf{x}}'\Delta\tilde{\mathbf{x}})^{-3/2}.$$

So the stationary points of $g_h(x)$ are given by the real roots of the $(3p-2)$th order polynomial function

$$\left(\frac{d\tilde{\mathbf{x}}'}{dx}\right) \mathbf{t}(\tilde{\mathbf{x}}'\Delta\tilde{\mathbf{x}}) - (\tilde{\mathbf{x}}'\mathbf{t}) \left(\frac{d\tilde{\mathbf{x}}'}{dx}\right) \Delta\tilde{\mathbf{x}},$$

which can easily be found numerically by using MATLAB® function `roots`. Denote all the real roots by $r_1,\cdots,r_q$ for some $0 \le q \le 3p-2$. It follows therefore

$$K_{1h}(\mathbf{t},\Delta,(a,b)) = \max\{g_h(a), g_h(b), g_h(r_1),\cdots,g_h(r_q)\}$$
$$K_{2h}(\mathbf{t},\Delta,(a,b)) = \max\{|g_h(a)|, |g_h(b)|, |g_h(r_1)|,\cdots,|g_h(r_q)|\}$$

from which $K_{1h}(\mathbf{t},\Delta,(a,b))$ and $K_{2h}(\mathbf{t},\Delta,(a,b))$ can easily be computed numerically.

E.2 Computation of $K_{2c}(\mathbf{t},(a,b))$ and $K_{1c}(\mathbf{t},(a,b))$

The definition of $K_{2c}(\mathbf{t},(a,b))$ is given in (7.11) by

$$K_{2c}(\mathbf{t},(a,b)) = \sup_{a<x<b} |\tilde{\mathbf{x}}'\mathbf{t}| \tag{E.3}$$

where $\mathbf{t}$ is a given $(p+1)$-vector, (a,b) a given interval, and $\tilde{\mathbf{x}} = (1,x,\cdots,x^p)'$. The definition of $K_{1c}(\mathbf{t},(a,b))$ is given in (7.13) by

$$K_{1c}(\mathbf{t},(a,b)) = \sup_{a<x<b} \tilde{\mathbf{x}}'\mathbf{t}. \tag{E.4}$$

Note that both the supremums in (E.3) and (E.4) can be attained only possibly at a, b and the stationary points of the function

$$g_c(x) = \tilde{\mathbf{x}}'\mathbf{t}.$$

Now

$$\frac{dg_c(x)}{dx} = \frac{d\tilde{\mathbf{x}}'}{dx}\mathbf{t} = (0,1,\cdots,px^{p-1})\mathbf{t}.$$

So the stationary points of $g_c(x)$ are given by the real roots of the $(p-1)$th order polynomial function $(0,1,\cdots,px^{p-1})\mathbf{t}$ which can easily be found numerically by using MATLAB function `roots`. Denote the real roots by $r_1,\cdots,r_q$ for some $0 \le q \le p-1$. It follows therefore

$$K_{1c}(\mathbf{t},(a,b)) = \max\{g_c(a), g_c(b), g_c(r_1),\cdots,g_c(r_q)\}$$
$$K_{2c}(\mathbf{t},(a,b)) = \max\{|g_c(a)|, |g_c(b)|, |g_c(r_1)|,\cdots,|g_c(r_q)|\}$$

from which $K_{1c}(\mathbf{t},(a,b))$ and $K_{2c}(\mathbf{t},(a,b))$ can easily be computed numerically.

Bibliography

Aitkin, M.A. (1973). Fixed-width confidence intervals in linear-regression with applications to Johnson-Neyman technique. *British Journal of Mathematical & Statistical Psychology*, **26**, 261-269.

Al-Saidy, O.M., Piegorsch, W.W., West, R.W. and Nitcheva, D.K. (2003). Confidence bands for low-dose risk estimation with quantal response data. *Biometrics*, **59**, 1056-1062.

Atkinson, A.C. and Donev, A.N. (1992). *Optimum Experimental Designs*. Oxford University Press.

Beran, R. (1988). Balanced simultaneous confidence sets. *Journal of the American Statistical Association*, **83**, 679-697.

Beran, R. (1990). Refining Bootstrap simultaneous confidence sets. *Journal of the American Statistical Association*, **85**, 417-426.

Berger, R.L. (1982). Multiparameter hypothesis testing and acceptance sampling. *Technometrics*, **24**, 295-300.

Berger, R.L. (1984). Testing whether one regression function is larger than another. *Communications in Statistics-Theory and Methods*, **13**, 1793-1810.

Berger, R.L. and Hsu, J.C. (1996). Bioequivalence trials, intersection-union tests and equivalence confidence sets. *Statistical Science*, **11**, 283-319.

Bhargava, P and Spurrier, J.D. (2004). Exact confidence bounds for comparing two regression lines with a control regression line on a fixed interval. *Biometrical Journal*, **46**, 720-730.

Bofinger, E. (1999). Homogeneity of two straight lines: equivalence approach using fitted regressions. *Australian & New Zealand Journal of Statistics*, **41**, 481-491.

Bofinger, E. and Bofinger, M. (1995). Equivalence with respect to a control – stepwise tests. *Journal of the Royal Statistical Society (B)*, **57**, 721-733.

Bofinger, E., Hayter, A.J., and Liu, W. (1993). The construction of upper confidence bounds on the range of several location parameters. *Journal of the American Statistical Association*, **88**, 906-11.

Bohrer, R. (1967). On sharpening Scheffé bounds. *Journal of the Royal Statistical Society (B)*, **29**, 110-114.

Bohrer, R. (1969). On one-sided confidence bounds for response surfaces. *Bulletin of the International Statistical Institute*, **43**, 255-7.

Bohrer, R. (1973a). An optimality property of Scheffé Bounds. *Annals of Statistics*, **1**, 766-772.

Bohrer, R. (1973b). A multivariate t probability integral. *Biometrika*, **60**, 647-654.

Bohrer, R. and Francis, G.K. (1972a). Sharp one-sided confidence bands for linear regression over intervals. *Biometrika*, **59**, 99-107.

Bohrer, R., and Francis, G.K. (1972b). Sharp one-sided confidence bounds over positive regions. *Annals of Mathematical Statistics*, **43**, 1541-1548.

Bowden, D.C. (1970). Simultaneous confidence bands for linear regression models. *Journal of the American Statistical Association*, **65**, 413-421.

Bowden, D.C. and Grabill, F.A. (1966). Confidence bands of uniform and proportional width for linear models. *Journal of the American Statistical Association*, **61**, 182-198.

Brand, R.J., Pinnock, D.E., and Jackson, K.L. (1973). Large sample confidence bands for the logistic response curve and its inverse. *American Statistician*, **27**, 157-160.

Bretz, F., Genz, A. and Hothorn, L.A. (2001). On the numeircal availability of multiple comparison procedures. *Biometrical Journal*, 43, 645-656.

Broemeling, L.D. (1985). *Bayesian Analysis of Linear Models*. Marcel Dekker.

Casella, G. and Berger, R.L. (1990). *Statistical Inference*. Duxbury Press.

Casella, G. and Strawderman, W.E. (1980). Confidence bands for linear-regression with restricted predictor variables. *Journal of the American Statistical Association*, **75**, 862-868.

Cheng, M.Y., Hall, P. and Tu, D. (2006). Confidence bands for hazard rates under random censorship. *Biometrika*, **93**, 357-366.

Cheng, R.C.H. (1987). Confidence bands for 2-stage design-problems. *Technometrics*, **29**, 301-309.

Chow, S.C. and Liu, J.P. (1992). *Design and Analysis of Bioavailability and Bioequivalence Studies*. Marcel Dekker.

Collett, D. (2003). *Modelling Binary Data, 2nd edition*. Chapman & Hall/CRC.

Cornell, J.A. (1990). *Experiments with Mixtures : Designs, Models, and the Analysis of Mixture Data, 2nd edition*. Wiley.

Cressie, N. and Read, T.R.C. (1989). Pearson's χ^2 and the likelihood ratio statistic G^2: a comparative review. *International Statistical Review*, **57**, 19-43.

Dalal, S.R. (1983). Exact simultaneous confidence bands for random intercept regression. *Technometrics*, **25**, 263-269.

Dalal, S.R. (1990). Simultaneous confidence bands for regression with unknown unequal variances. *Technometrics*, **32**, 173-186.

Dawson, D.V., Todorov, A.B. and Elston, R.C. (1980). Confidence bands for the growth of head circumference in achondroplastic children during the 1st year of life. *American Journal of Medical Genetics*, **7**, 529-536.

Dielman, T. (2001). *Applied Regression Analysis for Business and Economics.* Duxbury.

Dobson, A.J. (2002). *An Introduction to Generalized Linear Models, 2nd edition.* Chapman & Hall/CRC.

Donnelly, J. (2003). Simultaneous Confidence Bands in Linear Modelling. PhD Thesis, School of Mathematics, University of Southampton, UK.

Draper, N. and Smith, H. (1998). *Applied Linear Regression, 3rd edition.* Wiley.

Dunn, O.J. (1968). A note on confidence bands for a regression line over a finite range. *Journal of the American Statistical Association*, **63**, 1028-1033.

Dunnett, C.W. and Sobel, M. (1954). A bivariate generalization of Student's *t*-distribution, with tables for certain special cases. *Biometrika*, **41**, 153-169.

Dunnett, C.W. (1955a). A multiple comparison procedure for comparing several treatments with a control. *Journal of the American Statistical Association*, **50**, 1096-1121.

Dunnett, C.W. and Sobel, M. (1955b). Approximations to the probability integral and certain percentage points of a multivariate analogue of Student's *t*-distribution. *Biometrika*, **42**, 258-260.

Edwards, D. and Berry, J.J. (1987). The efficiency of simulation-based multiple comparisons. *Biometrics*, **43**, 913-928.

Efron, B. (1997). The length heuristic for simultaneous hypothesis tests. *Biometrika*, **84**, 143-157.

Efron, B. and Tibshirani, R.J. (1993). *An Introduction to the Bootstrap.* Chapman & Hall.

Elston, R.C. and Grizzle, J.E. (1962). Estimation of time-response curves and their confidence bands. *Biometrics*, **18**, 148-159.

Faraway, J.J. and Sun, J. (1995). Simultaneous confidence bands for linear regression with heteroscedastic errors. *Journal of the American Statistical Association*, **90**, 1094-1098.

Ferber, G. (2005). Statistical aspects of clinical QT safety assessment – recent developments surrounding the emerging ICH-E14. Presentation at the 3rd Annual Meeting on QT Prolongation and Safety Pharmacology, Paris, 23-24 February 2005.

Fletcher, R. (1987). *Practical Methods of Optimization*. Wiley.

Folks, L. and Antle, E. (1967). Straight line confidence regions for linear models. *Journal of the American Statistical Association*, **62**, 1365-1373.

Frey, J. (2008). Optimal distribution-free confidence bands for a distribution function. *Journal of Statistical Planning and Inference*, **138**, 3086-3098.

Gafarian, A.V. (1964). Confidence bands in straight line regression. *Journal of the American Statistical Association*, **59**, 182-213.

Gafarian, A.V. (1978). Confidence bands in multivariate polynomial regression. *Technometrics*, **20**, 141-149.

Gamerman, D. and Lopes, H.F. (2006). *Markov Chain Monte Carlo: Stochastic Simulation for Bayesian Inference*. Taylor & Francis.

Genovese, C. and Wasserman, L. (2008). Adaptive confidence bands. *Annals of Statistics*, **36**, 875-905.

Genz, A. and Bretz, F. (1999). Numerical computation of multivariate t-probabilities with application to power calculation of multiple contrasts. *Journal of Statistical Computation and Simulation*, **63**, 361-378.

Genz, A. and Bretz, F. (2002). Methods for the computation of multivariate t-probabilities. *Journal of Computational and Graphical Statistics*, **11**, 950-971.

Genz, A. and Bretz, F. (2009). *Computation of Multivariate Normal and t Probabilities*. Springer.

Gleser, L.J. (1973). On a theory of intersection-union tests. *Institute of Mathematical Statistics Bulletin*, **2**, 233.

Graybill, F. A. and Bowden, D. C. (1967). Linear segment confidence bands for simple linear regression models. *Journal of the American Statistical Association*, **62**, 403-408.

Gsteiger, S., Bretz, F. and Liu, W. (2010). Simultaneous confidence bands for population pharmacokinetic models. Manuscript, School of Mathematics, University of Southampton, UK.

Hall, P. (1992). *Edgeworth Expansion and Bootstrap*. Springer-Verlag.

Hall, P. (1993). On Edgeworth expansion and bootstrap confidence bands in nonparametric curve estimation. *Journal of the Royal Statistical Society (B)*, **55**, 291-304.

Hall, W.J. and Wellner, J.A. (1980). Confidence bands for survival curve from censored data. *Biometrika*, **67**, 133-143.

Halperin, M. (1963a). Confidence interval estimation in nonlinear regression. *Journal of the Royal Statistical Society (B)*, **25**, 330-333.

Halperin, M. (1963b). Interval estimation of nonlinear parametric functions. *Journal of the American Statistical Association*, **58**, 611-627.

Halperin, M. and Gurian, J. (1968). Confidence bands in linear regression with constraints on independent variables. *Journal of the American Statistical Association*, **63**, 1020-1027.

Halperin, M., and Rastogi, S.C., Ho, I., and Yang, Y.Y. (1967). Shorter confidence bands in linear regression. *Journal of the American Statistical Association*, **62**, 1050-1067.

Hanselman, D. and Littlefield, B. (2005). *Mastering MATLAB 7*. Prentice Hall.

Hauck, W.W. (1983). A note on confidence bands for the logistic response curve. *American Statistician*, **37**, 158-160.

Hayter, A.J. (1984). A proof of the conjecture that Tukey-Kramer multiple comparisons procedure is conservative. *Annals of Statistics*, **12**, 61-75.

Hayter, A.J. (1990). A one-sided studentized range test for testing against a simple order alternative. *Journal of the American Statistical Association*, **85**, 778-785.

Hayter, A.J., Kim, J. and Liu, W. (2008). Critical point computations for one-sided and two-sided pairwise comparisons of three treatment means. *Computational Statistics and Data Analysis*, **53**, 463-470.

Hayter, A.J. and Liu, W. (1996). On the exact calculation of the one-sided studentized range test. *Computational Statistics and Data Analysis*, **22**, 17-25.

Hayter, A.J., Liu, W., and Ah-Kine, P. (2009). A ray method of confidence band construction for multiple linear regression models. *Journal of Statistical Planning and Inferences*, **139**, 329-334.

Hayter, A.J., Liu, W., and Wynn, H.P. (2007). Easy-to-construct confidence bands for comparing two simple linear regression lines. *Journal of Statistical Planning and Inferences*, **137**, 1213-1225.

Hayter, A.J., Wynn, H.P. and Liu, W. (2006a). Slope modified confidence bands for a simple linear regression model. *Statistical Methodology*, **3**, 186-192.

Hayter, A.J., Wynn, H.P. and Liu, W. (2006b). Confidence bands for regression: the independence point method. Manuscript, School of Mathematics, University of Southampton, UK.

Hewett, J.E. and Lababidi, Z. (1980). Comparison of two populations with multivariate data. *Biometrics*, **36**, 671-675.

Hewett, J.E. and Lababidi, Z. (1982). Comparison of three regression lines over a finite interval. *Biometrics*, **38**, 837-841.

Hochberg, Y. and Marcus, R. (1978). On partitioning successive increments in means or ratios of variances in a chain of normal means. *Communications in Statistics – Theory and Methods*, **7**, 1501-1513.

Hochberg, Y. and Quade, D. (1975). One-sided simultaneous confidence bounds on regression surfaces with intercepts. *Journal of the American Statistical Association*, **70**, 889-891.

Hochberg, Y. and Tamhane, A. C. (1987). *Multiple Comparison Procedures*. Wiley.

Hoel, P.G. (1951). Confidence regions for linear regression. *Proceedings of the second Berkeley Symposium*, 79-81, University of California Press.

Hoel, P.G. (1954). Confidence bands for polynomial curves. *Annals of Mathematical Statistics*, **25**, 534-542.

Hollander, M., McKeague, I.W. and Yang, J. (1997). Likelihood ration-based confidence bands for survival function. *Journal of the American Statistical Association*, **92**, 215-226.

Horvath, L., Horvath, Z. and Zhou, W. (2008). Confidence bands for ROC curves. *Journal of Statistical Planning and Inference*, **138**, 1894-1904.

Hosmer, D.W. and Lemeshow, S. (1980). Goodness of fit tests for multiple logistic model. *Communications in Statistics – Theory and Methods*, **9**, 1043-1069.

Hotelling, H. (1939). Tubes and spheres in n-space and a class of statistical problems. *American Journal of Mathematics*, **61**, 440-460.

Hsu, J. C. (1996). *Multiple Comparisons – Theory and Methods*. Chapman & Hall.

Hsu, J.C., Hwang, J.T.G., Liu, H., and Ruberg, S.J. (1994). Confidence intervals associated with tests for bioequivalence. *Biometrika*, **81**, 103-14.

Hsu. P.L. (1941). Canonical reduction of the general regression problem. *Annals of Eugenics*, **11**, 42-46.

Hu, X. (1998). An exact algorithm for projection onto polyhedral cone. *Australia and New Zealand Journal of Statistics*, **2**, 165-170.

Jamshidian, M., Jenrich, R. and Liu, W. (2007). A study of partial F tests for multiple linear regression models. *Computational Statistics and Data Analysis*, **51**, 6269-6284.

Jamshidian, M., Liu, W. and Bretz, F. (2010). Simultaneous confidence bands for all contrasts of three or more simple linear regression models over an interval. *Computational Statistics and Data Analysis*, to appear.

Jamshidian, M., Liu, W., Zhang, Y. and Jamshidian, F. (2005). SimReg: a software including some new developments in multiple comparison and simultaneous confidence bands for linear regression models. *Journal of Statistical Software*, **12**, 1-22.

Johansen, S. and Johnstone, I. (1990). Hotelling's theorem on the volume of tubes: some illustrations in simultaneous inference and data analysis. *Annals of Statistics*, **18**, 652-684.

Johnstone, I. and Siegmund, D. (1989). On Hotelling's formula for the volume of tubes and Naiman's inequality. *Annals of Statistics*, **17**, 184-194.

Jones, R.A., Scholz, F.W., Ossiander, M. *et al.* (1985). Tolerance bounds for log gamma regression-models. *Technometrics*, **27**, 109-118.

Kabe, D.G. (1976). On confidence bands for quantiles of a normal population. *Journal of the American Statistical Association*, **71**, 417-419.

Kanoh, S. (1988). The reduction of the width of confidence bands in linear regression. *Journal of the American Statistical Association*, **83**, 116-122.

Kanoh, S., and Kusunoki, Y. (1984). One-sided simultaneous bounds in linear regression. *Journal of the American Statistical Association*, **79**, 715-719.

Khorasani, F. (1982). Simultaneous confidence bands for non-linear regression models. *Communications in Statistics – Theory and Methods*, **11**, 1241-1253.

Khorasani, F. and Milliken, G.A. (1979). On the exactness of confidence bands about a linear model. *Journal of the American Statistical Association*, **74**, 446-448.

Khuri, A.I. (2006). *Response Surface Methodology and Related Topics*. World Scientific.

Kleinbaum, D.G., Kupper, L.L., Muller, K.E. and Nizam, A. (1998). *Applied Regression Analysis and Other Multivariable Methods, 3rd Edition*. Duxbury Press.

Knafl, G., Sacks, J., and Ylvisaker, D. (1985). Confidence bands for regression-functions. *Journal of the American Statistical Association*, **80**, 683-691.

Knowles, M. and Siegmund, D. (1989). On Hotelling's approach to testing for a nonlinear parameter in regression. *International Statistical Review*, **57**, 205-220.

Kosorok, M.R., and Qu, R. (1999). Exact simultaneous confidence bands for a collection of univariate polynomials in regression analysis. *Statistics in Medicine*, **18**, 613-620.

Kotz, S. and Nadarajah, S. (2004). *Multivariate t Distributions and Their Applications*. Cambridge University Press, UK.

Lee, R.E. and Spurrier, J.D. (1995). Successive comparisons between ordered treatments. *Journal of Statistical Planning and Inference*, **43**, 323-330.

Li, L., Desai, M., Desta, Z. and Flockhart, D. (2004). QT analysis: a complex answer to a 'simple' problem. *Statistics in Medicine*, **23**, 2625-2643.

Lieberman, G.S. and Miller, R.G. (1963). Simultaneous tolerance intervals in regression. *Biometrika*, **50**, 155-163.

Limam, M.M.T. and Thomas, D.R. (1988). Simultaneous tolerance intervals for the linear-regression model. *Journal of the American Statistical Association*, **83**, 801-804.

Lin, D.Y. (1994). Confidence bands for survival curves under the proportional hazards model. *Biometrika*, **81**, 73-81.

Lin, S. (2008). Simultaneous Confidence Bands for Linear and Logistic Regression Models. PhD thesis, School of Mathematics, University of Southampton, UK.

Liu, W. and Ah-Kine, P. (2010). Optimal simultaneous confidence bands in simple linear regression. *Journal of Statistical Planning and Inference*, to appear.

Liu, W., Bretz, F., Hayter, A.J. and Wynn, H.P. (2009). Assessing non-superiority, non-inferiority or equivalence when comparing two regression models over a restricted covariate region. *Biometrics*, **65**, 1279-1287.

Liu, W. and Hayter, A.J. (2007). Minimum area confidence set optimality for confidence bands in simple linear regression. *Journal of the American Statistical Association*, **102**, 181-190.

Liu, W., Hayter, A.J., Piegorsch, W.W. and Ah-Kine, P. (2009). Comparison of hyperbolic and constant width simultaneous confidence bands in multiple linear regression under MVCS criterion. *Journal of Multivariate Analysis*, **100**, 1432-1439.

Liu, W., Hayter, A.J. and Wynn, H.P. (2007). Operability region equivalence: simultaneous confidence bands for the equivalence of two regression models over restricted regions. *Biometrical Journal*, **49**, 144-150.

Liu, W., Jamshidian, M., and Zhang, Y. (2004). Multiple comparison of several linear regression lines. *Journal of the American Statistical Association*, **99**, 395-403.

Liu, W., Jamshidian, M., Zhang, Y. and Bretz, F. (2005). Constant width simultaneous confidence bands in multiple linear regression with predictor variables constrained in intervals. *Journal of Statistical Computation and Simulation*, **75**, 425-436.

Liu, W., Jamshidian, M., Zhang, Y., Bretz, F. and Han, X. (2007a). Some new methods for the comparison of two linear regression models. *Journal of Statistical Planning and Inference*, **137**, 57-67.

Liu, W., Jamshidian, M., Zhang, Y., Bretz, F. and Han, X. (2007b). Pooling batches in drug stability study by using constant-width simultaneous confidence bands. *Statistics in Medicine*, **26**, 2759-2771.

Liu, W., Jamshidian, M., Zhang, Y., and Donnelly, J. (2005). Simulation-based simultaneous confidence bands in multiple linear regression with predictor variables constrained in intervals. *Journal of Computational and Graphical Statistics*, **14**, 459-484.

Liu, W. and Lin, S. (2009). Construction of exact simultaneous confidence bands in multiple linear regression with predictor variables constrained in an ellipsoidal region. *Statistica Sinica*, **19**, 213-232.

Liu,W., Lin, S. and Piegorsch, W.W. (2008). Construction of exact simultaneous confidence bands for a simple linear regression model. *International Statistical Review*, **76**, 39-57.

Liu, W., Miwa, T. and Hayter, A.J.(2000). Simultaneous confidence interval estimation for successive comparisons of ordered treatment effects. *Journal of Statistical Planning and Inference*, **88**, 75-86.

Liu, W., Wynn, H.P. and Hayter, A.J. (2008). Statistical inferences for linear regression models when the covariates have functional relationships: polynomial regression. *Journal of Statistical Computation and Simulation*, **78**, 315-324.

Loader, C. (2004). The volume-of-tube formula: computational methods and statistical applications. Manuscript, Department of Statistics, Case Western Reserve University, USA.

Loader, C., and Sun, J.Y. (1997). Robustness of tube formula based confidence bands. *Journal of Computational and Graphical Statistics*, **6**, 242-250.

Lu, X.J. and Chen, J.T. (2009). Exact simultaneous confidence segments for all contrast comparisons. *Journal of Statistical Planning and Inference*, **139**, 2816-2822.

Ma, G.Q. and Hall, W.J. (1993). Confidence bands for receiver operating characteristic curves. *Medical Decision Making*, **13**, 191-197.

Martinez, W.L. and Martinez, A.R. (2008). *Compuational Statistics Handbook with MATLAB, 2nd edition*. Chapman & Hall.

McCann, M. and Edwards, D. (1996). A path length inequality for the multivariate-t distribution, with applications to multiple comparisons. *Journal of the American Statistical Association*, **91**, 211-216.

McCullagh, P. and Nelder, J.A. (1989). *Generalized Linear Models, 2nd edition*. Chapman & Hall.

McCulloch, C.E. and Searle, S.R. (2001). *Generalized, Linear, and Mixed Models*. Wiley.

Mee, R.W., Eberhardt, K.R. and Reeve, C.P. (1991). Calibration and simultaneous tolerance intervals for regression. *Technometrics*, **33**, 211-219.

Merchant, A., McCann, M. and Edwards, D. (1998). Improved multiple comparisons with a control in response surface analysis. *Technometrics*, **40**, 297-303.

Miller, R.G. Jr. (1981). *Simultaneous Statistical Inference*. Springer-Verlag.

Myers, R.H., Montgomery, D.C. and Anderson-Cook, C.M. (2009). *Response Surface Methodology : Process and Product Optimization Using Designed Experiments*. Wiley.

Myers, R.H., Montgomery, D.C. and Vining, G.G. (2002). *Generalized Linear Models with Applications in Engineering and the Sciences*. Wiley.

Naiman, D.Q. (1983). Comparing Scheffé-type to constant-width confidence bounds in regression. *Journal of the American Statistical Association*, **78**, 906-912.

Naiman, D.Q. (1984a). Average width optimality of simultaneous confidence bands. *Annals of Statistics*, **12**, 1199-1214.

Naiman, D.Q. (1984b). Optimal simultaneous confidence bounds. *Annals of Statistics*, **12**, 702-715.

Naiman, D.Q. (1986). Conservative confidence bands in curvilinear regression. *Annals of Statistics*, **14**, 896-906.

Naiman, D.Q. (1987a). Simultaneous confidence-bounds in multiple-regression using predictor variable constraints. *Journal of the American Statistical Association*, **82**, 214-219.

Naiman, D.Q. (1987b). Minimax regret simultaneous confidence bands for multiple regression functions. *Journal of the American Statistical Association*, **82**, 894-901.

Naiman, D.Q. (1990). On volumes of tubular neighborhoods of spherical polyhedra and statistical inference. *Annals of Statistics*, **18**, 685-716.

Nitcheva, D.K., Piegorsch, W.W., West, R.W. and Kodell, R.L. (2005). Multiplicity-adjusted inferences in risk assessment: benchmark analysis with quantal response data. *Biometrics*, **61**, 277-286.

Odeh, R.E. and Mee, R.W. (1990). One-sided simultaneous tolerance limits for regression. *Communications in Statistics – Simulation and computation*, **19**, 663-680.

O'Hagan, A. and Forster, J. (2004). *Kendall's Advanced Theory of Statistics, Volume 2b: Bayesian Inference*. Arnold.

Pan, W., Piegorsch, W.W., and West, R.W. (2003). Exact one-sided simultaneous confidence bands via Uusipaikka's method. *Annals of the Institute of Statistical Mathematics*, **55**, 243-250.

Piegorsch, W.W. (1985a). Admissible and optimal confidence bands in simple linear regression. *Annals of Statistics*, **13**, 801-810.

Piegorsch, W.W. (1985b). Average-width optimality for confidence bands in simple linear-regression. *Journal of the American Statistical Association*, **80**, 692-697.

Piegorsch, W.W. (1986). Confidence bands for polynomial regression with fixed intercepts. *Technometrics*, **28**, 241-246.

Piegorsch, W.W. (1987a). On confidence bands and set estimators for the simple linear model. *Statistics & Probability Letters*, **5**, 409-413.

Piegorsch, W.W. (1987b). Model robustness for simultaneous confidence bands. *Journal of the American Statistical Association*, **82**, 879-885.

Piegorsch, W.W., and Casella, G. (1988). Confidence bands for logistic regression with restricted predictor variables. *Biometrics*, **44**, 739-750.

Piegorsch, W.W., West R.W., Al-Saidy O.M., and Bradley, K.D. (2000). Asymmetric confidence bands for simple linear regression over bounded intervals. *Computational Statistics & Data Analysis*, **34**, 193-217.

Piegorsch, W.W., West R.W., Pan, W. and Kodell, R. (2005). Low dose risk estimation via simultaneous statistical inferences. *Journal of the Royal Statistical Society (C)*, **54**, 245-258.

Pinheiro, J.C. and Bates, D.M. (2000). *Mixed-Effects Models in S and S-Plus.* Springer.

Plackett, R.L. (1960). *Principles of Regression Analysis.* Oxford University Press.

Rao, C.R. (1959). Some problems involving linear hypotheses in multivariate analysis. *Biometrika*, **46**, 49-58.

Ross, S. (1988). *A First Course in Probaility.* Macmillan Publishing Company.

Royen, T. (1995). On some central and noncentral multivariate chi-square distributions. *Statistica Sinica*, **5**, 373-397.

Ruberg, S.J., and Hsu, J.C. (1992). Multiple comparison procedures for pooling batches in stability studies. *Technometrics*, **34**, 465-472.

Ruberg, S.J., and Stegeman, J.W. (1991). Pooling data for stability studies: testing the equality of batch degradation slopes. *Biometrics*, **47**, 1059-1069.

Rychlik, I. (1992). Confidence bands for linear regressions. *Communications in Statistics – Simulation and Computation*, **21**, 333-352.

Sa, P. and Edwards, D. (1993). Multiple comparisons with a control in response-surface methodology. *Technometrics*, **35**, 436-445.

SAS Institute Inc. (1990). *SAS/STAT User's Guide, Version 6, Fourth edition, Volume 2.* SAS Institute Inc., Cary, NC.

Scheffé, H. (1953). A method for judging all contrasts in analysis of variance. *Biometrika*, **40**, 87-104.

Scheffé, H. (1959). *The Analysis of Variance.* Wiley.

Scheffé, H. (1973). A statistical theory of calibration. *Annals of Statistics*, **1**, 1-37.

Seber, G. (1977). *Linear Regression Analysis.* Wiley.

Selvin, S. (1998). *Modern Applied Biostatistical Methods Using S-Plus.* Oxford University Press.

Seppanen, E. and Uusipaikka, E. (1992). Confidence bands for linear-regression over restricted regions. *Scandinavia Journal of Statistics*, **19**, 73-81.

Serfling, R.J. (1980). *Approximation Theorems of Mathematical Statistics.* Wiley.

Shao, J. and Tu, D. (1995). *The Jackknife and Bootstrap.* Springer-Verlag.

Silverman, B. W. (1986). *Density Estimation for Statistics and Data Analysis.* Chapman & Hall.

Somerville, P. N. (1997). Multiple testing and simultaneous confidence intervals: calculation of constants. *Computational Statistics and Data Analysis*, **25**, 217-233.

Somerville, P.N. (1998). Numerical computation of multivariate normal and multivariate-t probabilities over convex regions. *Journal of Computational and Graphical Statistics*, **7**, 529-544.

Somerville, P.N. (1999). Critical values for multiple testing and comparisons: one step and step down procedures. *Journal of Statistical Planning and Inference*, **82**, 129-138.

Somerville, P.N. and Bretz, F. (2001). Obtaining critical values for simultaneous confidence intervals and multiple testing. *Biometrical Journal*, **43**, 657-663.

Spurrier, J.D. (1983). Exact uniform confidence bands for periodic regression. *Communications in Statistics – Theory and Methods*, **12**, 969-973.

Spurrier, J.D. (1992). Simultaneous confidence bands for quadratic regression over a finite interval. *Proceedings of the Physical and Engineering Sciences Section, American Statistical Association*, 124-129.

Spurrier, J.D. (1993). Comparison of simultaneous confidence bands for quadratic regression over a finite interval. *Technometrics*, **35**, 315-320.

Spurrier, J.D. (1999). Exact confidence bounds for all contrasts of three or more regression lines. *Journal of the American Statistical Association*, **94**, 483-88.

Spurrier, J.D. (2002). Exact multiple comparisons of three or more regression lines: pairwise comparisons and comparisons with a control. *Biometrical Journal*, **44**, 801-812.

Steihorst, R.K. and Bowden, D.C. (1971). Discrimination and confidence bands on percentiles. *Journal of the American Statistical Association*, **66**, 851-854.

Stewart, P.W. (1991a). Line-segment confidence bands for repeated measures. *Biometrics*, **43**, 629-640.

Stewart, P.W. (1991b). The graphical advantages of finite interval confidence band procedures. *Communications in Statistics – Theory and Methods*, **20**, 3975-3993.

Stuart, A. (1999). *Kendall's Advanced Theory of Statistics 2A, 6th edition*. Arnold.

Sun, J. (1993). Tail probabilities of maxima of Gaussian random fields. *Annals of Probability*, **21**, 34-71.

Sun, J. and Loader, C.R. (1994). Simultaneous confidence bands for linear regression and smoothing. *Annals of Statistics*, **22**, 1328-1346.

Sun, J., Loader, C.R., and McCormick, W.P. (2000). Confidence bands in generalized linear models. *Annals of Statistics*, **28**, 429-460.

Sun, J.Y., Raz, J., and Faraway, J.J. (1999). Confidence bands for growth and response curves. *Statistical Sinica*, **9**, 679-698.

Tamhankar, M.V. (1967). A characterization of normality. *Annals of Mathematical Statistics*, **38**, 1924-1927.

Thomas, D.L., and Thomas, D.R. (1986). Confidence bands for percentiles in the linear regression model. *Journal of the American Statistical Association*, **81**, 705-708.

Tong, Y.L. (1990). *Multivariate Normal Distribution*. Springer-Verlag.

Trout, J.R., and Chow, B. (1972). Table of percentage points of the trivariate t distribution with application to uniform confidence bands. *Technometrics*, **14**, 855-879.

Trout, J.R., and Chow, B. (1973). Uniform confidence bands for a quadratic model. *Technometrics*, **15**, 611-624.

Tseng, Y.L. (2002). Optimal confidence sets for testing average bioequivalence. *Test*, **11**, 127-141.

Tsutakawa, R.K. and Hewett, J.E. (1978). Comparison of two regression lines over a finite interval. *Biometrics*, **34**, 391-398.

Tukey, J.W. (1953). *The Problem of Multiple Comparisons*. Dittoed manuscript of 396 pages, Department of Statistics, Princeton University.

Turner, D.L., and Bowden, D.C. (1977). Simultaneous confidence bands for percentile lines in the general linear model. *Journal of the American Statistical Association*, **72**, 886-889.

Turner, D.L., and Bowden, D.C. (1979). Sharp confidence bands for percentile lines and tolerance bands for the simple linear model. *Journal of the American Statistical Association*, **74**, 885-888.

Uusipaikka, E. (1983). Exact confidence bands for linear-regression over intervals. *Journal of the American Statistical Association*, **78**, 638-644.

Weisberg, S. (2005). *Applied Linear Regression*. Wiley.

Wellek, S. (2003). *Testing Statistical Hypotheses of Equivalence*. Chapman & Hall/CRC.

Westfall, P.H., and Young, S.S. (1993). *Resampling-Based Multiple Testing: Examples and Methods for P-Value Adjustment*. Wiley.

Weyl, H. (1939). On the volume of tubes. *Journal of the American Mathematical Society*, **61**, 461-472.

Wilcox, R.R. (1987). Pairwise comparisons of J-independent regression lines over a finite interval, simultaneous pairwise comparison of their parameters, and the Johnson-Neyman procedure. *British Journal of Mathematical & Statistical Psychology*, **40**, 80-93.

Working, H. and Hotelling, H. (1929). Applications of the theory of error to the interpretation of trends. *Journal of the American Statistical Association*, **24**, 73-85.

Wynn, H.P. (1975). Integrals for one-sided confidence bounds: a general result. *Biometrika*, **62**, 393-6.

Wynn, H. P. (1984). An exact confidence band for one-dimensional polynomial regression. *Biometrika*, **71**, 375-9.

Wynn, H.P. and Bloomfield, P. (1971). Simultaneous confidence bands in regression analysis. *Journal of the Royal Statistical Society (B)*, **33**, 202-217.

Xu, X.Z., Ding, X.B. and Zhao, S.R. (2009). The reduction of the average width of confidence bands for an unknown continuous distribution function. *Journal of Statistical Computation and Simulation*, **79**, 335-347.

Zhao, Z.B. and Wu, W.B. (2008). Confidence bands in nonparametric time series regression. *Annals of Statistics*, **36**, 1854-1878.

Index